Birkhäuser

Trends in the History of Science

Trends in the History of Science is a series devoted to the publication of volumes arising from workshops and conferences in all areas of current research in the history of science, primarily with a focus on the history of mathematics, physics, and their applications. Its aim is to make current developments available to the community as rapidly as possible without compromising quality, and to archive those developments for reference purposes. Proposals for volumes can be submitted using the online book project submission form at our website www.birkhauser-science.com.

More information about this series at http://www.springer.com/series/11668

Volker R. Remmert · Martina R. Schneider
Henrik Kragh Sørensen
Editors

Historiography of Mathematics in the 19th and 20th Centuries

Birkhäuser

Editors
Volker R. Remmert
Interdisziplinäres Zentrum für Wissenschafts- und Technikforschung (IZWT)
Bergische Universität Wuppertal
Wuppertal
Germany

Henrik Kragh Sørensen
Department of Science Education
University of Copenhagen
Copenhagen
Denmark

Martina R. Schneider
Institut für Mathematik, AG Geschichte der Mathematik
Johannes Gutenberg-Universität Mainz
Mainz
Germany

ISSN 2297-2951 ISSN 2297-296X (electronic)
Trends in the History of Science
ISBN 978-3-319-39647-7 ISBN 978-3-319-39649-1 (eBook)
DOI 10.1007/978-3-319-39649-1

Library of Congress Control Number: 2016953001

2010 Mathematics Subject Classification Codes (AMS): 01A85 01A17 (Babylonian), 01A20 (Greek/Roman), 01A45 (17th C), 01A50 (18th C), 01A55 (19th C), 01A60 (20th C), 01A72 (Schools of mathematics)

Printed on acid-free paper

This book is published under the trade name Birkhäuser
The registered company is Springer International Publishing AG
The registered company address is: Gewerbestrasse 11, 6330 Cham, Switzerland
(www.birkhauser-science.com)

Contents

Contributors

David Aubin Institut de mathématiques de Jussieu-Paris rive Gauche, Sorbonne Universités, Université Pierre et Marie Curie (CNRS, Sorbonne-Paris-Cité, Sorbonne universités), Paris, France

Sonja Brentjes Max Planck Institut für Wissenschaftsgeschichte, Berlin, Germany

Maarten Bullynck Département de mathématiques et histoire des sciences, UFR 6 MITSIC & EA 1571 Centre de recherches historiques : Histoire des Pouvoirs, Savoirs et Sociétés Université Paris 8, St-Denis cédex, France

H. Floris Cohen Descartes Centre, Utrecht University, Utrecht, The Netherlands

Jeremy Gray Department of Mathematics and Statistics, The Open University, Milton Keynes, UK

Jens Høyrup Section for Philosophy and Science Studies, Roskilde University, Roskilde, Denmark

Anne-Sandrine Paumier Fondation Mathématique Jacques Hadamard, Institut Des Hautes Études Scientifiques, Bures-sur-Yvette, France

Volker R. Remmert Wissenschafts- und Technikgeschichte - Historisches Seminar, Fakultät für Geistes- und Kulturwissenschaften, Bergische Universität Wuppertal, Wuppertal, Germany

David E. Rowe Institut für Mathematik, AG Geschichte der Mathematik, Johannes Gutenberg-Universität Mainz, Mainz, Germany

Martina R. Schneider Institut für Mathematik, AG Geschichte der Mathematik, Johannes Gutenberg-Universität Mainz, Mainz, Germany; ERC-Project "Mathematical Sciences in the Ancient World" (SAW), CNRS; SPHERE, Université Paris 7 (Diderot), Paris, France

Reinhard Siegmund-Schultze Faculty of Engineering and Science, University of Agder, Kristiansand, Norway

Henrik Kragh Sørensen Department of Science Education, University of Copenhagen, Copenhagen, Denmark

Benjamin Wardhaugh All Souls College, University of Oxford, Oxford, UK

List of Figures

Pulling Harriot Out of Newton's Shadow: How the Norwegian Outsider Johannes Lohne Came to Contribute to Mainstream History of Mathematics

Introductory Remarks

Volker R. Remmert, Martina R. Schneider
and Henrik Kragh Sørensen

In this volume, we show through detailed case studies, how the historiography of mathematics has been influenced by the contexts and motivations of its practitioners.[1] The notion of historiography in the context of history of mathematics holds an important ambiguity: It can mean the history of mathematics or the methodology of the history of mathematics, and naturally these two are intimately related. As historians, we are situated in various discourses that shape the questions

[1]We would like to thank J. Barrow-Green, K. Chemla, and D. E. Rowe for their critical readings of earlier drafts of these introductory remarks. Part of the work by Martina R. Schneider leading to this volume has received funding from the European Research Council under the European Union's Seventh Framework Programme (FP7/2007-2013)/ERC Grant agreement n. 269804 "Mathematical Sciences in the Ancient World (SAW)".

V.R. Remmert
Wissenschafts- und Technikgeschichte - Historisches Seminar, Fakultät für Geistes- und Kulturwissenschaften, Bergische Universität Wuppertal, Gaußstraße 20, 42097 Wuppertal, Germany
e-mail: remmert@uni-wuppertal.de

M.R. Schneider (✉)
Institut für Mathematik, AG Geschichte der Mathematik,
Johannes Gutenberg-Universität Mainz, Staudinger Weg 9, 55099 Mainz, Germany
e-mail: mschneider@mathematik.uni-mainz.de

M.R. Schneider
ERC-Project "Mathematical Sciences in the Ancient World" (SAW),
CNRS; SPHERE, Université Paris 7 (Diderot), Paris, France

H.K. Sørensen
Department of Science Education, University of Copenhagen,
Øster Voldgade 3, 1350 København K, Denmark
e-mail: henrik.kragh@ind.ku.dk

V.R. Remmert et al. (eds.), *Historiography of Mathematics in the 19th and 20th Centuries*, Trends in the History of Science,
DOI 10.1007/978-3-319-39649-1_1

and methods involved in writing history of mathematics. Thus, in what follows, these two perspectives will frequently intersect and overlap.

Discussions about historiography of mathematics are necessarily multifaceted, and in this explorative volume, we cannot address all (or even most of) the pressing historiographical issues. Our interest lies in linking disciplinary and methodological changes in the history of mathematics to the wider cultural contexts of its practitioners, namely those doing history of mathematics during the nineteenth and twentieth centuries. We have chosen to restrict our cases to this period in order to focus on a transformation of the field during which it underwent the slow process of becoming an academic sub-discipline in its own right. Inevitably, our choice of focus has meant that we have had to leave out important discussions about mathematical contexts and practices studied during the past 200 years, that would also have fitted our purpose very well.

Important historiographical shifts during the nineteenth and twentieth centuries are reflected in shifts in the practitioners' academic backgrounds. In the nineteenth century specialists from a wide variety of different fields contributed to the history of mathematics. Some were philologists, philosophers, mathematicians, historians, teachers, and orientalists, others worked as missionaries and colonial administrators. In some cases they had profound knowledge in more than one of the fields involved. During the 20th century we still find contributions by members of some of these groups, but towards the end of the century the field is increasingly dominated by people who were either trained in mathematics as well as in the history of mathematics (or of science), hold academic positions designated to this field, or dedicate a lot of their time to research in the history of mathematics. This gradual shift and its nuances are reflected clearly in the contributions to our volume. For the purpose of these short introductory remarks, however, we have opted to refer to anyone engaged in the history of mathematics as a "historian of mathematics", in order to include any form of writing on the history of mathematics and to avoid simplistic divisions along disciplinary boundaries.

Over the past decades, historians of mathematics have taken an increasingly reflexive interest in their discipline and its methods. This tendency has been related to issues of professionalisation of the discipline and to a broadening of the research interests in the history of mathematics. This volume addresses anew some key aspects of the historiography of mathematics during the nineteenth and twentieth centuries with special attention paid to the cultural, social, and political contexts in which the history of mathematics was written.

During the period from roughly 1880 to 1940, mathematics developed in important ways, both concerning its content, its conditions for cultivation, and its identity. Increasing professionalisation and internationalisation of mathematics coincided with new trends in research in bringing about a new, more autonomous identity, in particular for pure mathematics. These developments in mathematics have all been the topics of extensive historical research, in particular over the past decades (see, for example, Gray 2008; Mehrtens 1990; see also Rowe 2013). As we shall see in this volume, sometimes the writing of the history of mathematics played an important role in the shaping of mathematical identity during the period in question.

Simultaneously with the development of mathematics, the history of mathematics gradually grew into a research field of its own with institutions such as journals, societies and academic positions (see, for instance, Dauben and Scriba 2002). Since the 1960s, the discipline has increasingly gained autonomy from mathematics and now occupies an interdisciplinary area with affiliations to mathematics, of course, but also, for instance, with specific linguistic and cultural studies, history of science, philosophy of mathematics, and with mathematics education.

Reflecting both a new professional identity and changes in the primary audience, various shifts of perspective in the ways of writing history of mathematics can be observed which continue to this day: From initially mostly concentrating on major internal, universal developments of certain sub-disciplines of mathematics, the field gradually expanded to incorporate the study of contexts of knowledge production involving individuals, local practices, problems, communities, and networks. Whereas this is true for the general scope of the field, variations do exist between subfields.

As is well known in historiography in general, the writing of history is conditioned and influenced by the context in which it is undertaken, although history of mathematics experienced a certain delay in coming to this realisation. However, during the past decade, the predominant philosophical understandings of mathematics, the goals, interests and interpretive skills in the history of mathematics, and the professional access to history of mathematics have all evolved in important and interrelated ways. Both professional historians of mathematics and research mathematicians, to name only two groups of practitioners, are deeply rooted in specific cultures, circumstances, and backgrounds which are sometimes difficult to discern analytically, but which our meta-historical approach can illuminate.

In order to introduce our perspectives and to illustrate our approach in this volume we will give three short examples. In the first example we discuss how history and biography have functioned as vehicles for the self-fashioning of mathematicians. We do so by considering the self-fashioning involved in Mittag-Leffler's biographical writings about his heroes Abel and Weierstrass. In the second example we highlight how the writing of history has been shaped by contemporaneous cultural discourses inside and outside mathematics. Here we analyse the historical interpretation of the discovery of incommensurability in ancient Greece in the light of the foundational crises of the late-nineteenth and early-twentieth centuries. The third example throws a light on the methodological development of the history of mathematics and its implications on its relation to mathematics, proper. We discuss the debate over who is competent to write history of mathematics, and how it should be written, as it unfolded between Unguru and Weil in the 1970s. In their different ways, each of these cases are instances of the well-known and fundamental phenomenon that those who write history appropriate the past for their specific contextual concerns which are more or less influenced by cultural, sociological and political factors. In this volume, our interest lies in analysing such contexts through the use and production of historical writing about mathematics.

During the first decades of the twentieth century, towards the end of his own active career, the Swedish mathematician G. Mittag-Leffler ventured into the historical, in particular the biographical, genre by preparing biographies of his two mathematical heroes N. H. Abel (published 1903/1907) and K. Weierstrass (published 1923). In these two biographies, Mittag-Leffler was not only presenting the lives of famous past mathematicians but also fashioning a particular role for himself and his own mathematical research (Sørensen 2012). Thus, he did not forget to allude to his own work in continuation of the ideas of Abel and Weierstrass. He offered his own interpretations of key events; interpretations that sometimes ran at odds with the standard narratives. For instance, Mittag-Leffler found fault with the frequent criticisms of the Norwegian state for not supporting Abel sufficiently; these he found provincial, which was of course contentious in a time when Norway was struggling for independence from Sweden. And he emphasised the necessity for international contacts and the importance of mathematical publishing, explicitly likening Abel's role in A.-L. Crelle's *Journal für die reine und angewandte Mathematik* to H. Poincaré's role in his own *Acta mathematica*. Thus, Abel's life and concerns were brought to bear on contemporary issues of internationalisation that were close to Mittag-Leffler's heart and agenda.

Later on, but in a more traditional genre, Mittag-Leffler's self-fashioning came to the fore when he delivered lectures on his efforts at internationalisation at Scandinavian and international conferences. In an example of retrospective actors' history, Mittag-Leffler recounted efforts to set up international meetings and collaboration, giving the events his own spin and emphasising his own contributions.

Thus, Mittag-Leffler here comes to represent the use and appropriation of history and biography of practising mathematicians for specific ends. As such, his historical and biographical writings become open to interpretation within the contemporary mathematical culture in which they were to serve, and we can get a richer picture of their author as a leading figure on the international mathematical scene and as an heir to Abel's and Weierstrass' mathematical legacies.

Whereas the above example illustrates how to use historical writings to discern the otherwise sometimes inaccessible appropriations, the history of mathematics is, of course, also influenced by the mathematical and cultural contexts in which it is written. For instance, when in 1936, O. Neugebauer was interviewed about the flourishing of mathematics in ancient Babylon, he pointed to the interactions between "different types of people". We can hardly consider this categorization neutral once we are aware of the fact that the author was an emigrant from Nazi oppression. On the more internal side, key developments within nineteenth-century mathematics, in particular the construction of the real numbers by R. Dedekind, Weierstrass and others, gave rise to a new line of historical interpretation of ancient, in particular pre-Euclidean mathematics, which has been the subject of debate since the mid-1960s.

This example points to two interesting perspectives on the impact of later historical and philosophical thought on the interpretation of the past. First, on the technical side, the new definitions of real numbers were observed to bear resemblances with the ancient theory of proportions that led some authors, like A.E. Taylor in the second quarter of the twentieth century, to search for traces of the modern concepts in obscure passages from ancient writers. Such approaches were later denounced as anachronistic: "these writers clearly betray a distortion of critical viewpoint owing to their awareness of the modern real-number concept" (Knorr 2001, 125).

Second, on the more contextual and philosophical side, the suggestion that the discovery of incommensurable quantities by the Pythagoreans must have led to a "veritable logical scandal" was argued by J. Tannery in 1887. Later, this was taken up by H. Hasse and H. Scholz who in 1928 argued that this discovery "must naturally have shaken the idea of the 'arithmetica universalis' of the Pythagoreans" (quoted from ibid., 127). Their argument was made with explicit reference to modern mathematical developments such as the arithmetization of analysis in the nineteenth-century and the contemporary Grundlagenkrise over the foundations of mathematics: "Just as in the past century and today, so also in the 2nd half of the 5th century, there was a severe foundations crisis" (cited from ibid., 127). This example illustrates that contemporary mathematical and philosophical discourses do influence the historical writing about ancient mathematics, and by extension, about any mathematical topic. The critical approaches of H. Freudenthal, W. Knorr and D. Fowler have served to highlight such inferences drawn from anachronistic positions.

The third example addresses questions of methodology with respect to the interdisciplinarity of history of mathematics. In the 1970s S. Unguru and A. Weil disputed over the qualifications needed to do research in the history of mathematics. They were not only fighting about professional legitimacy but also about the methodological underpinnings of the discipline. Unguru (1975) attacked the then-dominant historiography of ancient Greek mathematics as proposed by mathematicians such as, most prominently perhaps, B. L. van der Waerden. He criticized their approach and interpretation in general, which arose, in his view, from a Platonist perspective of mathematics. In particular, he discarded the by then well established concept of "geometric algebra" that, according to Unguru, drew on modern algebraic notations, tools and concepts foreign to ancient Greek mathematics, and had been created mainly by modern mathematicians.

A proper historical approach, Unguru argued, should refrain from using modern mathematical notations and concepts to explain old mathematics. Instead it should draw only on the (remaining) sources of the time to understand the texts and establish their contexts. Thus for Unguru, a training in today's mathematics seems not to be a prerequisite for doing history of mathematics, but rather an impediment since it could lead to anachronistic conclusions. Unguru called for a systematic historisation of ancient Greek mathematics to be done by historians who rely only on historical, textual, philological and philosophical evidence, and not on mathematical intuition. In this way, Unguru's attack can be seen to signal the beginning of a "textual turn" in the history of mathematics (see also Saito 1998) and it can be understood as being connected to processes of professionalisation in the fields of history of science and history of mathematics, as well as to power struggles between the sciences and the humanities in the United States.

Unguru's call for a radical change in the methodology of history of mathematics provoked a counterattack by Weil. In his response, Weil addressed the interdisciplinary nature of history of mathematics as situated between history and mathematics as problematic per se (Weil 1978). At the international congress of mathematicians in Helsinki in 1978 he took up the issue again in his address "History of mathematics: why and how" (Weil 1980). As mathematicians were, according to Weil, the main audience for history of mathematics, historians of mathematics required a thorough training in mathematics. This was also needed, he said, in order to trace "the early occurrence of concepts and methods destined to emerge only later into the conscious mind of mathematicians" (Weil 1980, 232)—which he felt was one of the most rewarding topics for mathematicians. However, he also conceded that proper historians had a raison d'être in the history of mathematics, and that fruitful cooperation was possible for historians and mathematicians in the field.

This kind of debate can be linked to discussions about internalist and externalist approaches in the history of mathematics and keeps resurfacing every now and then (see for example Blåsjö 2014; Fried 2014); it is also touched upon in some contributions to this volume. It shows underlying tensions between mathematicians and historians in a field which attracts experts from a variety of disciplines.

These three examples point to the fact that the study of history of mathematics can be—and has been—undertaken for a variety of purposes and with a broad spectrum of different interests and concerns. To show this diversity, we aim not only to focus on the interaction between history of mathematics and mathematics proper, but also to include the wider embedding of historiography in other contexts such as the history of science or social and political contexts.

By analysing the often hidden agendas of former producers of history of mathematics, we are led to reflect upon our own professional objectives as well as on the methods and tools we employ today. Thus, our aim for this volume is twofold: First, to present a number of individual cases from a variety of backgrounds to form an initial basis for comparison; second, to open up for methodological discussion of issues bearing on how to approach the contextualised "meta-history" of mathematics in the nineteenth and twentieth centuries. This volume brings together essays that reflect upon various aspects of the writing of history of mathematics during the past 200 years. As indicated above, this was a formative period for our discipline, and many of the issues were of a methodological kind that are still with us in our contemporary practice of history of mathematics. Yet, this volume is only indirectly concerned with explicit methodological discussions, although the theme will be recurring through the chapters.

Nevertheless, this volume is marked by the dualistic meaning of the term historiography alluded to above: On the one hand, it means the history of the history of mathematics, and on the other hand, it refers to the methodology of practicing history of mathematics. In the latter form, the history of mathematics has only recently begun to become as reflexive as its cousins in the history of science.

Despite their kinship, there are methodological and philosophical differences between the history of mathematics and the history of science that in part stem from similar differences in the disciplines themselves. Yet, for the history of mathematics, there is no equivalent to H. Kragh's *An Introduction to the Historiography of Science* (Kragh 1987, 1994), although Wardhaugh (2010) and Stedall (2012) provide short introductions to the questions and methodology of our discipline in forms that are also accessible to students. Instead, graduate training in history of mathematics has until recently been more by apprenticeship, and through imitation and inspiration from the literature. Thus, our ambition is also to provide a reflexive stance on these recent practices without aiming for the ambitious task of providing a more normative methodology.

With these introductory remarks, we hope to have stimulated the appetite for the detailed case studies presented in the chapters that follow. These case studies are ordered roughly chronologically by the principal historians of mathematics involved in order to link to the development of the historiography. Each chapter is preceeded by an abstract to allow quick orientation. It is our ambition, that eventually these cases and others will stimulate a more synthetic exposition of the shaping and continuous development of the historiography of mathematics. Clearly, the assumptions tacitly underlying these remarks are likely to betray our particular approaches, preoccupations, and perspectives but—as this volume shows–that is a necessity that we cannot avoid but only strive to critically reflect upon.

The contributions to the present volume stem from the workshop "Historiography of Mathematics in the 19th and 20th Centuries", initiated by Martina R. Schneider and Henrik Kragh Sørensen and organized together with Volker R. Remmert at the University of Wuppertal in March 2013. We wish to thank all participants in the workshop for their inspiring contributions.

Aarhus, Paris, Wuppertal in December 2015

References

Blåsjö, V. (2014). A critique of the modern consensus in the historiography of mathematics. *Journal of Humanistic Mathematics, 4*(2), 113–123.

Dauben, J. W., & Scriba, C. J. (Eds.). (2002). *Writing the history of mathematics: its historical development*, vol. 27 of *Science Networks. Historical Studies*. Basel, Boston, Berlin: Birkhäuser.

Fried, M. N. (2014). The discipline of history and the modern consensus in the historiography of mathematics. *Journal of Humanistic Mathematics, 4*(2), 124–136.

Gray, J. (2008). *Plato's ghost: The modernist transformation of mathematics*. Princeton, Oxford: Princeton University Press.

Knorr, W. R. (2001). The impact of modern mathematics on ancient mathematics. *Revue d'histoire des mathématiques, 7*, 121–135.

Kragh, H. (1987/1994). *An Introduction to the historiography of science*. Cambridge: Cambridge University Press.

Mehrtens, H. (1990). *Moderne Sprache Mathematik. Eine Geschichte des Streits um die Grundlagen der Disziplin und des Subjekts formaler Systeme*. Frankfurt am Main: Suhrkamp Verlag.

Rowe, D. E. (2013). *Review of Jeremy Gray, 'Plato's Ghost: The modernist transformation of mathematics'*, Princeton and Oxford: Princeton University Press, 2008. *Bulletin of the American Mathematical Society*, *50*(3), 513–521.
Saito, K. (1998). Mathematical reconstructions out, textual studies in: 30 years in the historiography of Greek mathematics. *Revue d'histoire des mathématiques*, *4*(1), 131–142.
Sørensen, H. K. (2012). Reading Mittag-Leffler's biography of Abel as an act of mathematical self-fashioning. In B. Wardhaugh (Ed.), *The history of the history of mathematics. Case studies for the seventeenth, eighteenth and nineteenth centuries* (pp. 115–144). Oxford: Peter Lang.
Stedall, J. (2012). *History of mathematics. A very short introduction*, (Vol. 305). Oxford: Oxford University Press.
Unguru, S. (1975). On the need to rewrite the history of Greek mathematics. *Archive for History of Exact Sciences, 15*(1), 67–114.
Wardhaugh, B. (2010). *How to read historical mathematics*. Princeton and Oxford: Princeton University Press.
Weil, A. (1978). Who betrayed Euclid? (extract from a letter to the editor). *Archive for History of Exact Sciences, 19*(2), 91–93.
Weil, A. (1980). History of mathematics: Why and how. In O. Letho (Ed.), *Proceedings of the International Congress of Mathematicians, Helsinki 1978* (vol. 1, pp. 227–236). Helsinki: Academia Scientiarum Fennica.

The History of Mathematics in the Progress of Mankind. Modifying the Narrative Around 1800

Maarten Bullynck

Abstract

The narrative patterning of historiography changed profoundly around 1800. Instead of the accumulative, encyclopaedic format typical of the 18th century, a historical narrative hinged on the progress of mankind became viable (Condorcet, Fichte etc.). This also had an impact on the writing of the history of mathematics. An interesting testimony of this transition can be found in Alexander von Humboldt's project on the origin and development of the decimal positional numeral system. What originally started as an ethnographic and encyclopaedic project became a hypothetical history of ideas, inspired by the new philologies and the new mathematics.

Keywords

Narrative formats of historiography · History of mathematical notation · Condorcet · Fichte · Alexander von Humboldt

Near the end of the 18th century a new historiographic scheme appeared "that starts from the idea that mankind will for ever and ever progress in his cultivation and amelioration" and, as a consequence, "all analogies between past and present disappear" as a contemporary notes.[1] This new format broke with mid 18th century

[1]"Man arbeitet heut zu Tage an historischen Systemen, und unter andern an einem, welches von dem Gedanken ausgeht: daß das Menschengeschlecht *immer und immer* in seiner Kultur und Verbesserung vorwärts schreite, u.s.w. Dieses hat besonders der französische Bürger Condorcet zu behaupten und zu beweisen gesucht […] Es fallen folglich alle analogischen Schlüsse weg, welche man von den alten Begebenheiten auf das machen kann, was unter unseren Augen vorgeht; denn wir sind mehr kultivirt, als man sonst war, haben mehr Gewandtheit der Kräfte u.s.w." (Laukhard 1796, 315–316).

M. Bullynck (✉)
Département de mathématiques et histoire des sciences, UFR 6 MITSIC & EA 1571 Centre de recherches historiques : Histoire des Pouvoirs, Savoirs et Sociétés Université Paris 8, 2 rue de la liberté, 93526 St-Denis cédex, France
e-mail: maarten.bullynck@univ-paris8.fr

V.R. Remmert et al. (eds.), *Historiography of Mathematics in the 19th and 20th Centuries*, Trends in the History of Science,
DOI 10.1007/978-3-319-39649-1_2

modes of writing as can be typically found in Pierre Bayle's *Dictionaire historique et critique* (1697), Christian Wolff's oeuvre or J.H. Zedler's *Universal-Lexikon* (1731–1754). These books tried to exhaust all knowledge accessible, either framed in a systematic and accumulative format (Wolff), or through an accretation of quotes old and new, larded with comments and references to books and voyages (Bayle, Zedler). Around 1800 this abundance of historical and geographical materials made way for a historiography that first made a selection of facts and figures and then assembled them into an account that tells of the systematic progression of ideas.[2]

1 Origins of the 'Progress of Mankind'-Narrative

1.1 Paris 1800

The first person to introduce convincingly this new historiographic format seems to have been the Marquis de Condorcet (1743–1794). In the introduction to his posthumously published *Esquisse d'un tableau historique des progrès de l'esprit humain* (1795) he wrote:

> we have to show, by reasoning and with facts, that there is no limit in the perfection of the human faculties; [...] that the progress of this perfection is now independent of every force that would try to stop it, and that it has no other limit than the duration that this world or nature gives us. Without any doubt, these progresses may follow a pace that is more or less fast, but it will never be retrograde.[3]

This idea of an infinite improvement resembles the classic religious eschatological scheme where mankind strives to perfection to ascend to God's kingdom, but in Condorcet's version it became secularised. Human reason and its main wings, science and technology, pushed mankind ever forwards to a better destination here on earth.

[2]It is interesting to note that a transition between genres of fiction occurred in parallel to this transition in historiography. Near the end of the 18th century the new literary genre 'Bildungsroman' appeared as a reaction to and transformation of the very popular mid-18th century genres of the picaresque novels, travel accounts and other forms of fiction that are essentially an accumulation of random events and experiences often held together by a loose narrative (Bakhtin 1986). The 'Bildungsroman' superimposes a teleological structure on the random sequence of events that the picaresque main character lives through. What at first sight appears a series of contingent occurrences, discloses itself at the end of the 'Bildungsroman' as a pre-arranged set of experiences that help to form (*bilden*) the hero's character, where the hero, as pars pro toto, stands for mankind.

[3]"il faut montrer, par le raisonnement et par les faits, qu'il n'a été marqué aucun terme au perfectionnement des facultés humaines; [...]; que les progrès de cette perfectibilité désormais indépendante de toute puissance qui voudroit les arrêter, n'ont d'autre terme que la durée du globe où la nature nous a jetés. Sans doute, ces progrès pourront suivre une marche plus ou moins rapide, mais jamais elle ne sera rétrograde" (marquis de Condorcet 1795, 4).

Condorcet's *Tableau historique* was certainly the most influential and most elaborate version of the idea of eternal progress of mankind. The *Tableau historique* pursued and altered earlier 18th century ideas on history.[4] Its most direct ancestor was a *discours* by Turgot (1727–1781) on the 'successive progresses of the human mind'. In order to rationalise and defend the religious scheme of progressing towards perfection, Turgot pointed out that science and the arts could bring about just that perfection. The sciences and technological innovations led by mathematics guaranteed that the chaos and arbitrariness of mankind's opinions and sentiments could finally be appeased to contribute to his progress. Mathematics stood out as a stable haven against a background of chaos and disorder and served as the 'torch' that helps discern truth from error. Reason, or *in casu* logic or mathematics, as a torch enlightening mankind was a stock element of Enlightenment discourse.

Whereas Turgot allowed for error and decadence and saw science mainly as the light that guides through the chaos of history, Condorcet consciously only painted the progresses of the human mind. Instead of the 18th century tendency towards encyclopaedic accretion, Condorcet sifted through the huge amount of information available, both diachronically (through books) and synchronically (through the travel accounts that were so popular in the 18th century). A new historiographic master narrative also implied a new methodology of lining up historical facts and figures:

> it is necessary to choose [the facts of history] among those of several peoples, to bring them together, to combine them so as to derive from them a hypothetical history of only one people and to form a table of its progresses.[5]

Condorcet advocated a selection and combination of facts of possibly chronological and/or topological heterogenous origin to assemble the "hypothetical" history of mankind and its progress.

This led to an epistemological periodisation of history.[6] Condorcet's subdivision of history into 10 epochs reflected the succession of ideas and innovations in technology and industry that drive mankind's progress. From tribes to agricultural settlements, from the invention of alphabetic writing to the invention of printing, from there to the Renaissance of the sciences and finally to the French Revolution.[7]

---

[4]A more complete overview of the roots of Condorcet's *Tableau historique* can be found in the critical edition of (Marquis de Condorcet 1795, 32–36).

[5]"il est nécessaire de choisir [les faits de l'histoire] dans celles de différens peuples, de les rapprocher, de les combiner, pour en tirer l'histoire hypothétique d'un peuple unique, et former le tableau de ses progrès" (Marquis de Condorcet 1795, 13).

[6]The term 'epistemological periodisation' is borrowed from the Condorcet (1795).

[7]Condorcet's epochs are: (1) Les hommes sont réunis en peuplades; (2) Les peuples pasteurs. Passage de cet état à celui des peuples agriculteurs; (3) Progrès des peuples agriculteurs jusqu'à l'invention de écriture alphabétique; (4) Progrès de l'esprit humain dans la Grèce jusqu'au temps de la division des sciences vers le siècle d'Alexandre; (5) Progrès des sciences depuis leur division jusqu'à leur décadence; (6) Décadence des lumières, jusqu'à leur restauration vers le temps des croisades; (7) Depuis les premiers progrès des sciences vers leur restauration dans l'Occident jusqu'à l'invention de l'imprimerie; (8) Depuis l'invention de l'imprimerie jusqu'au temps où les

In each epoch, mathematics is duly mentioned. The predilection of Socrates and Plato for mathematics that saved them from sophistry, the combination with observation that made ancient Greek mathematics so successful, the invention of logarithms and the renaissance of mathematics (mentioned immediately after the horrors of the religious wars of the 16th–17th centuries), the birth of algebra and analysis that made the age of reason possible.[8] In the discussion of the last epoch, the future, mathematics is invoked once more:

> The progress of science assures the progress of the art of teaching, and this in its turn accelerates the progress of science. This reciprocal influence [...] should be placed among the most active and powerful causes for the perfection of humankind. Today, a young man that leaves our schools knows more of mathematics than Newton had ever learned by profound studies or by genial discovery.[9]

Mathematics is also instrumental for the devices Condorcet proposes to accelerate progress in the future. These include a universal language, modeled after algebra, teaching of arithmetic in elementary schools or the application of mathematics for the organisation of a democratic state.[10]

Such an epistemological periodisation is also found in Bossut's contemporary history of mathematics, *Essai d'une histoire générale des mathématiques* (Bossut 1802). Charles Bossut (1730–1814) was a friend of Condorcet and they had both been involved in the redaction of the *Dictionnaire méthodique* in the 1780s. Instead of Montucla's (or Kästner's) 18th century histories of mathematics that are arranged according to the succession of the centuries, Bossut ordered his account after the succession of great ideas in mathematics: From the beginnings to the School of Alexandria, from the Arabs to the end of the 15th century (birth of symbolic algebra), from the 15th century to the invention of the calculus.[11] Instead of an absolute chronological timeframe, Condorcet and Bossut chose a relative timeframe

(Footnote 7 continued)
sciences et la philosophie secouèrent le joug de l'autorité; (9) Depuis Descartes jusqu'à la formation de la République Francoise; (10). Des progrès futurs de l'esprit humain.

[8]Pages 81–81; 102–107; 215–216; and 279–286 respectively.

[9]"Les progrès des sciences assurent les progrès de l'art d'instruire, qui eux-mêmes accélèrent ensuite ceux des sciences; et cette influence réciproque, dont l'action se renouvelle sans cesse, doit être placée au nombre des causes les plus actives, les plus puissantes du perfectionnement de l'espèce humaine. Aujourd'hui, un jeune homme, au sortir de nos écoles, sait en mathématiques, au-delà de ce que Newton avoit appris par de profondes études, ou découvert par son génie" (Marquis de Condorcet 1795, 372).

[10]See (Marquis de Condorcet 1795, 375–377), Condorcet, Condorcet's *Moyens d'apprendre à compter sûrement et avec facilité* (1800) and his *Tableau général de la science qui a pour objet l'application du calcul aux sciences politiques et morales* (1793) respectively. More generally on the role of mathematics in the plans of Turgot or Condorcet to organise and improve society in Brian (1994).

[11]The chapters in Bossut (1802) are: Etat des mathématiques depuis leur origine jusqu'à la destruction de l école d'Alexandrie; Etat des mathématiques depuis leur renouveau chez les Arabes jusque vers la fin du XVe siècle; Progrès des mathématiques depuis la fin du XVe siècle jusqu'à l'invention de l'Analise; Progrès des mathématiques depuis la découverte de l'Analise infinitésimale jusqu'à nos jours. Compare also with Novy (1996).

that follows and fits the internal logic of development of a field of knowledge. As a consequence, periods of decadence or lesser activity were mostly left out of the narrative. A further consequence was the disappearance of parallel narratives where, e.g., India and China would feature alongside of Europe, in favour of one master narrative that turns the rest into secondary or inferior plot developments.[12]

The new historiographic format would quickly become popular in many adaptations and versions. In France, people such as Lazare Carnot (1753–1823) or later Auguste Comte (1798–1857) with his positivism and corresponding philosophy of history would develop Condorcet's historiographic format further during the 19th century.

1.2 Berlin 1800

In the German-speaking states, Condorcet's *Tableau* was translated by E.L. Posselt (1763–1804) as early as 1796. The philosopher J.G. Fichte (1762–1814) provided one of the earliest and most influential adaptations of Condorcet's vision on history. In his *Grundzüge des gegenwärtigen Zeitalters* from 1804, Fichte wrote:

> The life of the human kind does not depend on blind arbitrariness, nor is it […] everywhere similar to itself […] but it goes along and it progresses always according to a fixed plan that *must* inevitably be realised. […] This plan is the following: The human kind has to form itself in this life with freedom into the pure image of reason. Their common life divises into five main epochs: a first, where reason rules as a blind instinct; a second where this instinct is transformed in an externally commanding authority; a third one where the rule of this authority, and therefore reason itself, is destroyed; a fourth one where reason and its laws are understood clearly and consciously; a fifth and final one where all relationships of the human kind will be ruled and ordered following those laws of reason.[13]

Fichte considered his own times as part of the third epoch in transition to the fourth one. Again, it is human reason that fronted mankind's progress and technology that played an important role. The invention of writing is capital for the second epoch, the invention of printing for the third, and general alphabetisation is, according to Fichte, one of the main instruments to pass over to the fourth epoch. Fichte also sharply criticised 18th century formats where the sciences were drawn hither and tither "by the blind tendency of the association between ideas" and where

---

[12]18th century European historiography often saw Asia as an 'equal partner' or at the very least as a valid point of comparison, see Osterhammel (1998).

[13]"Das Leben der menschlichen Gattung hängt nicht ab vom blinden Ohngefähr, noch ist es […] sich selbst allenthalben gleich […] sondern es geht einher und rückt vorwärts immer nach einem festen Plane, der nothwendig erreicht werden *muss*. […] Dieser Plan ist der: dass die Gattung in diesem Leben mit Freiheit sich zum reinen Abdruck der Vernunft ausbilde. Ihr gesammtes Leben zertheilt sich in fünf Hauptepochen: diejenige, da die Vernunft als blinder Instinct herrscht; diejenige, da dieser Instinct in eine äusserlich gebietende Autorität verwandelt wird; diejenige, da die Herrschaft dieser Autorität, und mit ihr der Vernunft selber zerstört wird: diejenige, da die Vernunft und ihre Gesetze mit klarem Bewusstseyn begriffen werden: endlich diejenige, da durch fertige Kunst alle Verhältnisse der Gattung nach jenen Gesetzen der Vernunft gerichtet und geordnet werden" (Fichte 1846a, 17).

the master idea for representing facts and figures was "after the sequence of letters in the alphabet", i.e. encyclopaedic or dictionary-like accretion.[14] Again, Antiquity's mathematics gets quoted to prove that the sciences need not be such sundries.

Although Fichte's scheme had much in common with Condorcet's (the idea of progress, the role of reason and technology), it was a decidedly more abstract blueprint that also integrated a new dialectical mechanism that both endangered and empowered the idea of progress. The third epoch constituted that moment of dialectic articulation where reason as the exterior force from the second epoch is destroyed to make way for the interiorisation of reason in the fourth epoch. This rather abstract idea may be concretized by an example. In the follow-up to the *Grundzüge*, the *Reden an die deutsche Nation* (1808), Fichte wrote about the sciences that were fixated in a dead (Latin) or foreign (French) language and therefore were but a "riven collection of arbitrary and inexplicable signs of equally arbitrary ideas" that one could only memorise but not develop.[15] If German would be adopted as a scientific language, however, it would be a living language where each sign was alive and sensible as part of language, culture and life, both in the past and in the present.[16] Whereas reason used to be written and expressed in an external language, it should now be transmuted in the language of a nation's inner life. Instead of following and even writing in foreign style and language, German science should develop its own German language and thereby overcome its enslavement to foreign manners. This is instance of the dialectic third epoch where reason is destroyed to be newly created again in a fourth epoch. The sciences old style should be replaced by sciences new style.

The dynamics of the words 'Begriff' (or 'Gedanken') and 'Zeichen' (or 'Symbol' or 'Sinnbild') is quite capital in Fichte's discourse. 'Begriff' stands for the living idea, but the 'Zeichen' (sign) that ports and communicates the living idea between people has an ambivalent role. It may either obscure the idea and make it inexplicable as part of arbitrary combinations, or it may clarify the essence of the idea by showing its embedding in a real live discourse. In our example, it may be expressed clumsily in a foreign language or clearly revelated in your mother tongue. In the first case, it would cause reason to have a fallback to the second epoch, in the second, it would progress to a fourth epoch. It is the task of philosophy to guide the sciences in their communication of ideas.

[14]"In Absicht seiner Meinungen über diese Gegenstände wird es durch den blinden Hang der Ideenassociation bald dahin bald dorthin gezogen werden […] Ein Meisterfund für die Darstellung eines solchen Zeitalters wäre es, wenn es darauf geriethe, die Wissenschaften nach der Folge der Buchstaben im Alphabete vorzutragen" (Fichte 1846a).

[15]"zerrissenen Sammlung willkürlicher und durchaus nicht weiter zu erklärender Zeichen ebenso willkürlicher Begriffe" (Fichte 1846b, 325).

[16]"Dieser übersinnliche Theil ist in einer immerfort lebendig gebliebenen Sprache sinnbildlich, zusammenfassend bei jedem Schritte das Ganze des sinnlichen und geistigen, in der Sprache niedergelegten Lebens der Nation in vollendeter Einheit, um einen, ebenfalls nicht willkürlichen, sondern aus dem ganzen bisherigen Leben der Nation nothwendig hervorgehenden Begriff zu bezeichnen, aus welchem, und seiner Bezeichnung, ein scharfes Auge die ganze Bildungsgeschichte der Nation rückwärtsschreitend wieder müsste herstellen können" (Fichte 1846b, 325).

Fichte's historiographic scheme and its dialectic moment would have a lasting influence on the philosophy of history in early 19th century Prussia, especially at the newly founded university of Berlin. For philology we could quote August Boeckh (1785–1867) who in his study of Antiquity saw mythology as a symbolisation of exterior ideas that gets replaced by an epoch of art that symbolises the inner ideas.[17] More important to our topic is G.W.F. Hegel (1770–1831) who turned the dialectic moment into a general principle that drives history itself.

Already in the introduction to the *Phänomenologie des Geistes* (1807) but more explicitly in his *Enzyklopädie der Wissenschaften* (1817) Hegel wrote against a mathematisation of philosophy, or against a primacy of mathematics over philosophy.[18] Again, an 'external' language, mathematics, is regarded as an unsuitable medium for communicating philosophical ideas, because the arbitrariness with which mathematical symbols can be combined disrespects the precise determination of an idea. In his polemic, Hegel took up the same dichotomic pair *Gedanke-Symbol* (idea/concept—symbol/figure):

> It would further be a superfluous and ungrateful effort to use such a recalcitrant and inadequate medium as are spatial figures and numbers for the expression of ideas […]. The first simple figures and numbers can adequately and without misunderstandings be used as symbols because of their simplicity, but they remain a heterogenous and beggarly expression for the idea. The first attempts at pure thought have used them as a makeshift, the Pythagorean number system is its most famous example. But for many concepts, this means is completely unsatisfactory, because the external combination and the arbitrariness of the concatenation of these symbols is inadequate for the nature of a concept, also, it becomes totally ambiguous which of the many relationships that are possible between combinations of numbers and figures should be apprehended. Furthermore, the fluidity of a concept evaporates in such an external medium where every determination dissolves in indifference. This ambiguity can only be lifted by explanation. The essential expression of an idea is this explanation, and such a symbolisation thus only an unsubstantial exuberance.[19]

---

[17]"Man könnte vielleicht sagen, die Kunst als Symbolisirung der Ideen sei später zu betrachten als die Mythologie, weil jene das Aeussere, diese das zu Grunde liegende Innere darstelle" (Boeckh 1877, 62).

[18]The mathematisation of philosophy had been very popular during the 18th century, typical exponents of this trend were Christian Wolff (1679–1754) or J.H. Lambert (1728–1777), see, e.g., Arndt (1971).

[19]"Es würde ferner eine überflüssige und undankbare Mühe sein, für den Ausdruck der Gedanken ein solches widerspenstiges und inadäquates Medium, als Raumfiguren und Zahlen sind, gebrauchen zu wollen [und dieselben gewaltsam zu diesem Behufe zu behandeln]. Die einfachen ersten Figuren und Zahlen eignen sich ihrer Einfachheit wegen ohne Missverstandnisse zu Symbolen, die jedoch immer für den Gedanken ein heterogener und kümmerlicher Ausdruck sind, angewendet zu werden. Die ersten Versuche des reinen Denkens haben zu diesem Nothbehelfe gegriffen; das pythagoreische Zahlensystem ist das berühmte Beispiel davon. Aber bei manchen Begriffen werden diese Mittel völlig ungenügend, da deren äusserliche Zusammensetzung und die Zufälligkeit der Verknüpfung überhaupt der Natur des Begriffs unangemessen ist, und es völlig zweideutig macht, welche der vielen Beziehungen, die an zusammengesetzte[n] Zahlen und Figuren möglich sind, festgehalten werden sollen. Ohnehin verfliegt das Flüssige des Begriffs in solchem äusserlichen Medium, worin jede Bestimmung in das gleichgültige Aussereinander fällt. Jene Zweideutigkeit könnte allein durch die Erklärung gehoben werden. Der wesentliche Ausdruck des Gedankens ist alsdann diese Erklärung, und jenes Symbolisiren ein gehaltloser Ueberfluss" (Hegel 1845, § 259).

Mathematics is thus surely not suited for expressing philosophical ideas, but worse still, according to Hegel, mathematics, by its use of symbols and signs, is not even adequate to express the essence of time and space. Therefore Hegel advocated a new mathematics, a philosophical mathematics. Signs might still be useful for calculating, but for comprehension and explication, philosophy would be needed.

2 A Case of Epistemological History: Alexander Von Humboldt's Project on Numeration Systems

Adapting the history of mathematics to the new format of a progression of ideas implied, as was clear in Bossut's *Essai*, that the facts of mathematics' history should be rearranged according to the inner logic of mathematics as a discipline. They should line up to provide a self-consistent development of fundamental ideas that made up 19th century mathematics. The historical genesis of mathematics should ideally repeat or foreshadow the individual's learning curve in mathematics, the order of learning mathematics anno 1800. Such a narrative sequences key moments in the history of mathematics in order of ascending difficulty as does a school curriculum. It could start with Greek mathematics (Euclid's *Elements*, geometry and elementary arithmetic), than pursue with symbolic algebra that dates back to the early 17th century and go on with the calculus that was invented in the late 17th century. One key invention, however, resisted the format: the origin of the decimal place-value notation for numbers. Whereas in the order of things mathematical, it should come before the others, it historically only appeared in Europe during the 8th to 13th century.

As noted in the famous *Rapport historique sur les progrès des sciences mathématiques* (1810) even Napoleon himself had pointed out that the fact that the ancient Greek used another numeral system than the modern one was "une lacune très remarquable dans l'histoire des mathématiques" (Delambre 1810, 37). The notation system for numbers did fit neither chronologically nor geographically into a nice derivation from Greek Antiquity to mathematics anno 1800. This spurred two new kinds of investigation of our numerals' origin. A first strand focussed upon the (Greek) practice of doing arithmetic, here we may quote J.B. Delambre's (1749–1822) "De l'arithmétique des Grecs" (1807) or Reimer's (1772–1832) additions in his translation of Bossut's *Essai* (1804). A second strand was nurtured by the influx of new information on various (oriental) languages and the emergence of modern philology.

In a project that was never completed, Alexander von Humboldt (1769–1859) tried to combine both approaches to get at the intellectual origins of the decimal place-value system. Both by his being in between France and Prussia and his being in between generations of ethnologists and philologists, his project documents changing methodologies in the beginning of the 19th century.[20] Alexander von Humboldt most literally concretised Condorcet's methodology to "choose, join, combine all facts of history and of different people to make a hypothetical history" (as Condorcet wrote) in his studies on our modern numeral system.

2.1 Hypothetical History

The very beginning of A. v. Humboldt's interest in numeral systems seems to have been of ethnographic origin. In his *Monumens des peuples indigènes de l'Amérique* (1816, 1824) Humboldt described a numeral system developed by the Muyscas people. They used words for numbers that have the same roots as words that indicate the phases of the moon. According to Humboldt, "this would be one of the most remarkable facts in the philosophical history of language", because in all other languages the roots for numerals are completely independent of roots for words that express objects from the physical world (von Humboldt 1824, 241–242). Indeed, in 18th century linguistics, it was thought that all words stood in one-to-one correspondence with real world sensations, except for function words such as pronouns, articles, numerals etc. It was, however, thought that all function words did descend from object words but that their original meaning and expression had faded over time. For most of the function words, etymological reconstructions were found or proposed, but the numerals proved to be the hardest case.

Between 1808 and 1827, when A. von Humboldt lived in Paris, his initial ethnographic interest slowly grew into a rather panoramic project to both describe the variety of numerals used throughout the world and derive the origins of our positional decimal system. It remained a project but he described the main lines of his project in two lectures: One at the Académie royale des sciences in Paris, 20th September 1819 (von Humboldt 1819) and one at the Akademie der Wissenschaften zu Berlin, 2nd March 1829 (von Humboldt 1829). These texts lay bare a hub of communication between France, Germany and England on matters mathematical and philological.

The 'natural milieu' in Paris for Humboldt's study on numerals was the circle of researchers that would found the Société Asiatique in 1822 (Silvestre de Sacy, Abel-Rémusat and others). This was complemented by his own network in the German states and by close communication with his brother Wilhelm in Berlin who was also interested in the project because of his interest in Sanscrit and the Kawi-language. These networks are representative of a more general trend, that of the renewal of the studies of language at the beginning of the 19th century. Silvestre de Sacy (1758–1838) had been appointed as professor of Persian at the Collège de France in 1806. In 1814 August Boeckh had founded the *Philologisches Seminar* at the Berlin university that would renew classical philology. In 1816 Franz Bopp

[20]An interdisciplary research group has recently been mounted at SPHERE (Paris) by M. Bullynck, A. Keller, I. Smadja and others to study the genesis and influence of A. von Humboldt's project.

(1791–1867) had published his *Über das Conjugationssystem der Sanskritsprache* that would found comparative linguistics. Also in 1816 Wilhelm von Humboldt (1767–1835) had secured Julius Klaproth (1783–1835) a pension from the Prussian state to pursue his Asiatic studies in Paris where Jean-Pierre Abel-Rémusat (1788–1832) held the newly founded chair of Chinese at the Collège de France.

Relying on those networks, A. von Humboldt managed to gather a variety of material on Arabic, Chinese, Indian, Ancient Egyptian, medieval Latin etc. use of numerals. He used these materials to derive an 'epistemological' reconstruction of the genesis of the decimal place-value system. The aim was to show how the Hindu-arabic notation could have been developed:

> I cannot historically develop that the origin of the Indian place-value system with 9 ciphers is really the one that I have indicated, but I believe I found a way in which the discovery could gradually have been made.[21]

To obtain this genesis of our numeral system, Humboldt proposed to abstract from the actual appearance of numeral signs and focus on the structural properties:

> In the studies of numerical signs, one has been occupied so far more with the characteristic physiognomy of the signs and their individual forms than with the idea of the methods [...] I have made it myself a rule in this article to use no other signs than our usual arithmetic and algebraic ones. In this way, the attention is focused more on the essence of things, on the idea of the method.[22]

Individual symbols stood in the way of the general ideas that developed through history, therefore they had to be transformed to see the inner structure of methods—by transcribing everything into modern algebraic and numerical signs. By destroying the original symbols with modern signs, the idea itself should become visible.

In Humboldt's hypothetical history of the Hindu-arabic numerals, as explained in the 1819 and 1829 lectures, there were basically three phases. A first epoch was that of juxtaposition, repetition of simple signs (group signs) such as dashes: III. The group signs might evolve and be replaced and/or supplemented by a fixed sequence of arbitrary signs such as the sequence of the letters of the alphabet or another series of conventional signs (such as our 1,2,3, ...). A second epoch advanced to the possible simplification of these group signs using an exponent to indicate repetition of the (group) sign, I^3. The 'diacritical' sign might be under (index) or over (exponent) the group sign or it might appear left of the group sign, as a coefficient, $3I$. In cases, the group sign might appear under or over the

[21] "Ich kann nicht historisch entwickeln, dass der Ursprung des indischen Stellenwerthes von 9 Ziffern wirklich der sei, welchen ich angegeben, aber ich glaube einen Weg gefunden zu haben, auf welchen allmählich die Entdeckung gemacht werden konnte" (von Humboldt 1829, 207).

[22] "Man hat sich bisher, in den Untersuchungen über die numerischen Zeichen [...] ernster mit der characteristischen Physiognomik der Zeichen und ihrer individuellen Gestaltung, als mit dem Geist der Methoden beschäftigt [...] Ich habe es mir in dieser Abhandlung zum Gesetz gemacht, keine anderen Zeichen, als die gewöhnlichsten arithmetischen und algebraischen zu gebrauchen. Die Aufmerksamkeit wird auf diese Weise mehr auf das Wesentliche, auf den Geist der Methode, gerichtet" (von Humboldt 1829, 204–214).

diacritical sign, 3^I reversing the positions of the group sign and the repetition sign. This last reversal was a critical transition point in Humboldt's reconstruction since it provided the link between juxtaposing numeral system (as the Chinese system) to a positional system such as the Indian number system. A final epoch dropped the (group) signs and kept only the exponents. Adding a 'zero' closed the deal.

2.2 Philologies

Three clusters of material were important 'cornerstones' for building Humboldt's argument (cf. Table 1). First, the discovery of the gobar numerals in the North of Africa by Silvestre de Sacy for the transition from I^3 to 3^I. Second, the book *The philosophy of arithmetic* (1817) by John Leslie (1766–1832), who basically compared the words and signs used to count in many cultures and introduced a distinction between palpable and figurative arithmetic. Palpable arithmetic meant doing arithmetic using moveable objects (from hands to calculating instruments), figurative arithmetics is characterised by the use of symbols to denote numbers and to operate on them. And finally, Henry Thomas Colebrooke's (1765–1837) work on Indian mathematics (that Alexander got via mediation of comparative philologist Franz Bopp) and his brother Wilhelm's studies on Sanscrit provided ample new information on numeral systems on the Indian subcontinent. These materials were key to Humboldt's history of transmission and transformation of numeral signs, providing the cultural substratum (either linguistic or material) that could accomodate for the transformations of numerals throughout their history (say, explain how it was transmitted and how it was transformed here rather than there).

The first question was of course: Where did the first impetus towards our numeral system come from before it got its present form in India? This was the most speculative aspect of Humboldt's paper and drew extensively on Leslie's differentiation between palpable and figurative arithmetic. Humboldt defended the idea that the Chinese suanpan (our abacus) might be at the origin of our number system.

> The usage of the suanpan made the people familiar with the idea of many ranks of groups; they showed an empty place (sifroun) where an intermediary group was lacking. The Chinese artifice that placed the unities as multiplicators over the group signs probably completed the discovery. It transplanted so to say the germ of the indian method from the domain of palpable arithmetic into the domain of figurative or graphical arithmetic.[23]

In a letter to Franz Bopp, A. v. Humboldt called this 'proof' of the Chinese origin of the Hindu-arabic numerals the main result of his 1819 paper.

---

[23]"L'usage du suanpan accoutumait les peuples à l'idée de plusieurs rangs de groupes; ils montraient un place vide (un sifroun) là où manquait un groupe intermédiaire. L'artifice chinois de placer des unités comme multiplicateurs au-dessus des signes de groupes acheva probablement la découverte. Il transplanta, pour ainsi dire, le germe de la méthode indienne du domaine de l'arithmétique palpable dans le domaine de l'arithmétique figurative ou graphique." (von Humboldt 1819, 100).

Table 1 Overview of some of A.v. Humboldt's sources with indication of the year(s) of the most important publication(s) and/or interaction(s)

Source	Year	Topic
A. v. Humboldt	1810	Muyscas-system
S. de Sacy	1810	Arabic gobar-notation
W. v. Humboldt	1810–1835	South-Indian and Malay numerals
A. Boeckh	1811–...	Greece and Rome
J. Leslie	1817	Palpable vs. figurative arithmetic
Th. Young	1814–1823	Egyptian numerals
J. Klaproth	1815–1835	Asiatic languages
H. Th. Colebrooke	1817	Indian numerals
Franz Bopp	1819–20	Indo-Germanic numerals
J.-F. Champollion	1822–1832	Egyptian numerals
Karl Otfried Müller	1820–1828	Rome and Etruscan
J.-P. Abel-Rémusat	1825	Chinese

A second development in the evolution and transmission of decimal positional number systems that Humboldt wanted to explain is: Why did it emerge in Sanscrit, in an Indian context? Humboldt had learned from Colebrooke that the Indians do and did not use only one system of notation, they used a variety of systems in parallel, both in writing and in speech.[24] They used a set of words for each numeral to be juxtaposed according to the metre, they used the same words but with position (without zero) also to fit in a metre; they used syllables for ciphers, indian ciphers, both from left to right as well as right to left, etc. Looking at the (non-indo-european) languages spoken on the Indian subcontinent, that his brother Wilhelm von Humboldt was studying, the variety became even more overwhelming. According to Humboldt, this was fertile ground once the Chinese way of notating numbers arrived in the Indian subcontinent. The rich variety of notations that were all linear and sequential proved the perfect substratum in which the new notation could settle, find its expression and slowly fixate the numeral system that would be ours. One thing changed, the direction of writing: "the order that had been established in a perpendicular writing must have been conservated in horizontal writing".[25]

The final transmission to be accounted for was the one between the Arabic world and the Indian world. Here, the gobar or dust writing provided the missing link. Silvestre de Sacy had signalled the gobar numerals in his *Grammaire arabe à l'usage des élèves de l'école spéciale des langues orientales vivantes* (1810) in a footnote. Borrowing the idea of diacritical marks often used in semitic languages to indicate vowels in an otherwise consonantic writing system, this notation used *n*

[24]For a modern overview, see Singh (1997).

[25]"l'ordre établi dans l'écriture perpendiculaire a dû être conservé dans l'écriture horizontale" (von Humboldt 1819, 101).

points above a cipher to indicate that the cipher should be multiplied with ten to the n-th power. According to Alexander von Humboldt's interpretation the gobar notation, used by the Mauretian customs, might be a variant of our numeral system with zero, though still displaying some older stage of its development. It was evidence of that transitional point where multiples of a number are expressed by marks placed above or under the signs and thus for the slow process of transmission between the Indian subcontinent and the Arabic world.

3 Beginnings of Modern History of Mathematics

From the year 1819 onwards the initial speculative enthusiasm present in A. v. Humboldt's first version of his 'Zahlzeichen'-project gradually dimmed. In the 1829 incarnation for Crelle's Journal and once again in its last appearance, as an endnote in Part 2 of the *Kosmos* (1847), the focus shifted. These shifts correspond to disciplinary evolutions during those years.

The 1820s were years of *rétablissement* in the German states after the wars with Napoleon. Not only the new universities of Berlin and Bonn were founded, but new professorships and new disciplines altogether were created. For instance, classical philology and Sanscrit studies began to flourish in the German states and this had an impact on the history of mathematics too. Alexander's brother Wilhelm had been working in the 1820s on his opus magnum *Ueber die Kawi-Sprache auf der Insel Java* (4 volumes, written 1830–1833, publ. 1836–1838) collecting many new sources on Sanscrit. Franz Bopp was working on his *Vergleichende Grammatik* (6 volumes, 1833-1852) of which a capital part was the comparison of the numerals in different languages. At Berlin university Boeckh himself had edited the texts of Philolaos (1819) and had published on ancient metrology (1817, 1838) and chronology (1855–6; 1862).[26] A number of his students, such as J.E. Nizze (1788–1872) and his colleague C.F.J. Hasenbalg would embark on historical critical editions of Greek mathematicians (Theodosius (1826), Hero (1826)). Others such as K.O. Müller (1797–1840) would study the Minean, Doric and Etruscan cultures (1820–1828). This influx of material from the Indo-European realm refocused A. v. Humboldt's project. It now featured lots of information on Sanscrit and other languages from the Indian subcontinent[27] and as a consequence the idea of palpable arithmetic and of Chinese influences disappeared from the main argument, though it was still mentioned. The main result of the 1829 paper was rather a summary of philologic findings and in this spirit Humboldt closed the paper with the wish that the philologists might find and exploit more material in a near future.

[26]On Humboldt and Boeckh, see also Knobloch (2011).

[27]The explanation is only briefly mentioned in 1819 (p. 101), but given more space in 1829 (pp. 212–213, 215–216, 219, 226–228).

In 1847 Humboldt's project had shrunk to a mere two-page endnote in his opus magnum *Kosmos*. By that time, a new generation of researchers, both competent in the new philologies and in mathematics, had begun to study historical sources of mathematics. Friedrich August Rosen (1805–1837), a student of Franz Bopp, went to work in London to edit Sanscrit sources and translated Al-Khwarizmi's *Algebra* in 1831. Franz Woepcke (1826–1864), another student of Berlin university, got a stipend through A. v. Humboldt to go work in Paris where he published and translated a variety of Arabic works on mathematics (1851–1863). G.H.F. Nesselmann (1811–1881), a student of C.G.J. Jacobi (1804–1851, himself once a student of Hegel and Boeckh in Berlin) wrote a *Versuch einer kritischen Geschichte der Algebra* (1842) that analysed the original texts and posited three phases in the history of algebra.[28] This had made Humboldt's own project nearly redundant. As (pure) mathematics had emancipated itself in the beginning of the 19th century to become a blossoming discipline in the 1850s, so it began to become self-conscious too. Using the new philologies that had matured in parallel, mathematics started to write its own histories.

References

Arndt, H. W. (1971). *Methodo scientifica pertractatum. Mos geometricus und Kalkülbegriff in der philosophischen Theorienbildung des 17. und 18. Jahrhunderts*. Berlin: De Gruyter.

Bakhtin, M. (1938/1986). The Bildungsroman and its significance in the history of realism. In C. Emerson & M. Holquist (Eds.), *Speech genres and other late essays* (pp. 10–59). Austin: University of Texas Press.

Boeckh, A. (1809/1877). *Encyclopädie und Methodologie der philologischen Wissenschaften*. In E. Bratuschek (Ed.), Teubner, Leipzig.

Bossut, C. (1802). *Essai d'une histoire générale des mathématiques*. Louis, Paris. German translation: N. Th. Reimer. 1804. *Carl Bossut's Versuch einer allgemeinen Geschichte der Mathematik*, 2 vols. Hamburg: Hoffman.

Brian, E. (1994). *La mesure de l'Etat. Administrateurs et géomètres au XVIIIe siècle*. Paris: Albin Michel.

Delambre, J. (1810). *Rapport historique sur le progrès des sciences mathématiques depuis 1799*. Paris: Imprimerie impériale.

Fichte, J. (1804(1845–46)a). Grundzüge des gegenwärtigen Zeitalters. In I. Fichte (Ed.), *Johann Gottlieb Fichtes sämmtliche Werke* (Vol. 7, pp. 3–256). Berlin: Veit und comp.

Fichte, J. (1808(1845–46)b). Reden an die deutsche Nation. In I. Fichte (Ed.), *Johann Gottlieb Fichtes sämmtliche Werke* (Vol. 7, pp. 257–502). Berlin: Veit und comp.

Hegel, G. W. F. (1817(1832–1845)). Enzyklopädie der philosophischen Wissenschaften im Grundrisse. In Hegel, G. W. F. (ed. by P. Marheineke et al., Vol. 8 and 9). Berlin: Duncker und Humblot 1832–1845.

[28]The first one is a rhetorical stage where everyday words are used for stating and solving a problem, the second stage is a syncopated stage of abbreviating words, and a final stage would be the symbolic stage that manipulates its elements with the utmost economy. This three-phase scheme has a close connection with Humboldt 1829-article, compare pp. 62–64 with p. 302 in Nesselmann (1842). More on these authors that mainly worked on Arabic sources can be found in Sonja Brentjes'contribution to this volume.

Humboldt, A. von (1816/1824). *Vues des Cordillères, et Monumes des Peuples indigènes de l'Amérique* (Vol. I & II). Librairie Grecque-Latine-Allemande: Paris.
Humboldt, A. von (1819). Considérations générales sur les signes numériques des peuples. *Annales de chymie et physique*, *12*, 93–101. (German translation in L. Oken's magazine *Isis*, IV, 190–195).
Humboldt, A. von (1829). Über die bei verschiedenen Völkern üblichen Systeme von Zahlzeichen und über den Ursprung des Stellenwerthes in den indischen Zahlen. *Journal für die reine und angewandte Mathematik*, *4*, 205–231. (French translation by Fr. Woepcke: De Humboldt A. 1851. "Des systèmes de chiffres en usage chez différents peuples, et de l'origine de la valeur de position des chiffres indiens.", *Nouvelles Annales Mathématiques*, *10*, 372–407).
Humboldt, A. von (1847). *Kosmos, Entwurf einer physischen Weltbeschreibung* (Vol. II). Cotta: Stuttgart.
Knobloch, E. (2011). Alexander von Humboldt und August Böckh über Zahlzeichen und Stellenwertsystem. In R. Werther (Eds.), *Alexander von Humboldt/August Böckh, Briefwechsel* (pp. 30–35). Berlin: Akademie-Verlag.
Laukhard, F. (1796). *Leben und Schicksale* (part II, vol. 3). Fleischer: Leipzig.
Marquis de Condorcet, J. (1795). *Esquisse d'un tableau historique des progrès de l'esprit humain.* Agasse, Paris. New and commented work edition: 2004. *Condorcet, Tableau historique de l'esprit humain. Projets, Esquisse, Fragments et Notes (1772-1794)*, ed. by the Groupe Condorcet. Paris: Ined.
Nesselmann, G. H. F. (1842). *Versuch einer kritischen Geschichte der Algebra*. Berlin: Reimer.
Novy, N. (1996). Les étapes ou les paradigmes des mathématiques? In E. Ausejo & M. Hormigón (Eds.), *Paradigms and mathematics* (pp. 148–168). Madrid: Siglo Veintiuno Editores.
Osterhammel, J. (1998). *Die Entzauberung Asiens: Europa und die asiatischen Reiche im 18. Jahrhundert*. München: C.H. Beck.
Singh, C. (1997). Histoire de la numération et de l'arithmétique indiennes des origines au douzième siècle. In *Actes du colloque 'L'océan Indien au carrefour des mathématiques arabes, chinoises, européennes et indiennes* (pp. 179–209). La Réunion: IUFM de la Réunion.

Practicing History of Mathematics in Islamicate Societies in 19th-Century Germany and France

Sonja Brentjes

Abstract

This paper discusses methodological and interpretive aspects of practices in the history of the mathematical sciences in Islamicate societies as they emerged in Germany and France during the nineteenth century. It argues that in the nineteenth century, those who practiced history of mathematics in Islamicate societies had a strong methodological commitment. They formulated three main research lines with clear methodological claims. Two of these approaches (a scientific history of mathematics and a serious investigation of primary sources) found general approval in history of mathematics at large. Thus, they continued to be followed in the historiographical and methodical practices during the twentieth century. The third (the integration of progress and source studies into a cultural and biographical narrative) was discarded as a methodological principle. Only under the impact of discussions in history of science and the humanities since the 1980/90s did approaches similar to, and at the same time more sophisticated than, this forgotten third way practiced in the late nineteenth century find new practitioners with a new methodological consciousness.

Keywords

Islamicate societies · History of mathematics · Germany · France · Nineteenth century · Methodological commitment

S. Brentjes (✉)
Max Planck Institut für Wissenschaftsgeschichte, Boltzmannstr. 22, 14195 Berlin, Germany
e-mail: brentjes@mpiwg-berlin.mpg.de

V.R. Remmert et al. (eds.), *Historiography of Mathematics in the 19th and 20th Centuries*, Trends in the History of Science,
DOI 10.1007/978-3-319-39649-1_3

1 Introduction

Interest in Arabic, Persian, and at times also Ottoman Turkish texts of the mathematical sciences existed in different parts of Europe since the Middle Ages. Until the nineteenth century, this interest was mostly of a contemporary scientific, but not of a historical nature. Hence, I will not consider these older activities in this paper. I will limit myself to discussing authors, their texts, and in a more limited manner their contexts that contributed to the formation of research on the history of these sciences in a variety of Islamicate societies. One comment needs to be made nonetheless on those older activities. Early on, the discussions of the value of mostly Latin translations of Arabic texts and of scholarly activities in the Ottoman Empire shaped attitudes and perspectives of first Christian and later also secular scholars in Europe to the intellectual achievements in Islamicate societies as a whole. The values that these discussions promoted were prevalently negative, even when respect was paid to the existence of numerous Arabic translations of Greek mathematical works. Latin translations of Arabic scientific, mathematical, medical, and philosophical texts were denounced from the 14th century onwards for their cumbersome Latin and their Arabisms. Arabic scholarly achievements were appreciated in the 16th and the first half of the 17th centuries, while Ottoman scholarship was talked of in derogatory terms. During the second half of the 17th century negative evaluations of Arabic scholarship increased, partly because expectations and attitudes changed as a result of the new forms of scholarly enterprise that developed in this period in parts of Europe and partly because some of the former enthusiasts of Oriental studies became disenchanted with their fruits of reading, traveling, and language learning.

As a consequence of these two strands and further political and cultural changes in Europe, North Africa, and western Asia two main positions emerged that affected thinking about the mathematical sciences in Islamicate societies profoundly and for a long time. A cultural and a temporal preference emerged that located worthwhile studies of past mathematical sciences in Arabic texts that were either translations from other languages, mostly ancient Greek, or new compositions executed between the later 8th and the early 13th centuries. Among those preferred carriers of knowledge the subset of texts that preserved ancient Greek works was particularly cherished. The thesis of independent new achievements by scholars from regions ruled by the Abbasid dynasty and its representatives like the Buyids and the Seljuqs, while formulated in clear manner already in the 1830s and supported by various writers in the course of that century, became generally acknowledged, supported, and respected by historians of mathematics only in the last third of the twentieth century. The reasons for this profound shift in evaluation and attitude lie in the rapid increase of professionalism in the second half of the 20th century, in cultural anti-colonial politics in India, Iran, Turkey, and the Arab world after WW I or WW II, as well as in immense migrations of people from western Asia and North Africa to Europe, Canada, and the US and their demands for cultural recognition and integration. Historians of science have shown a greater reluctance to integrate scholars from Islamicate societies and their activities and results into their canonical views of their profession. This applies in particular to the development of this field since the 1950s.

The second position that dominated evaluations and attitudes throughout the 19th century looked at such Arabic contributions from their potential to mediate between ancient Greece and medieval Latin Europe. This sandwiched position continues to be subscribed to by historians of science of other periods, regions, and cultures, in particular those who work on the Renaissance and the early modern period in parts of central and western Europe. Although historians of the mathematical sciences in Islamicate societies turned their back to it in the course of the second half of the 20th century, its impact on research orientations remains clearly visible. There was and is very little research on past mathematical sciences and their practitioners in regions like eastern Iran, Muslim India, or Sub-Saharan Africa. The same phenomenon is expressed by the dearth of studies on relationships between mathematical cultures in different Islamicate societies and their eastern or southern neighbors in Asia and Africa, when compared to such research on Graeco-Arabica or Arabo-Latina.

2 On Some of the Main Contextual Components

The contexts of new studies, historiographical positions, and interpretive stances with regard to the mathematical sciences in Islamicate societies during the nineteenth century have to be looked for in political, religious, social, and academic conflicts and changes. The specifics of each of these types of context differ between the three countries where the men came from who provided financial and intellectual support for these studies—Germany, France, and Italy.

In Germany, the Napoleonic wars and their impact on German boundaries, reforms, and political aspirations have to be seen as a major source of change. One of its consequences was Wilhelm von Humboldt's (1767–1835) reform of Prussian universities and the preference it gave to research (Dauben and Scriba 2002, 114). One outcome of this changed orientation of the university profile and the professional duties of university professors and lecturers was the rise of what was called pure mathematics to the detriment of the so-called applied mathematics. This led to a preference for theory over practice in many books and articles on history of mathematics during the nineteenth century. Folkerts, Scriba, and Wußing see Nesselmann as one of the first authors who embodied "the spirit of the educational reform" within history of mathematics, without, however, explaining why that is the case (Dauben and Scriba 2002, 114–5). One possible argument in favor of such a contextualization of Nesselmann's university career is his substantial involvement with novel research questions and methods.

In France, the French Revolution and its successors shaped the political fate of state and people through the different phases of the constitutional monarchy and its successor, the French Republic. In 1830, at the end of the restoration period,

Charles X invaded Algeria and started the second phase of French colonialism. This first French colonial war of the nineteenth century continued until 1848, when the second French revolution of the nineteenth century broke out and introduced the Second Republic. During those years, the first round of a major academic controversy over the value of scholarly contributions to the mathematical sciences from Islamicate societies and their relationships to ancient Greek scholarly achievements took place in Paris. Peiffer rightly highlighted the importance that the French conquest of North Africa has had upon the academic atmosphere of this period (Dauben and Scriba 2002, 14). The first chapter on Algeria of *Histoire pittoresque de l'Afrique française*, published in 1845 during the constitutional reign of Louis Philippe, duke of Orléans, begins with the claim, although the war would continue for another three years, that:

> Algeria is today a French province, or rather the new France. This region (will have) a future, (which is) one of the most prosperous and useful for our fatherland, which soon will draw from this addition incomparably greater advantages than those, (which came) from its long-ago colonies. The time is not far away, when France will be richly recompensed for all the sacrifices, which it has made for fifteen years in order to retain and organize that precious conquest. Today, the government seems determined not to neglect any means for installing forever the civilization, its laws, its mores, and the French industry in those regions where Turkish dominance and belief in the Muslim religion previously upheld barbarism.[1]

Before the anonymous propagator of French colonial glory turned to describing French military success, he wished to discuss the peaceful activities, since it was those, which "honor our arts, our sciences, our civilization, …".[2] Although the following pages treat more or less military activities, occasional comments clarify the involvement of civil, academic personnel, primarily geographers and archeologists, who drew up maps of the newly conquered terrains and pointed the military commanders to Roman ruins that they considered worthy of protection (Histoire pittoresque 1845, 44–5). This primary attention to historical objects, which were seen as part of the colonizing country's own historical legacy, is also visible in earlier descriptions of the conquest. The anonymous writer had also invested, on the other hand, much effort in reading Arabic geographical and historical sources about North Africa (Histoire pittoresque 1845, 46). This tension between the appreciation of ancient Greek and Roman cultures and their intellectual achievements and interest in Arabic scholarly texts continued during the entire nineteenth century. It

[1]"L'Algérie est aujourd'hui une province française, ou plutôt la nouvelle France. Cette contrée, est réservée à un avenir des plus prospères et de plus utiles à notre patrie, qui bientôt, retirera de cette adjonction, des avantages incomparablement plus grands que ceux résultants de ses colonies lointaines. L'époque n'est pas éloignée où la France, sera amplement dédomagée de tous les sacrifices quelle a faits, depuis quinze ans, pour se conserver et organiser cette précieuse conquète. Aujourd'hui, le gouvernement paraît décidé à ne négliger aucuns des moyens d'installer à jamais la civilisation, ses lois, ses mœurs et l'industrie française, dans ces contrées où la domination turque et la croyance de la religion musulmane, entretenaient auparavant la barbarie." (Histoire pittoresque 1845, 5).

[2]"… ils honorent nos arts, nos sciences, notre civilisation, …" (Histoire pittoresque 1845, 33).

contributed to the formation of French oriental studies on the one hand and two major public conflicts on the academic evaluation of scholarly contributions from Islamicate societies on the other.[3]

A further major political and cultural context of the historical study of the mathematical sciences in Islamicate societies concerns the relationship between attitudes towards Catholicism and the Roman Catholic Church on the one hand and modernization, secularization, and the formation of a "national state" on the other. In France, the turn towards science as a useful instrument for overcoming the destructions wrought by the French Revolution and the Napoleonic wars favored the rise of positivism as the dominant philosophy (Dauben and Scriba 2002, 14). The rejection of religion in general, Islam and the Roman Catholic Church's Inquisition in particular, was an important factor for Renan's wholly negative evaluation of the sciences in Islamicate societies in a public lecture at the Sorbonne in 1883. His anti-religious position reflects the Republican character of the governments of the Third Republic and their struggle against the Catholic Church and its orders in France. The republicans put part of the blame for the lost war against Prussia in 1870/71 on the clerics and their political allies, i.e. the monarchists, and their dominant role in the state institutions of the Second Empire, in particular the educational system. Modernization was identified with secular education and the separation between Church and state.

A similar situation prevailed in Italy. Mazzotti argues for Baldassare Boncompagni's (1821–1894) case that his approach to history of mathematics and his various publication projects were shaped by his fierce opposition towards the secular and anti-papal forces of the *Risorgimento* (1815–1870), which led to the end of the Papal state in 1870 with which Boncompagni was closely affiliated by social status, conviction, and academic activities (Mazzotti 2000, 260–69, 272–76). Mazzotti believes that the specific Roman conditions explain Boncompagni's interest in medieval Latin and Arabic mathematical texts and other features of his working practice, which included the generous patronage for Franz Woepcke (Mazzotti 2000, 269–82).[4]

In addition to these major political, religious, and cultural developments in Germany, France, and Italy during the nineteenth century, a number of more specific intellectual and institutional changes shaped the academic landscape in the humanities in this time. In addition to the impact of new trends in philosophy and history, highlighted by Folkerts, Scriba, and Wußing, the search for intellectual, linguistic, and other origins of European languages, peoples, and cultures, for instance in India or in ancient Greece, led to the formation of new academic disciplines such as Indology or comparative and historical linguistics, and to movements such as neo-humanism. Almost all scholars who worked in France and Germany on

[3]See below: the Sédillot-Biot controversy and the Renan-al-Afghani dispute.

[4]Mazzotti's analysis of Boncompagni's *Bullettino di bibliografia e storia delle scienze matematiche e fisiche* is, however, flawed. Moreover, he apparently lacks in understanding of the skills and working practice of a medievalist or Arabist. This lack caused several mistakes in his evaluation.

Arabic or Persian mathematical and astronomical texts also learned Sanskrit. They believed in a strong Indian background to the mathematical sciences in the Abbasid caliphate in addition to their ancient Greek legacy. The overwhelming dominance of so-called pure mathematics at German universities during the nineteenth century, an already mentioned outcome of Humboldt's educational reforms, was a further important factor that shaped in particular Woepcke's research and writing practice (Dauben and Scriba 2002, 124). Equally, Cantor's *Vorlesungen über Geschichte der Mathematik* are dedicated to the history of this type of mathematics.

In France, the philosophies of Jean Antoine Caritat de Condorcet (1772–1791) and Auguste Comte (1789–1857) created the contexts for a more diversified working practice (Dauben and Scriba 2002, 9, 14). The two authors propagated an image of the new sciences and their practitioners that demanded encyclopedic knowledge and cherished experimentation. This view asked for a combination of mathematics with the sciences, but included also a desire for the search for, collection, edition, translation, and analysis of ancient, medieval, and early modern works (Dauben and Scriba 2002, 14). Peiffer locates the precursor of French colonial attention to "Oriental" achievements in the mathematical sciences in the activities of early modern antiquarians such as Barthélémy d'Herbelot (1625–1695), astronomers and geographers of the Royal Academy such as Jean-Dominic Cassini (1625–1712) and Joseph-Nicolas Delisle (1688–1768), and the cooperation between these groups. This cooperative atmosphere also characterized much of the work undertaken in the nineteenth century, in particular its first half. As Peiffer shows, it was not limited to interpersonal exchanges, but included institutional forms and projects such as the translation of Ibn Yunus' (d. 1009) astronomical handbook by a commission of the *Academy of Sciences*, in which Jean-Baptiste Delambre (1749–1822), Pierre-Simon Laplace (1749–1827), and Jean-Jaques Caussin de Perceval (1759–1835) worked together, or the cooperation of Jean-Jacques Sédillot (1777–1832) with Delambre and Laplace at the *Bureau des Longitudes* (Dauben and Scriba 2002, 15–6). Unfortunately, she does not offer any further investigation of the impact of French colonialism on the study of the mathematical sciences in Islamicate societies, either in France or in the colonies (Dauben and Scriba 2002, 20).

The major specific context that shaped not only the research and debates on Islamicate societies and their sciences in Paris, but drew the attention of German and Italian scholars as well, is the fierce struggle about the value and content of the contributions of the practitioners of the mathematical sciences in those societies. The elder Sédillot publicly expressed his view of the "Arabs" as innovators and original scholars. His two collaborators Delambre and Laplace defended, despite their growing knowledge of "Oriental" astronomy, the older belief according to which the "Arabs" had merely preserved and transmitted Greek scientific texts without adding anything of value to this knowledge. Jean-Jacques Sédillot's son Louis-Amélie (1808–1875) continued to subscribe to his father's judgment. His life-long efforts to substantiate this view were partly the result of a mistake. In 1835, namely, he misinterpreted a passage in Abu l-Wafa''s (940–998) *Almagest* as already containing the inequality of the Moon's motion, then known as its "variation" and ascribed to Tycho Brahe (1546–1601) (Dauben and Scriba 2002, 17; Sédillot 1835). His main opponent was Jean-Baptiste Biot (1774–1862), a professor of mathematics and physics and member of the Academy of Sciences. Due to the early death of his son Edouard, who was a sinologist and whose unfinished book he prepared for print, Jean-Baptiste became interested in Chinese mathematics and astronomy and moved from there to Indian and Egyptian topics. His linguistic competence was, however, limited and did not include Arabic. Hence, his severe critique of the younger Sédillot's interpretation of Abu l-Wafa"s text relied on translations and prejudices (Dauben and Scriba 2002, 18).

The esteem of scholarly work in Islamicate societies was not limited to astronomy, algebra, arithmetic, and geometry. Before the two Sédillots, another French historian who investigated, among other themes, the Arabo-Latin translations of Aristotle's works, Amable Jourdain (1788–1818), had already declared that the medieval Occident owed "the Arabs" their knowledge of the most important Aristotelian texts and that "the philosophy of Fakhr-eddin, Algazel, Alfarabi, Avicenne, etc., had obtained a great success in the Orient, despite the censorship of (those doctors, who were most closely) attached to the purity of Islamism."[5]

As I will suggest below, the Sédillot-Biot controversy raised awareness of the issue at stake, namely the question of an adequate and fair interpretation of scientific achievements in non-Western cultures, an issue that continues to haunt us today. It also motivated serious research, even if on a quantitatively low scale when compared with other themes in history of mathematics that continued through the entire century. Young scholars of the last third of the nineteenth century carried the torch far into the twentieth century. The controversy in its general features found its repetition in the clash between Ernest Renan (1823–1893) and Jamal al-Din al-Afghani (1838–1897) in the 1880s. Without naming names, Renan referred to the Biot-Sédillot controversy and took sides against the latter.[6] The general tone of his lecture was racist and denigrating. He considered Islam as the most authoritarian of the three monotheistic religions and believed that it had always obstructed scientific and philosophical reflection, that nothing of what was called Arabic science and philosophy owed anything to this language or this religion, but everything to

[5]"La philosophie de Fakhr-eddin, d'Algazel, d'Alfarabi, d'Avicenne, etc., avait obtenu un grand succès en Orient, malgré les censures des docteurs les plus attachés à la pureté de l'islamisme." (Jourdain 1819, 111).

[6]Renan (1883): "Je n'ai point cherché, Messieurs, à diminuer le rôle de cette grande science dite arabe qui marque une étape si importante dans l'histoire de l'esprit humain. On en a exagéré l'originalité sur quelques points, notamment en ce qui touche l'astronomie; il ne faut pas verser dans l'autre excès, en la dépréciant outre mesure." http://fr.wikisource.org/wiki/L%27Islamisme_et_la_science (Accessed June 15, 2014).

ancient Greece.[7] He argued that "religious orthodoxy" had always and everywhere renounced such rationalism, persecuted its practitioners, and vilified its patrons.[8] In his view Ibn Rushd (d. 1198) was the last Muslim philosopher.[9] Renan's false claims dominated academic interpretive perspectives until the last third of the twentieth century. They can be found even today in writings of non-expert academics as well as popular literature. Al-Afghani, a religious reformer and political activist from Iran, engaged in a public debate with the French scholar when he was in Paris. In stark contrast to his rebuttal of Muslim educational reformers in India and their willingness to adopt British sciences as necessary for young Muslim men to master, in his speech to a French public he presented himself as a champion of modern sciences and philosophy and an opponent of all religion. In an almost ironic form, he repeated many of Renan's claims and opposed them, picking out the most ludicrous of them for his critique. He agreed with Renan about the negative impact of his religion on the sciences and philosophy, but rejected the latter's claim that Arabs (and other Semites) were by nature, language, and social form incapable of

[7]Renan (1883): "… ce grand ensemble philosophique, que l'on a coutume d'appeler arabe, parce qu'il est écrit en arabe, mais qui est en réalité gréco-sassanide. Il serait plus exact de dire grec; car l'élément vraiment fécond de tout cela venait de la Grèce. On valait, dans ces temps d'abaissement, en proportion de ce qu'on savait de la vieille Grèce. La Grèce était la source unique du savoir et de la droite pensée. La supériorité de la Syrie et de Bagdad sur l'Occident latin venait uniquement de ce qu'on y touchait de bien plus près la tradition grecque. Il était plus facile d'avoir un Euclide, un Ptolémée, un Aristote à Harran, à Bagdad qu'à Paris. Ah! si les Byzantins avaient voulu être gardiens moins jaloux des trésors qu'à ce moment ils ne lisaient guère; si, dès le huitième ou le neuvième siècle, il y avait eu des Bessarion et des Lascaris! On n'aurait pas eu besoin de ce détour étrange qui fit que la science grecque nous arriva au douzième siècle, en passant par la Syrie, par Bagdad, par Cordoue, par Tolède." http://fr.wikisource.org/wiki/L%27Islamisme_et_la_science (Accessed June 15, 2014).

[8]Renan (1883): "La philosophie avait toujours été persécutée au sein de l'islam, mais d'une façon qui n'avait pas réussi à la supprimer. A partir de 1200, la réaction théologique l'emporte tout à fait. La philosophie est abolie dans les pays musulmans. Les historiens et les polygraphes n'en parlent que comme d'un souvenir, et d'un mauvais souvenir. Les manuscrits philosophiques sont détruits et deviennent rares. L'astronomie n'est tolérée que pour la partie qui sert à déterminer la direction de la prière. Bientôt la race turque prendra l'hégémonie de l'islam, et fera prévaloir partout son manque total d'esprit philosophique et scientifique. A partir de ce moment, à quelques rares exceptions près, comme Ibn-Khaldoun, l'islam ne comptera plus aucun esprit large; il a tué la science et la philosophie dans son sein. … Entre la disparition de la civilisation antique, au sixième siècle, et la naissance du génie européen au douzième et au treizième, il y a eu ce qu'on peut appeler la période arabe, durant laquelle la tradition de l'esprit humain s'est faite par les régions conquises à l'islam. Cette science dite arabe, qu'a-t-elle d'arabe en réalité? La langue, rien que la langue." http://fr.wikisource.org/wiki/L%27Islamisme_et_la_science (Accessed June 15, 2014).

[9]Renan (1883): "Quand la science dite arabe a inoculé son germe de vie à l'Occident latin, elle disparaît. Pendant qu'Averroès arrive dans les écoles latines à une célébrité presque égale à celle d'Aristote, il est oublié chez ses coreligionnaires. Passé l'an 1200 à peu près, il n'y a plus un seul philosophe arabe de renom." http://fr.wikisource.org/wiki/L%27Islamisme_et_la_science (Accessed June 15, 2014).

scientific, philosophical, or metaphysical thought.[10] He emphasized that Christianity and in particular the Roman Catholic Church had also fought against scientific and philosophical theories and continued to do so in the 1880s.[11] Hence the differences between the two religions and the societies they permeated and controlled were not as great as Renan suggested. Islamicate societies were not incapable of modernization and Muslims could overcome the negative aspects of their religion.[12] Al-Afghani's texts against the so-called naturalists in India and against Renan's untenable evaluation of the sciences and philosophy in Islamicate societies of the past have seen a revival in the last decades in different academic and non-academic circles, whose method often consists in (very simplified) critiques not merely of the colonial period, but of contemporary policies of Western countries, in particular in the Middle East as well as today's sciences, social sciences, the humanities, and other cultural features of Western societies.

[10]"In truth, the Muslim religion has tried to stifle science and stop its progress. It has thus succeeded in halting the philosophical or intellectual movement and in turning minds from the search for scientific truth. … The Arabs, ignorant and barbaric as they were in origin, took up what had been abandoned by the civilized nations, rekindled the extinguished sciences, developed them and gave them a brilliance they had never had. Is not this the index and proof of their natural love for sciences? It is true that the Arabs took from the Greeks their philosophy as they stripped the Persians of what made their fame in antiquity; but these sciences, which they usurped by right of conquest, they developed, extended, clarified, perfected, completed, and coordinated with a perfect taste and a rare precision and exactitude." (Al-Afghani 1968, 176–7, https://disciplinas.stoa.usp.br/pluginfile.php/2004379/mod_resource/content/1/KEDDIE%2C%20Nikki.pdf) (Accessed Oct. 20, 2016).

[11]"A similar attempt, if I am not mistaken, was made by the Christian religion, and the venerated leaders of the Catholic Church have not yet disarmed, so far as I know. … Besides, the French, the Germans, and the English were not so far from Rome and Byzantium as were the Arabs, whose capital was Baghdad. It was therefore easier for the former to exploit the scientific treasures that were buried in these two great cities. They made no effort in this direction until Arab civilization lit up with its reflections the summits of the Pyrénées and poured its light and riches on the Occident. The Europeans welcomed Aristotle, who had emigrated and become Arab; but they did not think of him at all when he was Greek and their neighbor. Is there not in this another proof, no less evident, of the intellectual superiority of the Arabs and of their natural attachment to philosophy?" (Al-Afghani 1968, 177, https://disciplinas.stoa.usp.br/pluginfile.php/2004379/mod_resource/content/1/KEDDIE%2C%20Nikki.pdf) (Accessed Oct. 20, 2016).

[12]"If it is true that the Muslim religion is an obstacle to the development of sciences, can one affirm that this obstacle will not disappear someday? How does the Muslim religion differ on this point from other religions? … Religions, by whatever names they are called, all resemble each other. No agreement and no reconciliation are possible between these religions and philosophy. Religion imposes on man its faith and its belief, whereas philosophy frees him of it totally or in part." (Al-Afghani 1968, 177, https://disciplinas.stoa.usp.br/pluginfile.php/2004379/mod_resource/content/1/KEDDIE%2C%20Nikki.pdf) (Accessed Oct. 20, 2016).

3 Historiographical Approaches of Nineteenth-Century Writers to the Mathematical Sciences in Islamicate Societies

The main historiographical approaches to writing history of mathematics in Germany during the nineteenth century, shared by writers on every period and mathematical culture, included the study of primary sources, the search for a periodization of the history of mathematics based on philosophical concepts and principles, and the "kulturhistorische" approach (Dauben and Scriba 2002, 115–7, 122–4). As for Islamicate societies, the representatives of these three approaches were Georg Heinrich Ferdinand Nesselmann (1811–1881), Franz Woepcke (1826–1864), and Moritz Cantor (1829–1920). Nesselmann and Woepcke pursued at least two of them in their research practice, while Cantor employed all three when narrating history of mathematics in a universal style.

The most influential French author of treatises on Arabic astronomy, instruments, and mathematics, although not the first nor the last author of such texts in nineteenth-century France, was the already-mentioned Louis-Amélie Sédillot (Charette 1995, 93–4). He also focused on the study of primary source material and scientific progress as achieved by scholars from Islamicate societies. But, going beyond the themes and methods of Nesselmann and Woepcke, he used codicological methods and wrote about or, better, against Indian and Chinese contributions to scientific knowledge.

3.1 Nesselmann's Views on How to Do History of Algebra

In the dedication of his *Versuch einer kritischen Geschichte der Algebra*, Nesselmann characterized his turn to the history of mathematics, in particular algebra, as the result of the "spirited" teaching of mathematics by his high school teacher C.F. Buchner and one of his two university professors of mathematics, Carl Gustav Jacob Jacobi (1804–1851) (Nesselmann 1842, viii). Folkerts, Scriba, and Wußing identified explicit historiographical statements in the preface to this book as influenced by Johann Gottfried von Herder's (1744–1803) philosophical ideas on culture and the new German school of historicism, in particular the works of Barthold Georg Niebuhr (1776–1831) and Leopold von Ranke (1795–1886) (Dauben and Scriba 2002, 115). The book's title and the emphasis on the chosen critical approach to the past confirm Nesselmann's conceptual as well as rhetorical connection to historicism: "I intended, as the title says, to write a critical history; I did not wish to teach the history of algebra according to tradition, but as it results from a persevering and careful study of the sources and then present it accordingly with fidelity."[13] His historiographical commitment in this and other passages to the study of primary sources strengthens the bond with the new historical school in the realm of methods.

The first sentences in Chap. I offer a different kind of vocabulary, one that is indeed strongly reminiscent of Herder's philosophical doctrines:

> At the moment when the first seeds of historical consciousness awake in a people, the desire stirs towards a national history, which will bring this self-consciousness to completion. A people without history is a head that does not see the body on which it rests; it is a present that lacks the consciousness of a true past. In this (situation of an) onerous sentiment of emptiness and unconsciousness of a previous existence the historical element evolves at first in the form of the national epic among peoples, as soon as they reached a certain level of intellectual culture; from there it moves slowly to true history. ... This very same desire for history, which ensouls uprising peoples, can also be discovered among the admirers and propagators of the sciences and the arts. This is all the clearer and more definitive, when the material for further formation of the latter arises from within. Thus, the progress of theology, mathematics, and poetry was first worked out in a historical (form).[14]

But while positions of historicism permeated Nesselmann's research practice, Herder's philosophy remained a distant point of orientation that gave some rhetorical structure to the mass of details newly acquired from primary sources. The overall structure chosen by Nesselmann for presenting his critical history of algebra, namely, does not follow Herder's ideas of people and historical consciousness, but mathematical topics and methods that are ascribed to either peoples or periods (Nesselmann 1842, 30–4).

In addition to these two methodological and philosophical inspirations of Nesselmann's work a third motivation can be recognized in his desire to define a well-delineated historiographical working methodology with specific methods. The discussion of his predecessors in history of mathematics in Chap. I of *Versuch einer kritischen Geschichte der Algebra* leaves no doubt that Nesselmann perceived severe shortcomings in their works. These issues were of immense relevance to him. He decided to break with their approaches and methods and to make explicit what he

---

[13]"Ich wollte, wie der Titel besagt, eine kritische Geschichte schreiben, ich wollte die Geschichte der Algebra, nicht wie die Tradition sie lehrt, sondern wie sie sich aus dem ausdauernden und gewissenhaften Studium der Quellen ergiebt, erforschen und demgemäß treu darstellen." (Nesselmann 1842, x–xi).

[14]"Sobald in einem Volke die ersten Keime des Selbstbewußtseins erwachen, regt sich in ihm das Bedürfniß nach einer Nationalgeschichte, welche jenes aufstrebende Selbstbewußtsein zur Vollendung bringe. Ein Volk ohne Geschichte ist ein Kopf, welcher den Körper nicht sieht, auf dem er steht; es ist eine Gegenwart, der das Bewußtsein fehlt, daß sie auf einer thatsächlichen Vergangenheit beruht. In diesem drückenden Gefühle der Leere und der Bewußtlosigkeit früherer Existenz entwickelt sich bei den Völkern, sobald sie auf eine gewisse Stufe der geistigen Cultur getreten sind, zuerst das historische Element in der Form des Nationalepos und geht von da allmählig in wirkliche Geschichte über. ... Dieses nämliche Bedürfniß nach Geschichte, welches emporstrebende Völker beseelt, läßt sich auch unter den Verehrern und Fortbildnern der Wissenschaften und Künste wahrnehmen, und zwar um so deutlicher und bestimmter, je mehr den letzteren der Stoff zu ihrer Fortbildung von innen her zukommt; am frühesten sind daher die Fortschritte in der Theologie, der Mathematik und der Poesie geschichtlich bearbeitet worden." (Nesselmann 1842, 1).

considered a reliable and proper research as well as writing practice. He praises, for instance, Proclus (c. 412–485) for "dealing in detail with a subject, explaining fully the theoretical aspects of the most important theorems and doctrines, and presenting the content of mostly lost books in such precision and detail that the more recent mathematicians were enabled by this information to reconstruct those works," while chastising his immediate predecessors for mainly offering a catalogue of names and titles.[15] The mistakes committed by the latter are so numerous that Nesselmann saw himself obliged to instruct future writers to learn how to quote correctly, to quote only from a primary source, if one had read it, and to avoid substituting modern forms for older thoughts, methods, or results (Nesselmann 1842, 35, 37–38). Finally he argued that his predecessors had mistaken three different tasks for one: to write the biography of a mathematician; to compile a bibliography; to write in a scientific manner about the history of mathematics (Nesselmann 1842, 38). His declared intention was to write a scientific history of mathematics. That is why he asked his readers not to expect names, but things and not to demand a complete list of books, but to accept as sufficient the description of the content of those few books that had an impact on the progress of mathematics (Nesselmann 1842, 39). He did not believe to have invented this historiographical approach, a feat he ascribed to Jean-Étienne Montucla (1725–1799), but to have been the first who strictly adhered to these principles and worked hard to complete the entire project (Nesselmann 1842, 18, 39).

An important point to be raised is whether Nesselmann applied his historiographical ideas to his investigations of algebra and arithmetic in Islamicate societies, and, if so, how. This question can be discussed only with the utmost restriction since Nesselmann never published more than the first volume of his critical history of algebra. His translation of Baha' al-Din al-'Amili's *Essence of Arithmetic* contains only a few brief remarks about his goals in publishing the work, a justification of his use of a printed rather than a manuscript version, and a short biographical sketch of the Safavid author (Nesselmann 1842, iii–v, 74–6). In his discussions of the names of algebra, the various number systems, and their symbolic representations he works indeed with Arabic and Persian primary sources in manuscript as well as printed form and presents in his footnotes the translated quotes in their original language, if they are more than a mere phrase or between parentheses in the text, if they are single words or short expressions (Nesselmann 1843, 42–3, 45–51, 73, 78–9 etc.). The scarcity of his source material though is remarkable. In addition to al-'Amili's *Essence of Arithmetic*, a Persian translation with commentary by Rawshan 'Ali Jawnpuri (19th century), and a didactic poem by some Najm al-Din 'Ali Khan, like Jawnpuri a Muslim scholar from India, Friedrich August Rosen's (1805–1837) publication and translation of al-Khwarizmi's algebra, and apparently Thabit b. Qurra's (d. 901) translation of Nicomachus's (2nd c)

[15]"(er begnügt sich aber nicht, wie so viele neuere Historiker, bloße Namen und Büchertitel zu nennen,) sondern er geht ausführlich in die Sache ein, führt die wichtigsten Sätze und Lehren theoretisch vollständig aus, und giebt den Inhalt von vielen, jetzt größtentheils verloren gegangenen Büchern so genau und ausführlich an, daß es neueren Mathematikern möglich geworden ist, nach diesen Angaben jene Werke wiederherzustellen." (Nesselmann 1842, 7).

Introduction to Arithmetic, Nesselmann refers only to a very small number of secondary sources such as Silvestre de Sacy's (1758–1831) Arabic grammar.

Despite the brevity of Nesselmann's own remarks in his reprint and translation of al-'Amili's *Essence of Arithmetic*, its preface is relevant to the discussion of his historiographical stance. It makes clear that he thought that editing a text meant working from manuscripts and using more than a single one, if possible (Nesselmann 1843, iii). The guiding model of classical studies is clearly recognizable. Secondly, Nesselmann justified his publication of the Safavid text primarily because of its information on what scholars in Islamicate societies had done with their intellectual heritage since Muhammad b. Musa al-Khwarizmi had written the first algebraic text in Arabic in the early ninth century (Nesselmann 1843, iii–iv). Only at the end, in his biographical excursus, did he also point to the treatise's function as a textbook and its dominant place in Indian madrasas (Nesselmann 1843, 75). This perspective corresponds well with his goal of producing a scientific history of mathematics that looks for progress and not for socio-cultural roles and functions of knowledge or any other property such knowledge might have. In a sense it also sets the tone that was to dominate research and interpretation of the mathematical sciences in Islamicate societies until very recently.

The reputation that Nesselmann's work acquired among his contemporaries is nicely summarized in the short biography by Moritz Cantor:

> Scientific fame came to him, however, first in a work of the year 1842, his "Algebra of the Greeks". It was the first work in German since Kästner that treated mathematical-historical things on the basis of proper research; but in erudition, critical insight, and comparative investigative power it ranks far above all German predecessors. Only Chasles, "History of Geometry" (1837) and Libri, "History of Mathematics in Italy" (1838—1841) can be compared with Nesselmann's "Algebra of the Greeks" and form with it the three exemplary models that all successors on the same domain learned from and were inspired by.[16]

3.2 Louis-Amélie Sédillot

The main interpretive issues that were discussed in the nineteenth century concerned the value of mathematical and astronomical contributions by scholars from Islamicate societies, the relative importance of ancient Greek and ancient or early medieval Sanskrit texts for these contributions, and the sciences in Arabic in general. Contributions in Persian or Ottoman Turkish received much less attention.

---

[16]"Wissenschaftliche Berühmtheit verschaffte ihm dagegen zuerst ein Werk aus dem Jahre 1842, seine "Algebra der Griechen". Es war seit Kästner das erste Werk in deutscher Sprache, welches am Grund eigener Forschung mathematisch-geschichtliche Dinge behandeltea, ber es steht dabei an Gelehrsamkeit, an kritischer Einsicht, an vergleichender Spürkraft weit über allen deutschen Vorgängenr. Nur Chasles, Geschichte der Geometrie (1837) und Libri, Geschichte der mathematik in Italien (1838–1841) lassen sich mit Nesselmann's Algebra der Griechen vergleichenund bilden mit diesem die drei Musterwerke, aus welchen alle Nachfolger auf dem gleichen Gebiete gelernt und geschöpft haben." (Cantor. 1886. Nesselmann, Georg Heinrich Ferdinand. In Allgemeine Deutsche Biographie. https://www.deutsche-biographie.de/pnd117550469.html) (Accessed June 15, 2014).

The discussions about values and evaluation did not begin in the nineteenth century, but have a history that reaches centuries back. The fierceness of the conflicts that arose in particular in France was, however, more intense than ever, with the exception perhaps of the sixteenth century.

Louis-Amélie Sédillot, like his father, spoke out in favor of Arabic ingenuity and was happy to support his views with the authority of Alexander von Humboldt, with whom he also shared his research results. He expressed with verve a position that found supporters as well as opponents throughout the century (Charette 1995, 105–6, 121–39). Among the opponents were scholars such as Guillaume Libri (1803–1869) and Jean-Baptiste Biot (1774–1862). In general terms, Sédillot described his beliefs as follows in the preface to his *Matériaux pour servir à l'histoire des sciences mathématiques chez les Grecs et les Orientaux*:

> … from the ninth to the thirteenth century, one sees the formation of one of the most vast literatures that exist; manifold productions, precious inventions attest to the marvelous (intellectual activities) and making felt their action on Christian Europe, they seem to justify the opinion that *the Arabs were in all matters our masters*.[17]

Given the recent French colonial conquests of Arab territories in North Africa and the strong belief held by many French and other European academics in the exclusive superiority of ancient Greek sciences, such a bold statement was bound to encounter scorn and rejection among some factions of academia. While in its generality it was certainly more correct than the opposite position that limited the historical role of "Arabic" scholars to the translation of Greek mathematical and scientific works and their preservation until their translation into Latin in the twelfth and thirteenth centuries, one of Sédillot's main specific arguments, the invention of the lunar variation by Abu l-Wafa', was certainly false.

Convinced of the righteousness of his approach, and based on his father's as well as his own results, Sédillot proposed to prove in his book that "… the School of Baghdad was able to surpass the Schools of Athens and Alexandria."[18] The main content of the first volume consists in Sédillot's effort to rebuff those who did not accept his interpretation of Abu l-Wafa''s statement. But much more was at stake than the correct or incorrect interpretation of a few lines in an Arabic text of the tenth century. The quarrel was not limited to a comparative up- or downgrading of ancient Greeks, medieval Arabs, and modern Europeans. It also included ancient and medieval Indians and Chinese. The questions, simply formulated, were who had created a scientific astronomy and who had contributed what to such a science? Sédillot ardently defended the right of the "Arabs" to be considered as major contributors to science, while denying such a right to the Chinese and the Indians, and diminishing that of the Greeks. Biot defended the scientific valor of the Greeks, recognized Chinese astronomy as scientific, and argued vehemently against the

[17]"… du neuvième au treizième siècle, on voit se former une des plus vastes littératures qui existent; des productions multipliées, des précieuses inventions attestent l'activité merveilleuse des esprits, et faisant sentir leur action sur l'Europe chrétienne, semblent justifier l'opinion que *les Arabes ont été en tout nos maîtres*." (Sédillot 1845–49, vol. 1, iii).

[18]"… l'École de Bagdad a su dépasser les Écoles d'Athène et d'Alexandrie." (Sédillot 1845–49, vol. 1, v).

"Arabs", whom he described with a slightly altered and, according to Sédillot, anonymous quote from Cervantes's *Don Quijote* as people of whom one could not expect a single truth, because they were embezzlers, falsifiers, and liars.[19] This usage of literary fantasy, which in addition ignores (of course) the various expressions of appreciation for Islamicate cultures and their people found in Cervantes's works, as a scholarly argument indicates the expanse of what was considered feasible in discussing a specific and limited piece of astronomical knowledge. Emotions obviously ran high in French academia in the middle of the nineteenth century.

Doubts and suspicions also characterize Libri's more academic objections against Sédillot's "discovery". He doubted that Abu l-Wafa' had undertaken celestial observations (Sédillot 1845–49, vol. 1, 100). He asked why no later astronomer had mentioned the third inequality, if Abu l-Wafa' truly had discovered it (Sédillot 1845–49, vol. 1, 51).[20] Others wondered why the geometrical methods used by Abu l-Wafa' and Tycho Brahe were so close to each other. They suggested that this either meant that Brahe had known of Abu l-Wafa''s discovery or that Abu l-Wafa''s text had been modified or even falsified by a scribe in the seventeenth century (Sédillot 1845–49, vol. 1, 51, 54, 66–7, 87, 91, 93, 100, 115). Sédillot's answers to these points came from three practices: (1) a careful study of the manuscript he worked with and its para-textual information about copying dates, scribal hands, ownership marks, seals, and material features like binding and paper as well as other manuscripts, which offered elements for rebutting the objections; (2) the study of new publications on various points of Islamicate history and art objects from Islamicate societies, which were related to different features of the manuscript; (3) general historical arguments about the relationship between ancient Greek, medieval Arabic, and early modern European astronomy and geometry (Sédillot 1845–49, vol. 1, 52–8, 69, 72–3, 101–6, 116–7 et al.). Current codicologists and, in a more limited manner, also historians of the mathematical sciences in Islamicate societies check the same kind of items when they wish to determine age, provenance, or pathways of any given manuscript through different libraries. Hence, while Sédillot certainly erred and went too far in his efforts to defend his father's reputation and to establish the recognition of the scholarly value of medieval authors of Arabic astronomical and mathematical treatises, he also was one of the pioneers in applying basic codicological skills to the evaluation of the content of such texts. The various points made in this debate indicate the difficulties the participants faced when trying to interpret and evaluate different kinds of data. Prejudices and presuppositions played an important role in these processes as did the lack of technical skills, codicological and other experiences as well as historiographical methodologies (Sédillot 1845–49, vol. 1, 56–7, 67–9, 71–2, 76, 112, 115).

In addition to the efforts undertaken by Sédillot to establish a fundament for a history of the mathematical sciences in Islamicate societies that rested on philology, codicology, history, art history, the sciences, the study of the materiality of manuscripts and other relevant objects, and bibliography, he also formulated a series of

[19]"… de los moros no se puede esperar verdad alguna, porque todos son embelecadores, falsarios y chimeristas." (Sédillot 1845–49, vol. 1, 119).

[20]For a rejection of Sédillot's interpretation of the Arabic text see (Carra de Vaux 1892).

research tasks that were taken up in the late nineteenth and in particular in the second half of the twentieth centuries, although some of them still await their students: edition, translation, and analysis of major Arabic astronomical texts and tables, in particular those by al-Battani and Ibn Yunus; edition, translation, and analysis of the mathematical works of "the School of Baghdad" (a term used repeatedly by Sédillot) and the astronomical observations carried out in the Abbasid capital during the ninth and tenth centuries; study of regional contributions to astronomy and mathematics, in particular in al-Andalus, the Maghrib, Egypt, Iran, and Transoxania; study of tables and mechanical, astronomical, and mathematical instruments (Sédillot 1845–49, vol. 1, 134–9, 274–364). On the other hand, he also gave voice, as did other French students of Arabic, Persian, or Ottoman manuscripts, to strong and often declarative, i.e. little investigated, historiographical positions, which dominated approaches during the entire twentieth century. According to him, the imagined interpolation of the third inequality into the eleventh-century copy of Abu l-Wafa''s text could not have been executed by an "Arab of the seventeenth century". His argument was the lack of awareness of the "Turks and the Arabs subjugated by them" of the scientific knowledge of the Europeans that made "most of them even today believe in the immobility of the earth" (Sédillot 1845–49, vol. 1, 56). In Sédillot's view, the end of scientific curiosity and activity occurred in the fifteenth century with the death of Ulugh Beg (Sédillot 1845–49, vol. 1, 270). As for the beginning of the mathematical sciences, in particular astronomy, Sédillot locates it in translations of Sanskrit texts, but believes that the shift to Greek theories, parameters, and methods was a quick and easy process, finished already around 820, due to the inferiority of the Indian material (Sédillot 1845–49, vol. 2, 440). He believed, like many historians of the mathematical sciences and scholars of Oriental matters, in the power of progress and scientific truth.

3.3 Woepcke's Goals, Claims, and Methods

Franz Woepcke pursued a different set of goals and ideas than Nesselmann (Narducci 1869). As I will show below, Woepcke was not interested in progress as such, but in the contributions of scholars from Islamicate societies to the mathematical sciences. This nuance might appear insignificant, but was of importance for Woepcke's historiographical practice. It enabled him to look critically at various interpretations of ancient Greek mathematical achievements proposed by his contemporaries. He introduced new questions, worked consequently with primary sources, and systematically pursued the question of what scholars from Islamicate societies had contributed to the mathematical sciences without trying to equate their achievements with nineteenth-century developments. Woepcke's historiographical approach, as expressed in his texts, reflected three elements: (1) the severe conflicts that shook Paris in the first half of the nineteenth century in regard to the history of astronomy and mathematics; (2) Alexander von Humboldt's interests in the achievements of scholars from Islamicate societies in general and the genesis of number and calculation systems in particular; (3) beliefs and values of mathematicians in Germany and France in the middle of the nineteenth century.

By contrast, Woepcke's texts show no clear or substantial relationship with the historiographical positions of his last major patron, Baldassarre Boncompagni (1821–1894), for whom he wrote several treatises and who published and republished a number of his works. Boncompagni's interests in the history of mathematics had shifted in this period primarily "to reconstructing the chronology and the channels of the transmission of mathematical knowledge from the Arabic world to Christian Europe." (Mazzotti 2000, 260) Woepcke was one of Boncompagni's main collaborators with regard to Arabic mathematical manuscripts and the appropriation of Arabic and Indian contributions. The two exchanged some 164 letters, which so far have not been analyzed for the historiographical similarities and differences between the two men.

Given Mazzotti's ascription of Boncompagni's rejection of a synthetic history of mathematics, of evaluations and judgments, and his conscious limitation to library studies, editions, and translations of mathematical texts and a problem-oriented research to his conservative political and religious convictions, the question arises in how far Woepcke's views and working practice had similar features and were due to comparable political, religious, and academic beliefs.

The obituary written by Hippolyte-Adolphe Taine (1828–1893), according to his own testimony a close friend of Woepcke, paints a picture of Woepcke's historiographical convictions that goes beyond the positions expressed by the scholar himself in his papers. It agrees in several points with the attitudes attributed by Mazzotti to Boncompagni: the rejection of a synthetic history of mathematic as unattainable, the insistence on the need to restrict research for a long time to the study of manuscript texts and the accumulation of factual knowledge, and severe critique of speculations and superficial judgments.[21] Taine situates these points, however, exclusively in

[21] "… il ne s'était engagé dans les recherches limitées et dans les questions particulières que par une aversion naturelle pour les considérations vagues, et parce qu'il regardait ces travaux bornés et concentrés comme la meilleure discipline de l'esprit. … Il entrevoyait dans l'avenir, pour la fin de sa vie, une histoire générale des mathématiques, du moins depuis leurs origines dans l'Inde jusqu'à la renaissance. Mais il n'y comptait guère: "On se donne cette espérance à soi-même, me disait-il, c'est pour s'encourager. Mais c'est là une illusion d'esprit; le travail est trop grand, et la vie d'un homme est sujette (sic) à trop de chances." - "Je ferais bien un système, ajoutait-il une autre fois, il n'y faudrait qu'un peu d'invention, et peut-être en suis-je capable comme un autre; mais à quoi bon, puisque mon système ne serait pas prouvé, et pourquoi perdrais-je mon temps à me duper moi-même avec des phrases?" Il pensait que les jugements d'ensemble sur l'ancienne histoire des mathématiques et sur le passage des sciences anciennes aux sciences modernes doivent demeurer encore en suspens pour un ou deux siècles. Il comparait les connaissances que nous avons aujourd'hui sur la science et la civilization arabes aux celles que nous avions au seizième siècle sur la science et la civilization grecques, et croyait que pendant bien longtemps tout travail fructueux doit se réduire comme au siècle de Casaubon et de Scaliger, à la publication des manuscrits. … Le trait le plus marquant de son esprit était la haine du charlatanisme; il ne devenait moqueur et caustique que sur ce point; et quand il mettait le doigt sur les prétensions et l'insuffisance de quelques contemporains, ses petits exposés de faits, si exacts et d'apparence si sèche, arrivaient au plus haut comique. Pour ce qui est de lui-même, il était toujours prêt à se réduire, même à se rebaisser. … Son plus vif désir était de n'être jamais dupe de lui-même; il tenait toujours dans sa main une balance pour peser ses opinions; il ne voulait rien admettre que de vrai et de prouvé, et préférait l'ignorance aux conjectures. Il avait un sentiment profond de l'imperfection de nos sciences, des limites de chaque esprit, des bornes du sien entre tous les autres." (Taine 1866, 385–7, 389).

Woepcke's character traits without even the smallest hint at any other reason behind these beliefs. Moreover, Taine denies that these beliefs signified a strong rejection of synthesis and theoretical reflection by Woepcke. On the contrary, Taine forcefully stresses Woepcke's interest in and adoration of, as he calls it, metaphysical conceptions of history.[22] Hence, the question must remain unanswered and a task for future research as to what shaped Woepcke's historiographical views beyond the three factors that Woepcke explicitly expressed in his publications.

The Impact of Louis-Amélie Sédillot's Appreciation of Abu l-Wafa' and Other Scholars from Islamicate Societies on Woepcke's Methodological Stance

Woepcke's historiographical practice cannot be isolated from the positions formulated and contested in Paris during the first half of the nineteenth century. The impact of the harsh debates on whether "the Arabs" had contributed anything to human scientific progress beyond the "preservation" of ancient Greek texts cannot be missed in the publications of French scholars during Woepcke's stay in Paris. As his first publication shows, Woepcke's approach was a continuation and extension of that proposed by Sédillot. In this work on 'Umar Khayyam's (d. ca. 1123) algebra, Woepcke states:

> The works of the illustrious mathematicians, whom Greece produced during a period of six centuries, have been, practically without interruption, the object of scholarly works. Since the beginning of the Middle Ages until our days, they have been translated, commented on, published, often by geometers who themselves were very famous. It suffices here to remember the names of Nassir eddin al Thusi, of Bachet de Mezériac, of Halley. The important discoveries during a similarly long period through which the Arab genius has enriched the same science have not been as fortunate in attracting an equal attention. Some have even gone so far as to claim that the Arabs in general have invented nothing or almost nothing beyond that which they scooped from Greek authors translated into Arabic since the time of the caliphs Haroun Alrachid and Almamoun. Careful and extended research will probably lead to markedly different results.[23]

---

[22] "Quoiqu'il eût aimé passionnément la métaphysique, il l'avait laissée derrière lui et la considérait seulement comme une façon commode de grouper les faits, comme un systèmes provisoire, utile pour tirer l'esprit des recherches spéciales et pour le diriger vers les vues d'ensemble. … Non qu'il fût sèchement positiviste; il suivait avec intérêt et sympathie les hautes constructions idéales que l'on essaye d'élever sur ces rares soutiens; et il estimait que chacun doit essayer ou esquisser la sienne, et il jugeait qu'après tout le plus noble emploi de science est de fournir matières à ses divinations grandioses par lesquelles, en dépit de nos erreurs et de nos doutes, nous prennons part aux contentements et à l'œuvre des siècles qui nous suivront." (Taine 1866, 390).

[23] "Les œuvres des illustres mathématiciens, que la Grèce a produits pendant l'espace de six siècles, ont été presque continuellement l'objet de travaux savans. Dès le commencement du moyen age, jusqu'à nos jours, elles ont été traduites, commentées, publiées, souvent par des géomètres, qui eux-mêmes avaient une haute célébrité. Il suffira ici de rappeler les noms de Nassir eddin al Thusi, de Bachet de Mezériac, d'Halley, Les découvertes importantes par lesquelles le génie Arabe pendant une période d'une semblable durée, a enrichi la même science, n'ont pas été assez heureuses pour s'attirer une pareille attention. On est même allé jusqu'à soutenir, que les Arabes n'avaient en général rien inventé, ou presque rien au-delà de ce qu'ils avaient puisé des auteurs Grécs, traduits en Arabe depuis le temps des khalifes Haroun Alrachid et Almamoun. Des recherches soigneuses et étendues meneront probablement à des résultats fort différens." (Woepcke 1850, 160).

Woepcke hoped to offer clear evidence for his claim. Comparing Khayyam's work to those of Muhammad b. Musa al-Khwarizmi's algebra and Baha' al-Din al-'Amili's arithmetic, the only two Arabic treatises with chapters on algebra published in a European language during the first half of the nineteenth century, Woepcke left no doubt as to Khayyam's more advanced mathematical level.[24] In his view, Khayyam's higher scholarly rank is not only expressed in the more complicated mathematical problems and methods, but also in his reliance on Aristotelian philosophy in the definitions of his new theory and in his methodology.[25] The latter Woepcke summarizes as focusing on true difficulties, while ignoring questions of lesser relevance and quoting contemporary scholars whose errors he occasionally corrected.[26] Woepcke's main praise for Khayyam's treatise consists in acknowledging that it pushed back the boundaries of a science.[27] Here we find major themes of research that have continued to stimulate research interests until the end of the twentieth century.

Woepcke also spoke out against premature and inappropriate evaluations of the intellectual achievements of scholars in Islamicate societies during those centuries offered by his contemporaries.

> By comparing the treatises of Mohammed Ben Moûçâ and Behâ Eddìn, Colebrooke arrived at the conclusion (Algebra of the Hindus, Dissertation, p. LXXIX) that algebra had remained almost stationary in the hands of the Musulmans. Would it not be the same to put into doubt the discoveries of Apollonius, Archimedes, Diophant, because neither Euclid's Elements nor Marcianus Capella's "Nuptials of philology and Mercury" gives us knowledge of the most beautiful monuments, which Greek geometry has left (to us)? No, the mathematical sciences did not remain stationary in the Orient in the time from Mohammed Ben Moûçâ until Behâ Eddìn; they took, in an intermediary epoch, an upsurge and a development worthy of true admiration.[28]

---

[24]"Je remarque qu'en général on doit accorder à Alkhàyàmî un rang supérieur à celui de Mohammed Ben Mousa ou de Behà-Eddìn, vue que les triatès de ceux-ci n'ont pour but que l'instruction des commençans, tadnis que celui d'Akhayàmî porte un caractère plus élevé, …. " (Woepcke 1850, 161). Compare also (Woepcke 1851, xix).

[25]"… contenant les définitions des notions fondamentales de cette science. Ces définitions sont assez intéressantes, parcequ'elles font voir combien la philosophie d'*Aristote* a influé sur la science Arabe; …" (Woepcke 1850, 161).

[26]"… effleurant seulement les questions d'une portée inférieure; appuyant sur les difficultés réelles, citans les travaux contemporains, corrigeant patfois leurs erreurs, …" (Woepcke 1850, 161–2).

[27]"Ce n'est pas un des livres, qui reproduisent ce qu'on sait dans une science, mais un de ceux qui en reculent les bornes." (Woepcke 1850, 162).

[28]"En comparant entre eux les traités de Mohammed Ben Moûçâ et de Behâ Eddìn, Colebrook était arrivé à la conclusion (Algebra of the Hindus. Dissertation. P. LXXIX), que l'algèbre était resté à peu près stationnaire entre les mains des musulmans. Ne serait-on pas également fondé à mettre en doute les découvertes d'Apollonius, d'Archimède, de Diophante, parce que ni les Éléments d'Euclide, ni les "Noces de la philologie et der Mercure" de Marcianus Capella, ne nous font connaître les plus beaux monuments qu'ait laissés la géométrie grecque? Non, les mathématiques ne sont pas restées stationnaires en Orient depuis Mohammed Ben Moûçâ jusqu'à Behâ Eddìn; elles ont pris, à une époque intermédiaire, un essor et un développment dignes d'une véritable admiration." (Woepcke 1851, xix).

Humboldt's Questions as Sources of Inspiration for Sédillot and Woepcke

In Louis-Amélie Sédillot's view, Humboldt had acknowledged explicitly in the *Kosmos*, "a book that France, in a period of political agitation, has not yet fully appreciated, … the services that the Arabs have rendered to civilization and at the same time gave an indication of all that which one should expect from future studies (if they) are capably undertaken".[29] Humboldt encouraged Woepcke to study Arabic in order to bring to light these expected achievements. Summarizing the most important ideas, problems, methods, and results was one of the strategies by which Woepcke tried to fulfill this task in several of his most extensive articles. He stressed their programmatic interconnectedness and his goal of proving that scholars from Islamicate societies went beyond the ancient Greek writers in their mathematical activities.

Woepcke expressed his appreciation for and devotion to Humboldt very early in his career by translating the latter's essay *Über die bei den verschiedenen Völkern üblichen Systeme von Zahlzeichen und über den Ursprung des Stellenwertes in den indischen Zahlen* (1829) into French (Woepcke 1851). Six of Woepcke's articles discuss subjects related to Humboldt's treatise, although Boncompagni's patronage was essential for at least half of them (Woepcke 1853, 1854, 1855, 1863, 1865–66).

Humboldt's patronage of Woepcke and his demands for uncovering innovative results in Arabic texts were of an undeniably great stimulus for the young researcher. However, these demands also produced undesirable side effects. In particular the refusal to edit and translate Arabic or Persian texts as a whole and preference for extracts and summaries led to a focus on the mathematical content only and its quick identification with 'modern' formulas. Woepcke's working practice was oriented towards excavating highlights to the detriment of any effort to study the many texts available to him in the Imperial Library from at least a limited historical perspective, for instance as groups of interrelated works and authors.

The Impact of Values and Beliefs of Nineteenth-Century Mathematicians on Woepcke's Approach to Arabic Texts

Woepcke considered mathematicians and perhaps school teachers as a main part of his public. It is to such readers that he offers his discussion of Arabic texts. He tries to please them, adapts his writing style and explanations to them, and apologizes when he thinks they might be displeased. This is not limited to his publications in mathematical journals, but also in those of Oriental matters like

[29]"… a livre que la France, à une époque d'agitations politiques, n'a pas encore suffisament apprécié, … (le *Cosmos* de M. de Humboldt trace un tableau impartial des) services que les Arabes ont rendus à la civilisation, et laisse en même temps pressentir tout ce qu'on doit attendre de recherches ultérieures habilement dirigées." (Sédillot 1845–49, vol. 2, ii–iii).

the *Journal Asiatique*.[30] Typical of Woepcke's argumentation is his reliance on mathematical issues for explaining or criticizing al-Khayyam's concepts, methods, and failures.[31] In his comparative studies, Woepcke exclusively focuses on the mathematical content of a problem or the kind of a method in order to determine whether they are of the same style, similar, or profoundly different. Hence, comparisons of texts, even if they are of a later period than the Arabic or Persian one he analyzed, are at the heart of his working practice. Questions of how such problems or methods were disseminated, in case they resembled each other, and which historical conditions favored such a transfer of knowledge, to name only two issues of relevance to his comparative studies, played no role in his work.

This relative narrow set up of historical comparison does not mean, however, that Woepcke drew facile, superficial conclusions from his mathematical observations, although he always stressed the need to recognize the intellectual productivity of the scholars whose texts he analyzed.[32] His main argument for the Indian origin of two of Abu l-Wafa''s problems and solutions is that "they are quite noticeably detached from the spirit of the Arabic geometry, which is always faithful, with regard to the form, to its Greek models."[33] His second argument is the judgment of Chasles, who had declared in 1837 that a certain method in Bhaskara's arithmetic was "indeed of Indian origin", but without any comparison with Abu l-Wafa''s text, which was not yet made accessible.[34] Woepcke's praise for Chasles leaves no doubt about his devotion to the French mathematician. It also shows that both found it completely acceptable to make sweeping evaluations on a very thin basis of evidence.

This kind of macro-historical judgment does not necessarily yield wrong results. But it ignores too many questions that need to be studied if we wish to understand processes of knowledge transfer beyond the simple comparison of mathematical

---

[30]An example for this apologetic style is the following: "Comme un reproduction de ces démonstrations aurait décuplé l'étendue de cette notice, j'ai dû me borner à ne donner que les énoncés des propositions, vu le peu d'espace que ce Journal peut accorder à des publications de ce genre. Mais pour satisfaire les géomètres, j'ai placé en note des démonstrations de ces propositions en me servant de la notation algébrique moderne, où le plus souvent la démonstration se réduit à la simple inspection d'une identité." (Woepcke 1852, 422).

[31]"Toutefois, il est très-suprenant qu'Alkhayyâmi, en construisant les équations du troisième degré, n'ait pas remarqué l'existence des racines négatives. … C'est la vicieuse habitude de ne tracer que des demi-cercles, des demi-paraboles, et une seule branche des hyperboles, qui a fait manquer au géomètre arabe cette belle découverte. … Les Arabes savaient déjà qu'il existait une certaine équation du second degré à deux racines …; si donc Alkhayyâmî avair remarqué que pareillement une équation cubique admettait, en certains cas, trois solutions, il est difficile à croire que cette coincidence entre le degré du problème et le nombre des solutions ne l'eût pas frappé et conduit à des réflexions, et peut-être à des découvertes, ultérieures." (Woepcke 1851, xvi).

[32]An example is Woepcke's comparison between several constructions of Abu l-Wafa' and Bhaskara (Woepcke 1855, 219–20).

[33]"… elles s'éloignent très-sensiblement de l'esprit de la géométrie arabe, toujours fidèle, sous le rapport de la forme, à ses modèles grecs, …" (Woepcke 1855, 230).

[34]"La seconde est tout-à-fait d'origine indienne; …" (Chasles 1837, 454).

forms. First and foremost, such fast and facile evaluations reflect the lack of historical training of this generation of writers on the history of mathematics. This interpretive approach is on a par with the dominance of mathematical evaluations of works of the past in Woepcke's research practice. Both features highlight the manner in which he understood his obligations as a historian of mathematics. Being very honest about his beliefs and working methods, Woepcke formulated his position clearly:

> Namely, in the discussion of scientific borrowings made by one people from another one, the criterion, which needs to stand in the first place and which contributes much beyond all the others, is the conformity or the difference of the spirit of the methods; in the actual case, this criterion decides, as we have seen, in favor of the Indian origin of the two constructions of Abu l-Wafa'.[35]

These two core practices had a tremendous impact on the practice of historians of mathematics until recently. Macro-historical judgments and abstention from historicizing mathematical practices and their products continued to dominate the practice of historians of mathematics in Islamicate societies during the entire twentieth century.

4 Cantor's Strategies for Representing the Cultures of the Mathematical Sciences

Cantor's general history of mathematics follows other historiographical principles and goals than those favored by the specialists. First, Cantor wrote explicitly for only one audience—mathematicians. Second, he had no knowledge of Arabic, Persian, or Ottoman Turkish and could thus only work with translations or research done by specialists. It is here where his main relevance lies for a discussion on historiographical practices in regard to the mathematical sciences in Islamicate societies. Third, as Folkerts, Scriba, and Wussing explained, he was influenced in his approaches to the history of mathematics by the philosophical ideas of Arthur Arneth (1802–1858), who was a high school teacher in Heidelberg and a lecturer (Privatdozent) at the city's university (Dauben and Scriba 2002, 116, 123). Folkerts, Scriba, and Wussing considered two points in Arneth's publications as being of particular relevance to the way in which Cantor conceived of how to do history of mathematics. Arneth's belief that the development of mathematical knowledge was closely intertwined with general intellectual and cultural history, or as he wrote, the history of the human mind, furnished a strong stimulus for Cantor to look beyond

[35]"Or, dans la discussion des emprunts scientifiques faits d'un peuple à un autre, le critérium, qui doit figurer en première ligne, et qui l'emporte de beaucoup sur tous les autres, est la conformité ou la différence de l'esprit des méthodes, et dans le cas actuel, ce critérium décide, comme nous venons de le voir, en faveur de l'origine indienne des deux constructions d'Aboûl Wafâ." (Woepcke 1855, 237).

the confines of mathematics proper. Second, Arneth's engagement with the debate on the relationship between Greek and Indian mathematics that prevailed in France during much of the nineteenth century as mentioned already in the discussion of Woepcke's studies motivated a special attention to this issue (Dauben and Scriba 2002, 112, 116).

Moritz Cantor published the first volume of his *Vorlesungen über Geschichte der Mathematik*, which covered the period "from the oldest times until the year 1200", in 1880. No expert of the mathematical sciences in any Islamicate society himself, he could rely on a body of studies in German and French published by authors mentioned repeatedly in this paper and a few others like Jean Jacques Emmanuel Sédillot (1777–1832), Friedrich August Rosen (1805–1837), or Adolf Hochheim (1840–1898). He also worked with other secondary literature, in particular surveys. The most important books of this type for Cantor were Hermann Hankel's (1839–1873) *Zur Geschichte der Mathematik in Althertum und Mittelalter*, Gustav Weil's (1808–1889) *Geschichte der islamitischen Völker von Mohammed bis zur Zeit des Sultan Selim übersichtlich dargestellt* (Weil 1866), and Alfred von Kremer's (1828–1889) *Kulturgeschichte des Orients unter den Chalifen* (von Kremer 1877). Moreover, he read biographical and bibliographical material as provided, for instance, by Johann G. Wenrich (1787–1847) in his *De auctorum Graecorum versionibus et commentariis Syriacis, Arabicis, Armeniacis Persicisque* or Ferdinand Wüstenfeld (1808–1899) with his *Geschichte der arabischen Aerzte und Naturforscher* (Cantor 1880, vol. 1, 593–700). In addition, Cantor worked with literature from the early nineteenth century, publications of primary sources as early as the seventeenth century, and the most recent publications on history of ancient and medieval, European and non-European mathematics, Arabic literary and political history, and catalogues of cross-cultural translations, commentaries, and pseudepigraphic works. Here his preferred authors in addition to Woepcke, Sédillot, and Hankel were Moritz Steinschneider (1816–1907), Adolf Hochheim (1840–1898), Aristide Marre (1823–1918), Joseph von Hammer-Purgstall (1774–1856), und Friedrich Hultsch (1833–1906).

As his reading practice went beyond the mere mathematical and bio-bibliographical, his writing practice aimed at combining the political, cultural, and scientific in a manner that did not separate too strictly the one from the other. Although Cantor devoted many pages to the description of purely mathematical content, his introduction to Chap. XXXII clearly aims at explaining the emergence of highly successful scholars interested in the mathematical sciences as well as the particularities of the scholarly communities and their specific interests on the basis of what he extracted from the surveys on Islamicate history and culture. Similar efforts characterize Chaps. XXXIV–XXXVII (Cantor 1880, vol. 1, 629–30, 633–4, 636–7, 642, 649–50, 653, 668–9, 680–2).

Cantor follows two different pathways by discussing political and cultural history in its most general forms on the one hand and treating individual scholars not as mere repositories of old or new mathematical knowledge, but as people with a life and stories to be told. In this sense, his "history of mathematics of the Arabs" was conceptually more than the "scientific history of mathematics" of Nesselmann and the study of primary sources for discovering the Arabic achievements by Sédillot or Woepcke. Cantor certainly gave these approaches prominence in his survey, but embedded them in a political and cultural history, which he mainly borrowed from his Orientalist colleagues in Austria and Germany. In my view this effort possesses two impressive features. One is its existence; the second is its relative comprehensiveness, at least for the so-called classical period. Almost all major themes for explaining what today is called the translation movement can already be found in Cantor's summary: the role of Syriac-speaking Christian communities and their centers in Antioch, Edessa, Emessa, and Nisibis; the function of Christian communities in Sasanian Iran and of Middle Persian texts; the impact of Sanskrit texts and the (alleged?) visit of an Indian embassy to the Abbasid court in 773; the enormous thematic breadth, cultural diversity, and number of the translated texts; the role of the administrators, in particular the Barmakids; the relevance of trade, finances, and patronage (Cantor 1880, vol. 1, 594–602). Major translators and scholars of the mathematical sciences of the ninth and tenth centuries are mentioned by name and activities (Cantor 1880, vol. 1, 602–3, 627, 629). In addition, Cantor points to two further themes as if unproblematic: the division of Islamicate societies and their mathematical sciences in Eastern (Morgenland) and Western (afrikanische Nordküste, Spanien und Sicilien) and the emergence of a somehow impermeable separation (Scheidewand) between these two regions; and the fact that translating was often carried out in combination with commenting, although Cantor discusses these two activities separately from each other, since to him only the latter included "original thinking" (Cantor 1880, vol. 1, 605–606).[36]

All the themes that Cantor raised shaped the historical practices of scholars in the second half of the twentieth century and their historiographical debates. By contrast, the function of diagrams and diagram letters as carriers of historical information about practices and cross-cultural transformations, which Cantor emphasizes in his reflections on how to determine the cultural origins of al-Khwarizmi's algebra, has received solid attention among recent historians of mathematics only in the last decade or two (Cantor 1880, vol. 1, 621, 630.). Cantor is also conscious of two important methodological points that should guide the practice of a historian of mathematics (or any other field), but did not do so during the twentieth century: historical events or products are rarely the result of a single cause; speculations about possible motivations, factors, or causes that brought forth such events or products need to be based on in-depth and broadly conceived investigations of texts and other historical material. Hence, in sometimes perhaps tedious excursions, he discusses all possible reasons or interpretations of a method, a text, or an approach, including not only the standard search for whether something is of Greek or Indian origin, but also regional customs, consequences of theological

[36]The last point is expressed in the following statement: "Die Uebersetzungsthätigkeit war auch von einer vielfach commentirenden begleitet, auf die wir aber, da sie immerhin einige Ansprüche an das Selbstdenken des Commentators erhebt, bei den Originalarbeiten zu reden kommen." (Cantor 1880, vol. 1, 605).

debates, and conflicts with regard to mathematical themes and methods and their evaluation by medieval scholars (Cantor 1880, vol. 1, 616–26, 653–7).

Finally, it is not surprising that the sharp debates about interpretation left their traces in Cantor's depiction of Arabic mathematical texts. He sided with Sédillot and Woepcke by acknowledging original achievements made by Arabic writing scholars from Islamicate societies. He also sided with Biot, Hankel, and others by privileging the impact of Greek scientific works on the mathematical sciences in those later societies and by recognizing as the only other relevant source of inspiration Indian texts.

Despite refusing to follow Hankel in his mostly negative evaluations of the contributions to mathematical progress by Arabic writing scholars, Cantor takes up his predecessor's historiographical perspective in one point: decline. Chapter XXXVI is titled: The Decline of the eastern Arabic mathematics. Egyptian mathematicians.[37] The men we encounter in the first part of this chapter are: Nasir al-Din Tusi (1201–1274), Ghiyath al-Din Kashi (d. perhaps in 1429) and Baha' al-Din al-'Amili from Iran, Qadizade al-Rumi (d. after 1440) from Anatolia, and Ulugh Beg (d. 1449) from Transoxania. They all are portrayed as the last mathematicians and astronomers and somehow the representatives of decline (Cantor 1880, vol. 1, 668–72). Cantor presents as his understanding of decline the notion "that each progress ends".[38] The men listed as examples do not fit such a claim, as we know today.

A second of Cantor's criterion for decline is the disappearance of "receptivity in the domain of mathematics".[39] Here, his only example is Baha' al-Din al-'Amili's *Essence of Arithmetic* published by Nesselmann (Cantor 1880, vol. 1, 672–5). With this formulation, he did not mean to say that there was no appropriation of new mathematical knowledge from other than Islamicate societies. This theme does not belong to those that Cantor addressed in a serious manner. He simply means the fact that the compiler himself does not offer anything that Cantor could recognize as relevant for the progress of the mathematical sciences (Cantor 1880, vol. 1, 674–5).

Cantor's cultural and biographical approach to the history of mathematics in Islamicate societies was limited to macro-history. He had no strong interest in understanding local contexts and their impact on the use and representation of mathematical knowledge. Afterall, his question was that of most historians of mathematics of the nineteenth and twentieth centuries whose education had focused on mathematics and the sciences: who had contributed what to mathematical progress?

[37]"Der Niedergang der ostarabischen Mathematik. Aegyptische Mathematiker." (Cantor 1880, vol. 1, 668).

[38]"Mit den Männern, welche wir zuletzt genannt haben, hört jeder Fortschritt bei den Einen auf, während er bei den Anderen zu immer rascherer Gangart sich gestaltet." (Cantor 1880, vol. 1, 672).

[39]"Und auch die Empfänglichkeit der Araber auf mathematischem Gebiet war dahin." (Cantor 1880, vol. 1, 672).

5 Concluding Remarks

These limited explorations in the historiography of mathematics in Islamicate societies as practiced in the nineteenth century in France and Germany illustrate the profoundly ideological underpinnings of all these efforts, independent of their methodological orientations and technical tools in philology, mathematics, philosophy, or historiography. This should encourage us more than ever to reflect about our own ideological commitments and their impact upon our approaches to primary sources and their interpretation. The second result of this brief dip into some of the publications of nineteenth-century writers on subjects of the history of mathematics is the recognition that these writers formulated the basic approaches that have dominated our own research during the last decades. Their methodological perspectives and technical arsenal were, despite its factual limitations, broader than what we accept as respectable tools and questions. This came as an unexpected surprise to me and shows how productive an analysis of working practices of earlier generations of historians of mathematics can be. An important conclusion from my interrogations of the works discussed and the practices they represent is thus to encourage further studies of past working practices, their theoretical underpinnings and analytical tools. Such studies, if done with a sympathetic, but critical eye, can only enrich our own practices and improve our own capabilities of investigating and interpreting past mathematical works and their historical meanings. A third point has to be made with regard to interpretive practices. The authors of the nineteenth century were willing, too readily as I think, to generalize on a very thin basis of evidence and to create pictures of centuries of mathematical activities in a broad range of societies and scholarly communities that by now are so deeply anchored in our collective view of Islamicate societies that it takes great efforts to question, check, or deconstruct them. Macro-historical reconstructions with their immanent judgments have proven dangerous, if not detrimental, to a fair understanding of past processes and products. It is a major trap from which we have not yet liberated ourselves in our own practices.

The last comment to be made is not based directly on the material discussed in my paper. It results from a conversation I had recently with Lorraine Daston (Max Planck Institute for the History of Science, Berlin) on the topic of when and why the sciences and technology became identifiers of cultural progress of civilizations. She pointed out that a major shift occurred during the nineteenth century, which is not well studied yet and thus only badly understood. It replaced the concept of people by the idea that different peoples constituted civilizations apart from other such groupings and that such big clusters possessed an identifying culture whose values in a universal history of humanity could be measured in terms of their scientific and technological progress. Colonialism and its accompanying "civilizational mission", which was invented for legitimizing conquest and appropriation of large territories with everything that existed and lived there and the subjugation of "nature and culture" to the unconstrained exploitation by the conquerors, explain parts of this profound shift in perspective and explanatory approach to culture and its values.

But other changes in those societies that struggled for global domination must have occurred to call for this new grand narrative of civilizational superiority. One further social change that most likely contributed to its emergence were the internal struggles for or against modernization, secularization, and republicanism. Some of the consequences of those struggles were pointed out in my paper in the description of Boncompagni's patronage of history of mathematics and the questions and methods he favored. His conservative approach to society and politics was certainly not the only reflection of the impact that the fierce struggles within Italian and French societies for their further socio-political directions exercised on the practices of historians of mathematics. Hence, this larger perspective on the interpretive frameworks used by nineteenth-century writers about historical subjects also calls for more studies of the practices of our predecessors, their motivations and engagements with past mathematical cultures not only qua historians of mathematics, but as members of a living society with a broad spectrum of problems that necessitated them to take a position and stance through their choices of research topics, approaches, and tools.

References

Al-Afghani, J.-D. (1968). *An Islamic response to imperialism* (pp. 175–187) (N.R. Keddie, Trans.). Berkeley: University of California Press.

Cantor, M. (1880). *Vorlesungen über Geschichte der Mathematik* (Vol. 1). Leipzig.

Carra de Vaux, B. (1892). L'Almageste d'Abū ʾlWéfa alBūzdjāni. *Journal Asiatique, sér., 8*(19), 408–471.

Charette, F. (1995). *Orientalisme et histoire des sciences: l'historiographie européenne des sciences islamiques et hindoues, 1784–1900*. Thèse de M.A: Université de Montréal.

Chasles, M. (1837). *Aperçu historique sur l'origine et le développement des méthodes en géométrie*. Brussels: M. Hayez.

Dauben, J. W., & Scriba, C. J. (Eds.). (2002). *Writing the history of mathematics: Its historical development*. Basel: Birkhäuser.

Mazzotti, M. (2000). For science and for the Pope-King: Writing the history of the exact sciences in nineteenth-century Rome. *BJHS, 33*, 257–282.

Narducci, E. (1869). Intorno alla vita ed agli scritti di Francesco Woepcke. *Bullettino di bibliografia e storia delle scienze matematiche e fisiche*, 2, 119–152.

Nesselmann, G.H.F. (1842). *Versuch einer kritischen Geschichte der Algebra*. Berlin: G. Reimer. (Reprint Frankfurt am Main 1969).

Nesselmann, G.H.F. (1843). *Essenz der Rechenkunst von Mohammed Beha-eddin ben Alhossain aus Amul, arabisch und deutsch*. Berlin: G. Reimer.

Renan, E. (1883). *L'Islamisme et la science*. 29 mars Conférence. Université de Sorbonne, Paris.

Sédillot, L. A. (1835). Découverte de la variation, par Aboul-Wafā, astronome du Xe siècle. *Journal Asiatique, sér.*, *2*(16), 420–438.

Sédillot, L.P.E.A. (1845–1849). *Matériaux pour servir à l'histoire comparée des Sciences mathématiques chez les Grecs et les Orientaux* (Vol. 1). Paris: Firmin Didot Frères.

Taine, H.-A. (1866). *Nouveaux essais de critique d'histoire* (2e éd.). Paris: Librairie de L. Hachette et Cie.

Weil, G. (1866). *Geschichte der islamitischen Völker von Mohammed bis zur Zeit des Sultan Selim übersichtlich dargestellt*. Stuttgart: Rieger.

Woepcke, F. (1850). Notice sur un manuscrit Arabe d'un traité d'algèbre par Aboul Fath Omar Ben Ibrāhīm Alkhayāmī, contenant la construction géomètrique des équations cubiques. *Journal für die reine und angewandte Mathematik*, *40*, 160–172.

Woepcke, F. (1851). *L'algèbre d'Omar Alkhayyāmī, publiée, traduite et accompagnée d'extraits de manuscrits inédits*. Paris: Benjamin Duprat.

Woepcke, F. (851). Des systèmes de chiffres en usage chez différents peuples, et de l'origine de la valeur de position des chiffres indiens. *Nouvelles Annales des Mathématiques. Journal des Candidats aux Ecoles Polytechnique et Normale*, *10*, 372–207.

Woepcke, F. (1852). Notice sur une théorie ajoutée par Thābit Ben Korrah à l'arithmétique spéculative des Grecs. *Journal Asiatique, sér.*, *4*(20), 420–429.

Woepcke, F. (1853). *Extrait du Fakhrī, traitéd'algèbre par Aboū Bekr Mohammed Ben Alhaçan Alkarkhī, précédé d'un mémoire sur l'algèbre indéterminée chez les Arabe*. Paris: Imprimerie Impériale.

Woepcke, F. (1854). Recherches sur l'histoire des sciences mathématiques chez les orientaux d'après des traités inédits arabes et persans. Premier article. Notice sur des notations algébriques employées par les Arabes. *Journal Asiatique, sér.*, *5*(4), 348–384.

Woepcke, F. (1855). Sur une donnée historique relative à l'emploi des chiffres indiens par les Arabes. *Annali di scienze matematiche e fisiche compilati da Barnaba Tortolini*, *6*, 321–323.

Woepcke, F. (1863). Mémoire sur la propagation des chiffres indiens. *Journal Asiatique, sér.*, *6*(1), 27–79, 234–290, 442–529.

Woepcke, F. (1865–66). Introduction au calcul gobārī et hawāī. *Atti dell'accademia de'Nuovi Lincei* , *19*, 365–383.

Mesopotamian Mathematics, Seen "from the Inside" (by Assyriologists) and "from the Outside" (by Historians of Mathematics)

Jens Høyrup

Abstract

Since the 1950s, "Babylonian mathematics" has often served to open expositions of the general history of mathematics. Since it is written in a language and a script which only specialists understand, it has always been dealt with differently by the "insiders", the Assyriologists who approached the texts where it manifests itself as philologists and historians of Mesopotamian culture, and by "outsiders", historians of mathematics who had to rely on second-hand understanding of the material (actually, of as much of this material as they wanted to take into account), but who saw it as a constituent of the history of mathematics. The article deals with how these different approaches have looked in various periods: pre-decipherment speculations; the early period of deciphering, 1847–1929; the "golden decade", 1929–1938, where workers with double competence (primarily Neugebauer and Thureau-Dangin) attacked the corpus and demonstrated the Babylonians to have possessed unexpectedly sophisticated mathematical knowledge; and the ensuing four decades, where some mopping-up without change of perspective was all that was done by a handful of Assyriologists and Assyriologically competent historians of mathematics, while most Assyriologists lost interest completely, and historians of mathematics believed to possess the definitive truth about the topic in Neugebauer's popularizations.

Keywords

Cuneiform script, decipherment · Mesopotamian mathematics, historiography · Hincks, Edward · Rawlinson, Henry · Oppert, Jules · Thureau-Dangin, François · Neugebauer, Otto

J. Høyrup (✉)
Section for Philosophy and Science Studies, Roskilde University,
Box 260, DK-4000 Roskilde, Denmark
e-mail: jensh@ruc.dk

V.R. Remmert et al. (eds.), *Historiography of Mathematics in the 19th and 20th Centuries*, Trends in the History of Science,
DOI 10.1007/978-3-319-39649-1_4

1 "Through a Glass, Darkly": Historians of Mathematics Before Assyriology

Until 1850, historians of mathematics had no other way to know about pre-classical Near Eastern mathematics[1] than using the information they could draw from classical authors, at best submitted to historical and epistemological common sense—whence the quotations from 1 Corinthians 13:12 in the above headline (which entails no promise like that of St. Paul that in the end we shall see "face to face"). This, for instance, is Jean-Étienne Montucla's (1725–1799) account, in Enlightenment spirit, about "what is told" about the birth of arithmetic (Montucla 1758: I, 46f)[2]:

> The Phoenicians, some say, were the first and the most able merchants of the world; but Arithmetic is nowhere more useful and more necessary than in trade: these people must therefore also have been the first Arithmeticians. Strabon[3] relates this as the accepted opinion of his age; and even if we should believe a historian,[4] Phoenix son of Agenor wrote as first an arithmetic in Phoenician language. On the other hand Egypt boasts of having been the cradle of this art[5]; and since a human intelligence hardly seemed to suffice for so useful an invention, one devised the pious fable that a god was its author, and had communicated it to mankind.[6] At least it was the general opinion, according to Socrates or Plato[7] that *Theut* was the inventor of numbers, calculation and geometry; and it is quite likely that the Greeks took from here the idea to attribute to their Mercury, with whom *Theut* or the Egyptian *Hermes* has a conspicuous connection, the jurisdiction of trade and arithmetic.[8]
>
> But I shall insist no further on these fabulous or risky lines; who wants to discuss the origin of our knowledge somewhat philosophically will see that Arithmetic must have preceded everything else. The first civilized societies could not do without it; it suffices to

[1]A conceptual clarification: The "Near East" encompasses Egypt, the Palestino-Syrian area, Arabia and Mesopotamia—sometimes other neighbouring areas are included as well. Mesopotamia largely coincides with present-day Iraq. Its northern third is Assyria, and the remainder is Babylonia. Chaldea strictly speaking is the southern third (in the third millennium BCE Sumer), but often in the quotations that follow it stands for the whole of Babylonia.

[2]My translation, as everywhere below where nothing else is stated. All translated quotations can be found in original language in the preprint version of the article, on http://rudar.ruc.dk//bitstream/1800/10613/1/Hoyrup_2013_c_Mesopotamian_mathematics_from_the_inside_and_from_the_outside_S.pdf. In cases where the titles of publications have been translated, the genuine titles can be found in the bibliography. The notes to the quotation are due to Montucla, my additions are in square brackets. Similarly below for notes within quotations.

[3]*Geograph.*lib.XVII.

[4]Cedrenus [an 11th-century Byzantine historian].

[5]Diog. Laer. *in proemio*. [Hicks 1925: I, 12].

[6]*In Phædro*. p. 1240 ed. 1602. [274c].

[7]Ibid.

[8][At this point, the astronomer Joseph-Jérôme Lalande (1732–1807) adds the following in the second edition—the earliest reasoned reference to *Babylonian* mathematics (Montucla 1799: I, 43f): "It is even quite difficult not to affiliate them with the Chaldeans. They, indeed, present us with the first traces of astronomical knowledge, very advanced at that. How would they, without that tool, have been able to discover several astronomical periods, knowledge of which has come down to us!" Apart from that, Montucla's passage is unchanged].

> possess something for being forced to use numbers, and even the first men, if only they had to count days, years, their age, that is enough for saying that they knew Arithmetic. Admittedly, richer or more trading societies may have expanded the limits of this natural Arithmetic by inventing perhaps shortened ways or procedures; and in this sense Strabon has said nothing contrary to reason. As regards *Josephus'* account[9] indicating Abraham as the first of Arithmeticians and making him teach the Egyptians the first elements of Arithmetic, it is easy to see that this historian wanted to adorn the first father of his nation with part of that knowledge which he saw honoured among foreigners. This is one of those pictures that will be favourably received only by some compiler deprived of critical sense and reasoning.

The last line could be directed at Petrus Ramus (1515–1572), in whose *Scholae mathematicae* (1569: 2) this story is taken for a fact (yet with a correct reference to chapter 8 of Josephus).[10]

Abraham Gotthelf Kästner (1719–1800) has no more sources than Montucla and is even more cautious in his *Geschichte der Mathematik* (1796: I, 2):

> To us, the oldest teachers of mathematics were the Greeks. What *they* may have learned from the Orient we only know from their own confessions, and how far their teachers have continued on their own, that they did not consider it necessary to write down. [...].

These two quotations, with the addition quoted in note 8, illustrate how much could be known about the mathematics of Mesopotamia and neighbouring areas until the birth of Assyriology.

2 The Beginnings of Assyriology

The earliest dead languages and writing systems to be deciphered were Aramaic dialects—first Palmyrene in 1754, then in 1764 and 1768 Phoenician and Egyptian Aramaic (Daniels 1988, 431); all three scripts were alphabetic, and the basis was provided by bilingual texts containing proper names, which were skilfully exploited by Jean-Jacques Barthélemy (1716–1795).

Much more famous is Jean-François Champollion's (1790–1832) use of the Rosetta Stone in the decipherment of the hieroglyphs and the Demotic script (1824), proving the mixed alphabetic-ideographic character of the former as well as the existence of homophones in the alphabet.

---

[9]*Ant. Jud.* liv. I c. 9. [Actually chapter 8].

[10]Abraham is at least absent from Giuseppe Biancani's (1566–1624) *Clarorum mathematicorum chronologia* (1615, 39), and also from Gerardus Vossius's (1577–1649) *De universae mathesios natura et constitutione liber* and *Chronologia mathematicorum* (1650), while Polydorus Vergilius (c. 1470–1555) (1546, 59f) has no chapter reference. Since Montucla does not abstain from identifying Ramus by name when chiding him for following "the inclination of the mob toward everything that seems marvellous" (p. 450), the present reference is most likely at least not to be to Ramus alone.

The decipherment of the cuneiform scriptures was a more involved affair—a short description will illustrate how much more involved. It will make it clear why even understanding of cuneiform *mathematics* had to be made in very small and slow steps.

Initially, everything was based on the trilingual inscriptions from Persepolis, which Pietro della Valle (1586–1652) had seen in 1621 to be written from left to right.[11] The development until around 1800 is described by Fossey as on the whole a "period of groping and of hazardous and contradictory hypotheses" (p. 90). Noteworthy positive contributions were, firstly, Carsten Niebuhr's (1733–1813) new and more precise copies of the Persepolis inscriptions—his discovery that three different scripts are involved—and his confirmation of the writing direction (1774: II, 138f, pl. XXIII, XXIV, XXXI); and secondly, at the very close of the period, Friedrich Münter's (1761–1830) dating of the inscriptions to the Achaemenid era (1798, published in Danish in 1800)—his confirmation that three scripts are used—and his arguments that the first of these is alphabetic, the second apparently mixed alphabetic-syllabic and the third perhaps mixed alphabetic-logographic (Münter 1802, 83–86)—his identification of a few signs from the alphabetic script as vowels (ibid., 104–109)—and his identification of its language as Old Iranian (more precisely he suggests Zend). Also of importance was Münter's verification that the Persepolis writing type had also been used in Babylon, and that it had probably originated in Mesopotamia (*ibid*, 129–144).

In 1802, Georg Friedrich Grotefend (1775–1853) presented a memoir to the Göttingen Academy[12] which is habitually taken as the stumbling beginning of decipherment proper. He came to the same conclusions as Münter (whose work only appeared in German during the same year, and which Grotefend may not have known). He went further on three decisive points: showing that all inscriptions were linked to Darius and Xerxes; finding the royal names mentioned in the inscriptions as well as the word for king; and using this to identify a number of letters (he claimed identification of 29 letters of the alphabetic script, 12 of which were later confirmed).

Over the next four decades or so, a number of scholars extended and corrected Grotefend's work, removing false values and adding new ones (not always correctly at first), and identifying the language as an Old Persian dialect distinct from Zend (adding also new inscriptions to the corpus) (Fossey 1904, 112–146). However, all of this concerned the alphabetic script, which was certainly derived from the cuneiform script of Mesopotamia but had a totally different character (and moreover concerned matters without the slightest relation to mathematics).

[11]What follows about work done before 1860 is drawn, when no original sources are referred to, from Charles Fossey's (1869–1946) very detailed exposition of (good and bad) arguments and results (Fossey 1904, 85–220).

[12]Grotefend (1802) was published only in full in (Meyer 1893), for which reason I build on Fossey's account (1904, 102–111) of the arguments that circulated.

Decipherment of the second script (Elamite), using about one hundred signs and being in a language with no known kin, made some but little progress during the same period, and is anyhow irrelevant for the present purpose. Grotefend made some attempts at the third script, which is Akkadian (the language of which Babylonian and Assyrian are dialects). His firm belief that the language had to be an Iranian dialect was one of the reasons he had no success—but until the second half of the 1840s nobody else did much better. In the meantime, excavations had begun, and a much larger, geographically wider and chronologically deeper text corpus was now available.

From 1845 onward, a large number of workers took up discussion and competition about the third script, from which some 300 signs were known: Isidore Löwenstern (1810–1858; 1859; pertinent publications 1845 and onward); Henry Rawlinson (1810–1895; 1846 and onward); Paul-Émile Botta (1802–1870; 1847 and onward); Edward Hincks (1792–1866; 1846 and onward); Félicien de Saulcy (1807–1880; 1847 and onward); Henry de Longpérier (1816–1882; 1847); Charles William Wall (1780–1862; 1848); and Moriz Abraham Stern (1807–1884; 1850)—of whom Rawlinson, Botta and Hincks were by far the most important. Before 1855 it was known that the language of the third script was that of Babylonia and Assyria; that this language (Akkadian) was a Semitic language, and thus a cognate of Arabic and Hebrew; that the same sign might have (mostly several) phonetic and (often several) logographic values, and even function as a semantic determinative (an unexpected function which Champollion had discovered in Hieroglyphics); and that the original shape of the signs had been pictographic. Moreover, Hincks had shown early on that the inventors of the script must have spoken a non-Semitic language. This is all summarized in a letter written by the young Jules Oppert (1825–1905) in 1855 (published as (Oppert 1856)), together with observations and hypotheses of his own. So, from now on large-scale reading of documents could begin—and we may speak of the birth of Assyriology. In (1859), Oppert himself was to stabilize the field—in his obituary of Oppert, Léon Heuzey (1831–1922) was eventually to write as follows (Heuzey 1906, 7):

> After some works on ancient Persian, Oppert concentrated his principal effort on the Assyrian inscriptions. Having been charged by Fresnel with a mission into Babylonian territory, he published at his return, in 1859, a volume, the second tome (in date actually the first) of his *Expédition en Méslopotamie* [sic] in which, using recently discovered sign lists or syllabaries, he established the central rules of decipherment. This volume, Oppert's *chef d'œuvre*, indicates a turning-point; it put an end to gropings and instituted Assyriology definitively.

3 Assyriologists' History of Mathematics, 1847–1930

On one account Oppert says nothing in his letter from 1855, even though this was to be one of the things that occupied him during his later brilliant career: mathematics.

However, already in a paper read in 1847 (published as (Hincks 1848)), Hincks had described the "non-scholarly" number system correctly.[13] In comparison, of the 76 syllabic values identified in this early paper only 18 turned out eventually to be correct or almost correct, while 46 had the right consonant but erred in the vowel, and 12 were wholly wrong (Fossey 1904, 185)—which however was already a significant step forward. The discovery of the place-value system followed soon. It was also due to Hincks (1854a, 232), who detected it in a tabulated "estimate of the magnitude of the illuminated portion of the lunar disk on each of the thirty days of the month".[14] A slightly later publication dealing with the numbers associated with the gods (Hincks 1854b, 406f) refers to the "use of the different numbers to express sixty times what they would most naturally do" on the tablet just mentioned; there, 240 is indeed written as iv (Hincks uses Roman numerals for the cuneiform numbers), while "iii.xxviii, iii.xii, ii.lvi, ii.xl, etc." stand for "208, 192, 176, 160, etc.".

Rawlinson also contributed to the topic in (1855) (already communicated to Hincks when the second paper of the latter was in print, in December 1854). A five-page footnote (pp. 217–221) within an article on "The Early History of Babylonia" points out that the values ascribed by Berossos (Cory 1832, 32) to σάρος (*šār*), νήρος (*nēru*) and σώσσος (*šūši*), respectively 3600, 600 and 60 years, are "abundantly proved by the monuments" (p. 217). As further confirmation Rawlinson presents an extract of "a table of squares, which extends in due order from 1 to 60" (pp. 218–219), in which the place-value character of the notation is obvious but only claimed indirectly by Rawlinson. The note goes on as follows:

> while I am now discussing the notation of the Babylonians, I may as well give the phonetic reading of the numbers, as they are found in the Assyrian vocabularies.

All three of "cuneiform's 'holy' triad", as Rawlinson, Hincks and Oppert were called by Samuel Noah Kramer (1897–1990) (1963, 15), were indeed quite aware that numbers and what had to do with them was important for understanding Mesopotamian history and culture.[15]

The reason that this was so is reflected further on in Heuzey's obituary:

> Oppert's scientific activity followed many directions: historical texts and religious texts, bilingual (Sumero-Assyrian) texts and purely Sumerian texts, juridical texts and divination texts, Persian texts and neo-Susian texts, there is almost no branch of the vast literature of

[13]This system is sexagesimal but not positional until 100, after which it is combined with word-signs for 100 and 1000.

[14]Archibald Henry Sayce (1845–1933), when returning to the text in (1875, 490; cf. Sayce 1887, 337–340), reinterprets the topic as a table of lunar longitudes. Geometrically, the two interpretations are equivalent, but the final verb of the lines (DU, "to go") suggested this new understanding.

[15]In contrast, the just published Blackwell *Encyclopedia of Ancient History* planned the same number of pages for Mesopotamian mathematics and Mesopotamian hairstyles. It should be added that those who planned the volume had little idea about Mesopotamia (nor were they very interested in receiving advice, however).

the cuneiform inscriptions he has not explored. The most special questions, juridical, metrological, chronological, attracted his curiosity [...].

Evidently, administrative, economical and historiographic documents could—and can—only be understood if numeration and metrology were/are understood. Reversely, such documents, in particular administrative and economical records, are and were important sources for understanding numeration and metrology. Oppert's observations on "the notation for capacity measures in cuneiform juridical documents" from (1886) offer an example.

For a long time, however, they were far from being the only sources for knowledge and assumptions. Already the decipherment of the scripts had drawn much on sources from classical Antiquity (how else could the names of the Achaemenid kings have been known?) and on comparison with Zend, Hebrew and Arabic. Similarly, known or supposedly known metrologies and numerical writings from classical Antiquity were drawn upon—sometimes with success (Rawlinson's use of Berossos is an example), sometimes with exaggerated confidence in the stability and uniformity of metrologies. Didactical materials such as bilingual lexical lists and tables also played their role (as they had done in the decipherment); so did astronomical texts (as already for Hincks and Rawlinson in 1854–55).

Three illustrative examples are (Norris 1856, Smith 1872, Oppert 1872). Edwin Norris (1795–1872) not only draws much on Biblical material in his dubious article but also reads the cuneiform signs on Mesopotamian weight standards as Hebrew characters ("I thought the first word looked like מנה"—p. 215). George Smith (1840–1876) makes use of metrological lists in order to establish the sequence of length units and their mutual relations, and of "lion" and "duck weights" (that is, stone sculptures of these animals on which their weight is inscribed) and of written documents in order to reach a similar understanding of the weight system (which turns out to be contradictory).

Oppert makes use of similar material. But he also believes in a shared stable "ancient" metrology,[16] and draws in particular on Hebrew parallels (and on Hebrew measures which he assumes *must* have had a parallel[17]). Friberg (1982, 2) justly characterizes the outcome as "somewhat premature", even in comparison with other publications from the period.

The use of the place-value principle not only for integers but also for fractions was established in Johann Strassmaier's (1846–1920), Josef Epping's (1835–1894) and Franz Xaver Kugler's (1862–1929) analysis of the Late Babylonian astronomical texts, beginning with (Strassmaier and Epping 1881); so far, however, it

[16]We may see this belief as the last scholarly and pseudo-scholarly survivor of the Renaissance faith in ancient *prisca sapientia*. Paradoxically, Oppert had pointed out already in (1886, 90) that there were "in Assyria and Chaldea, as everywhere else, ceaseless variations in the measures", which should have warned him against the dangers inherent in the comparative method.

[17]See for example p. 427f on the postulated unit "hair", which leads him to rather far-fetched hypotheses (presented "with all reserve", it is true).

was understood only in analogy with the use of sexagesimal fractions in Ptolemaic and modern astronomy.[18]

In a way, Hermann Hilprecht's (1859–1925) *Mathematical, Metrological and Chronological Tablets from the Temple Library of Nippur* (Hilprecht 1906) constitutes a decisive step. As we have seen, tables of squares and metrological lists had already been used in the early period by Rawlinson (1855) and Smith (1872). Hilprecht, however, put at the disposal of Assyriologists a large number of arithmetical and metrological tables. Unfortunately, his failing understanding of the floating-point character of the place-value system; the still strong conviction that the classical authors could provide interpretations of Mesopotamian texts; and a belief that everything Babylonian had to be read in a mystico-religious key[19] caused him not only to read very large numbers into the texts but also to understand a division of 1;10 (or 70) by 1 as 195,955,200,000,000 ÷ 216,000 (p. 27), where the denominator was then explained from a (dubious) interpretation of the passage about the "nuptial number" in Plato's *Republic* VIII, 546B–D (pp. 29–34) and coupled to postulated cosmological speculations:

> How can this number influence or determine the birth and future of a child? The correct solution of the problem is closely connected with the Babylonian conception of the world, which stands in the centre of the Babylonian religion. The Universe and everything within, whether great or small, are created and sustained by the same fundamental laws. The same powers and principles, therefore, which rule in the world at large, the macrocosm, are valid in the life of man, the microcosm.

So, while Hilprecht's publication represented a material step forward, his approach remained that of the nineteenth century.

Franz Heinrich Weißbach's (1865–1944) article "about the Babylonian, Assyrian and Old Persian weights" (Weißbach 1907), on the other hand, inaugurated a new trend. As formulated by Powell (1971, 188), "the study of Mesopotamian weight norms can be divided into two eras: the pre-Weissbach and the post-Weissbach eras". Weißbach discarded the comparative method, concentrating (like George Smith) on what could be derived from Mesopotamian sources and artefacts. He did not convince those who were committed to the "comparativist paradigm"; instead, the process confirms the observation made by Max Planck (1950, 33) (concerning Ludwig Boltzmann) and famously quoted by Thomas Kuhn (1970, 151), namely that

[18]Basing himself on indirect evidence and on Greek writings, Johannes Brandis (1830–1873) had already claimed that the unending sexagesimal fraction system of the Greek astronomers had to be of Chaldean origin "even if we never find direct or indirect testimony ascribing it to them" (Brandis 1866, 18).

[19]Hilprecht quotes this passage from Carl Bezold's (1859–1922) "concise survey of the Babylonian-Assyrian literature" (Bezold 1886, 225): "As far as we know by now, Babylonian-Assyrian mathematics primarily served astronomy, and this on its part a pseudo-science, astrology, which probably arose in Mesopotamia, propagated from there into the Gnostic writings, and was inherited by the Middle Ages, although we are not yet able to reconstruct this whole chain, the links of which are often broken from each other".

a new scientific truth does not triumph by convincing its opponents and making them see the light, but rather because its opponents eventually die, and a new generation grows up that is familiar with it.

The following generations of Assyriologists, indeed, less trained in Hebrew and classical scholarship but familiar with the results of a mature discipline, followed the model set by Weißbach and by the immensely influential François Thureau-Dangin (1872–1944). The earliest work on metrology and mathematical techniques of the latter had been published in 1897 (a sophisticated interpretation of the intricate calculations on a field plan from the outgoing third millennium BCE); he was going to publish on metrological questions for decades to come.[20]

So far, Assyriologists had been concerned with *mathematics in use*, namely in use in non-mathematical documents (including astronomy and texts serving in elementary mathematical training). The first to come to grips with what became known among historians of mathematics as "Babylonian mathematics" from the 1930s onward—namely mathematics which was complicated enough to be counted as mathematics by those same historians in the twentieth century—was the 25-years old Ernst Weidner (1891–1976) in (1916) (he had already published a volume on Babylonian astronomy in 1911 (Jaritz 1993, 15)). Weidner's article begins with the observation that

As regards the knowledge of the Akkadians in the area of mathematics we are still quite badly informed.[21] Beyond a few tables containing square and cube numbers and rather many multiplication tables we have nothing real except building- and field-plans, which already in quite early time makes us presuppose an Akkadian ability also to accomplish rather difficult calculations.

which summarizes the situation perfectly. Two difficult texts had been published in 1900, Weidner says[22] (only in cuneiform, with neither transliteration nor translation, which presupposes no understanding and conveys none); but these texts are then characterized as "probably the most difficult transmitted in cuneiform", which explains that nothing had been done on them. In the Berlin Museum he had now seen other texts of the same type, and he analyses two problems from one of them (VAT 6598): two different approximate calculations of the diagonal of a rectangle (a first and a corrupt second approximation, see (Høyrup 2002, 268–272)).

Compared to the analysis of the same text offered by Otto Neugebauer (1899–1990) in (1935), Weidner's interpretation contains some important mistakes, for which reason it can certainly be characterized as premature. However, Weidner's short paper, together with the commentaries of Heinrich Zimmern (1861–1931)

[20]Outside Assyriology, in particular among natural scientists taking interest in Antiquity and its mysteries, the comparativist trend is still alive and kicking—see (Berriman 1953; Rottländer 2006; Lelgemann 2004).

[21][A footnote refers to Moritz Cantor's *Vorlesungen* I, on which below.].

[22]Now known as BM 85194 and BM 85210.

(1916) and Arthur Ungnad (1879–1945) (1916, 1918) provided the first understanding of Babylonian mathematical *terminology*.[23]

In (1922), Cyril John Gadd (1893–1969) published a text dealing with subdivided squares, and added some further important terms (not least those for square, triangle and circle). Since the text in question contains no calculations, only terms for mathematical objects occur, none for operations. In (1928), finally, Carl Frank (1881–1945) published the collection of *Straßburger Keilschrifttexte*—a spelling that adequately reflects his working situation: he had made the copies before the War, when Strasbourg was Strassburg, and only received his own material in 1925, with no possibility of collating. None the less, Frank's book added another batch of terms. Because Frank's texts are even more difficult than the short ones dealt with by Weidner, Zimmern and Ungnad, and because Frank translated all sexagesimal place value numbers into modern numbers (repeatedly choosing a wrong order of magnitude), his understanding of the texts was rather defective.

This is how far Assyriologists went in the exploration of cuneiform mathematics until 1930—when Assyriology was half of its present age.

4 Historians of Mathematics Until c. 1930

On the whole, historians of mathematics depended during the same period not only on the material put a their disposal by Assyriologists but also on their interpretations.

In the posthumous (Hankel 1874), Hermann Hankel (1839–1873) dealt with "the Babylonians" (once, p. 65, accompanied by the Assyrians) on scattered pages of his discussion of the "pre-scientific period". Given the difficulties of Assyriologist with not only absolute but also relative chronologies until (Hommel 1885), it is no wonder that Hankel's observations are messy on this account. Substantially, he speaks about the sexagesimal divisions of metrologies (pp. 48f; not mentioning that not all subdivisions are sexagesimal, which was known at least since (Smith 1872)); a hunch of sexagesimal fractions (pp. 63, 65; but only to one place, and understood as written with a denominator which is "usually omitted"); the existence of tables of squares and astronomical tables (the two texts used by Hincks and Rawlinson in 1854–55), from which the hypothesis is derived that the Babylonians were interested in arithmetical series (p. 67); and a low level of geometry, concluded on the basis of the "building art deprived of style" (p. 73). Iamblichos's claim that Pythagoras had his knowledge of the harmonic proportion from the Babylonians is mentioned but explicitly not endorsed (p. 105).

[23]Only the terms for (what can approximately be translated as) square and cube roots were known since Moritz Cantor's use of Hilprecht's material in (1908). Quite a few of Weidner's readings later turned out to be philologically wrong while their technical interpretation was adequate. What was correct, however, was important later on, and some of the philological errors were still taken over in Neugebauer's early interpretations without great damage.

In the first edition of volume I of his *Vorlesungen* from (1880), Moritz Cantor (1829–1920) dedicates separate chapters to the Egyptians and the Babylonians—the latter on pp. 67–94. He is much better informed than Hankel—in part, it must be admitted, from publications that had appeared too late to be taken into account by Hankel, such as (Oppert 1872).[24] He offers an orderly exposition of the numerals and the "natural fractions" 1/6, 1/3, 1/2, 2/3 and 5/6. Further, he describes the tables of squares and cubes (which in (1908) he was going to see as tables of the corresponding roots), and he discusses the sexagesimal place value principle in connection with astronomy. Geometry is dealt with on the basis of geometric decorations, Herodotos and other Greek authors, and the Old Testament. Babylonian numerology is also discussed, in particular the ascription of numbers to the gods.

In the third edition from (1907), Cantor deals with the Babylonians before the Egyptians (pp. 19–51). The five extra pages allow him to tell the historiography of the field, but apart from a suggestion of that reinterpretation of Rawlinson's tables of squares as tables of square roots which he was to publish in full in (1908), nothing substantial is changed in the account of Babylonian mathematics. There are, however, some remarks about the material published by Hilprecht in (1906), with faithful adoption of his immense numbers (pp. 28f).

Hieronymus Georg Zeuthen's (1839–1920) *Geschichte der Mathematik im Altertum und im Mittelalter* (1896) dedicates a chapter (pp. 8–13) to what the Egyptians and the Babylonians knew in mathematics at the moment they came into touch with the Greeks, and which the Greeks might possibly have taken over from them (thus pp. 8f). Of the six pages, 26 *lines* deal with the Babylonians. 21 of these lines refer to astronomy and the division of the circle into 360°, and 5 to the possibility that Greek numerology was in debt to Babylonians and Chaldeans.

Johannes Tropfke (1866–1939) follows Hankel's pattern in the first volume of his *Geschichte der Elementarmathematik* (1902), mentioning the Babylonians now and then but not treating Babylonian mathematics per se—the obvious choice, given his full title "history of elementary mathematics presented systematically". But he only speaks about the sexagesimal system (mentioning Rawlinson's "square table" but without describing it). Only on two (quite dubious) points does he go beyond Hankel: he considers Iamblichos a certain source, and he claims (p. 304) that the Babylonians knew the solution 3–4–5 to the "Pythagorean equation"; he gives no source, and would have been unable to, since no pertinent text was known at the time. Most likely, he misremembers Cantor's idea (1880, 56) ("admittedly, for the moment without any foundation", thus Cantor) that *the Egyptians* might have used 3–4–5 triangles on ropes to construct right angles.

[24]Already Cantor's "mathematical contributions to the cultural life of the nations" had contained a chapter on the Babylonians (Cantor 1863, 22–38). At the time, however, he had only been able to speak about the decipherment; about "Oriental" culture in general; and about the writing system, about integer numerals and about the possible use of some kind of abacus (a hypothesis which he repeats in the *Vorlesungen*).

In general, historians of mathematics were not interested in Babylonian matters during the period. Inspection of 21 of the first 26 volumes of the series *Abhandlungen zur Geschichte der Mathematik*[25] (1877–1907) reveals no single article on the subject. For good reasons, as revealed by what Hankel and Cantor had been able to say about it—what Assyriologists had succeeded in finding out was still so tentative and so incoherent that it invited more to speculation than to solid work. The other possible explanation—that the authors should have been interested only in the higher level of mathematics—can be safely disregarded for the period before 1914, witness the many articles on elementary topics published in the same series.

5 The Long 1930s—Neugebauer, Struve, Thureau-Dangin, and Others

Beginning in 1929, the distinction between Assyriologists and historians of mathematics becomes irrelevant (for a while). This is the period when the advanced level of Old Babylonian and Seleucid mathematics was deciphered for good, after the modest but decisive beginnings made by Weidner, Zimmern, Ungnad, Gadd and Frank.

Neugebauer, it is true, is normally counted as a historian of mathematics. If anything, historian of astronomy would be the correct denomination—as we shall see, mathematics only occupied a rather short stretch of his life. But he had also been trained in Assyriology by none less than Anton Deimel (1865–1954), as he tells with gratitude in (1927, 5). Vasilij Vasil'evič Struve (1889–1965) was an Egyptologist but had also been trained in Assyriology, which was soon to become his main field. Thureau-Dangin was one of the most eminent Assyriologists of his (and all) times, but the contrast between his works from the 1920s (and before) and those from the 1930s demonstrates how much the discussions (and competition) with Neugebauer and the perspective of the history of mathematics had changed his approach. Hans-Siegfried Schuster (1910–2002), who made an important contribution in c. 1929, became an Assyriologist but participated in Neugebauer's seminar in Göttingen (Kurt Vogel, personal information; (Neugebauer 1929, 80)); Heinz Waschow (1914–?) studied not only Oriental philology (including Assyriology) from 1930 until 1934 but also applied mathematics (Waschow 1936, unpaginated CV). Albert Schott (1901–1945), the last of Neugebauer's contacts, had a strong interest in astronomy (but the numerous references to his assistance in (Neugebauer 1935–1937) all refer to strictly philological matters). Kurt Vogel (1888–1985), who sometimes took part in the discussion, was a mathematician and historian of mathematics but also trained in Egyptian (as well as Greek, and later also medieval Italian and German) philology—not to speak of his war experience as a military engineer.

[25] The exceptions are vols 2, 16, 19–20 and 25, to which I have no access; there is no reason to believe they should change the general picture.

Since Neugebauer's person was all-important for what happened in the 1930s, some words about his background may be fitting. His *Doktorarbeit* from (1926) had dealt with the Egyptian fraction system, but already while working at it he had become interested in the mathematics of the Sumerian cultural orbit as a parallel that might throw light on Egyptian thought, and been convinced (with due reference to Thureau-Dangin) that metrology was all-important for the development of early mathematics (Neugebauer 1927, 5).

In 1929, he launched *Quellen und Studien zur Geschichte der Mathematik, Astronomie und Physik* together with Julius Stenzel (1883–1935) and Otto Toeplitz (1881–1940). Since the co-editors were 17 respectively 19 years older than Neugebauer, there can be little doubt that the initiative was his. He may not have written the *Geleitwort* ("accompanying observations"), but if not he must at least have agreed with it (Neugebauer et al. 1929, 1–2):[26]

> [...] Through the title "Sources and Studies" we want to express that we see in the constant reference to original sources the necessary condition for all historical research that is to be taken seriously. It will therefore be our first aim to makes *sources* accessible, that is, in as far as possible to present them in a form that can meet the requests of modern philology but also, through translation and commentary, enable the non-philologist to always convince himself of the precise words of the original. To really meet the legitimate wishes of *both* groups, philologists and mathematicians, will only be possible if a genuine *collaboration* between them is brought about. To prepare that will be one of the most important tasks of our enterprise.
>
> The technical realization of this programme we intend to achieve through the publication of two irregular publication series. One, A, will accommodate the major proper editions, containing the text in its original language, philological apparatus and commentary and a translation that is as literal as possible, and which will make the contents of the text as easily accessible also to the non-philologist as can reasonably be done. Each issue of these "Sources" is to be a self-contained piece. The issues of Section B, "Studies", are to contain each a collection of articles having a closer or more distant relation to the material coming from the Sources.
>
> The "Sources and Studies" are to provide contributions to the *history* of mathematics. But they do not address specialists of the history of science alone. They certainly intend to present their material in such a form that it can *also* be useful for specialists. But they also address all those who think that mathematics and mathematical thinking is not only the concern of a particular science but also deeply connected with our total culture and its historical development, and that the attention to the *historical* genesis of mathematical thinking can provide a bridge between the so-called "sciences of the spirit" [*Geisteswissenschaften*] and the seemingly so a-historical "exact sciences". Our final aim is to participate in the building of such a bridge. [...]

So, a common endeavour between philologists and historians of mathematics was aimed at, for the benefit of both groups as well as a broader educated public. That those publications about Babylonian mathematics that appeared in the journal did not cast much light on the role of mathematics in general culture was not a result of failing will; as Neugebauer had to point out in (1934, 204), one should "not

[26]Some stylistic features do point to Neugebauer as the writer. As revealed by the quotation in note 41, at least Toeplitz also had an attitude in conflict with the present text.

forget that we still know practically nothing about the whole setting of Babylonian mathematics within the context of general culture".

In the first issue, Neugebauer and Struve (1929) published an article "about the geometry of the circle in Babylonia" (actually also about other geometrical objects). Among the results is the identification of a technical term for the height of plane or solid geometric figures. The explanation is philologically mistaken, but as in the case of Weidner's similar errors this is not decisive, as pointed out by Thureau-Dangin (1932a, 80) in the note where he gives the correction.

The preceding article in the same issue is by Neugebauer alone (1929). It offers a new analysis of some of Frank's texts, and manages to elucidate much which had remained in the dark for Frank. Neugebauer's main tool is of astonishing simplicity: he retains the sexagesimal shape of numbers, while Frank, in order to get something more familiar to a modern mathematical eye, had translated them into decimal numbers (and often translated them into a wrong order of magnitude, as observed above). Beyond that, Neugebauer offers a number of improved readings.[27] Some of the problems, it turns out, contain questions of the second degree. Neugebauer concludes (pp. 79f) in these words:

> One may legitimately say that the present text presents us with a piece of Babylonian mathematics that enriches our all too meagre knowledge of this field with essential features. Even if we forget about the use of formulas for triangle and trapezium, we see that complex linear equation systems were drawn up and solved, and that the Babylonians drew up systematically problems of *quadratic* character and certainly also knew to solve them – all of it with a computational technique that is wholly equivalent with ours. If this was the situation already in Old Babylonian times, hereafter even the later development will have to be looked at with different eyes.

A note added after the proofs had been finished then reveals that a text has been found which *solves* mixed second-degree problems, referring to the essential role played Schuster for understanding this, while an article written by Schuster (1930) and appearing in the second issue analyses the solution of four such problems in a Seleucid text.

The conclusion just quoted announces the approach which was to be that of the 1930s. Since the meanings of terms for mathematical operations were derived from the numbers that resulted from their use, the operations were almost by necessity understood as arithmetical operations; as a rather natural consequence, problems were understood as (arithmetical) equations and equation systems. And of course Neugebauer, as everybody else, expressed amazement that complicated matters such as second-degree equations were dealt with correctly.

Neugebauer knew very well that Old Babylonian (1800–1600 BCE, according to the "middle chronology") and Seleucid (third-second century BCE) mathematics were formulated in different terminologies. But he believed that the difference was one of terminology and implicitly supposed, as we see, that there must have been steady progress of knowledge from the early to the late period.

---

[27]"Moreover, even the readings themselves can be considerably improved, once the substantial contents has been elucidated" (p. 67).

A number of publications from Neugebauer's hand (and three from that of Waschow (1932a, b, c)) followed in *Quellen und Studien* B until 1936 (in vol. 4, from 1937–38, Neugebauer has turned completely to astronomy). In 1935–37, Neugebauer also published in *Quellen und Studien* A the monumental *Mathematische Keilschrifttexte* (MKT) (Neugebauer 1935–1937). They can be said to bring to completion the interpretation of his (1929)-paper; but they also make clear that Neugebauer had not left behind his interest in metrology and other simple matters—he was not looking merely after matters that might be seen as analogous to modern equation algebra. The conclusion of volume III (Neugebauer 1935–1937, III, 79f) gives two warnings to the reader. Firstly, that MKT is a *source edition*—"It does not belong among the tasks that I have proposed for myself in this edition to develop the consequences which can be drawn from this text material". Secondly,

> Since our knowledge of these things is of relatively recent date, and current datings had to be pushed considerably, there is an obvious danger to overestimate the mathematics of the Babylonians. In order to somehow gloss over the lack of a basis in sources, many familiar books change elementary mathematical things into "propositions" and "discoveries" that must be ascribed to great men. It seems to me that we should not stamp the Babylonians as such discoverers. What is often overlooked and cannot be sufficiently emphasized is the terrible difficulty and slowness of the development of the very simplest fundamental mathematical concepts, first of all of a genuine computational technique. This, however, is not the achievement of a single person; it can only be understood within a historical process, inextricably attached to the emergence of a whole culture. Once this stage has been reached, then there is nothing in Babylonian mathematics that must be seen as an unexpected brilliant performance.

The last sentence refers to Neugebauer's hypothesis (which he considers an established fact) (Neugebauer 1935–1937, III, 79),

> that Babylonian mathematics first grew out of the numerical methods of sexagesimal calculation, the practical advantage of which was fully understood, and then, decisively sustained by the possibilities offered by ideographic writing, soon reached a strongly "algebraic"[28] treatment of purely mathematical problems that were of or could be reduced to linear or quadratic character.

Thureau-Dangin, as we have seen, had been interested in metrology and mathematical techniques since (1897). He started dialogue with Neugebauer in (1931) (making a philological correction that also concerns Frank, whom he does not mention). His weighty *Esquisse d'une histoire du système sexagésimal* (1932c), however, is rather a crown on his work from the 1920s, describing both the sexagesimal place-value system and the non-positional system and non-sexagesimal fractions, together with their uses.[29] But very soon, Thureau-Dangin moved from

[28][The quotes around the word *algebraic* indicate that Neugebauer refuses to make hypotheses about which kind of algebraic thought is involved in the texts. The many algebraic formulas in his commentary are not meant to map the thinking of the authors of the texts; they show why the calculations are pertinent (or, rarely, why they are not)].

[29]This booklet had no strong impact—it drowned in the fury surrounding the new discoveries of the time. However, a revised English translation (including much about the Babylonian "algebra") appeared in *Osiris* in (1939) on George Sarton's initiative (p. 99).

purely philological emendations and addenda to the publication of new mathematical texts and to considerations of their mathematical substance—for example in (1932b, 1934, 1936)—and to a synthesis about "the method of false position and the origin of algebra" (1938a) along with the source edition *Textes mathématiques babyloniens* (1938b).[30] In several of these works Thureau-Dangin can be seen to be much less wary than Neugebauer when speaking of the algebraic thinking of the Babylonians. He also shows himself familiar with a very wide range of later mathematical sources, from Diophantos, Ptolemy and al-Khwārizmī to Stevin and Wallis.

Then, in 1937–38, this "heroic period" ended abruptly. In 1945, it is true, Neugebauer and Abraham Sachs published *Mathematical Cuneiform Texts* (Neugebauer and Sachs 1945),[31] an edition of texts from American collections that had not been included in MKT, and Neugebauer's popularization *The Exact Sciences in Antiquity* from (1951) (revised in 1957) contains a chapter on the topic; but apart from that Neugebauer only published two or three small items on Babylonian mathematics after 1937, dedicating instead himself wholly to the history of astronomy (and to the launching of the *Mathematical Reviews*, after the National Socialists had seized power over his earlier creation *Zentralblatt für Mathematik*). Schuster published nothing in the area after 1930 (he is better known as a Hittitologist), while Waschow entered the army in 1934, writing at the same time a dissertation on Kassite letters (1936).[32] In 1938 he published a book (*4000 Jahre Kampf um die Mauer*) about siege techniques since Old Babylonian times, after which I have been unable to find information about his fate (I would guess that as

[30]This is what von Soden (1939, 144) tells about the purpose of this parallel edition: "This new work is not meant to replace Neugebauer's MKT; indeed, the phototypes and autographs are not repeated, nor are all texts treated anew. Th.-D.s aim was instead, leaving the arithmetical tables completely aside (only the introduction speaks briefly about them) to make those problem texts that are sufficiently well preserved to allow at least a generally satisfying understanding available to as many researchers as possible in a cheaper edition, since the exorbitant price of the MKT unfortunately hampers its wider diffusion." But further: "While thus the specialist researcher will also in future not be able to give up Neugebauer's MKT as the complete source collection, with the just mentioned exception [two small texts from Susa with area calculations published by Vincent Scheil in 1938], then precisely he will also not be able to pass over Th.-D.'s new edition, as nobody will be able to digest in brief the large number of corrected readings and the immensely weighty lexical, grammatical and substantial observations, masterly concise though they are." In Thureau-Dangin's own words (TMB, xl): "The present volume contains no text which has not been published elsewhere in its original form [that is, without a translation of ideograms into syllabic Akkadian]. The main task I have set myself while preparing it has been to make documents accessible to the historians of mathematical thought.".

[31]Curiously enough, (Neugebauer and Sachs 1945) is much less afraid of ascribing modern mathematical concepts to the Babylonians than Neugebauer had been in the 1930s—such as logarithms, p. 35, cf. (Neugebauer (1935–1937), I, 363–365). Whether this is due to Sachs's influence or Neugebauer himself had been convinced by what others had read into (Neugebauer 1935–1937) I am unable to say.

[32]The edition of one long Seleucid text (BM 34568) in (Neugebauer 1935–1937, III, 14–22) is also, according to Neugebauer, "apart from a few trifles due to Herrn Dr. Waschow". This work must be dated between 1935 and 1937.

an officer he lost his life during the war). Albert Schott concentrated on astronomy, while Kurt Vogel's *Habilitationsschrift* (1936) dealt with Greek logistics. Thureau-Dangin returned to other Assyriological questions.

In (1961), Evert Bruins (1909–1990) and Marguerite Rutten (1898–1984) published a volume with mathematical texts from Susa. They had started work around 1938, and Bruins was very proud of having been trained by Thureau-Dangin.[33] No wonder that the volume is wholly in the style of the 1930s—yet on a much lower philological level than what had been published during this epoch, and full of groundless speculations and misreadings (with interspersed good ideas, it should be added).

6 Assyriologists, 1940–1980

After 1940, Assyriologists would usually put aside any tablet containing too many numbers in place-value notation as "a matter for Neugebauer" (thus Hans Nissen, at one of the Berlin workshops on "Concept Formation in Mesopotamian Mathematics" in the 1980s). In consequence, very few new texts (apart from the batch from Susa) were published during the following four decades.

There is one important exception to this generalization (and a few other less important ones). Between 1950 and 1962 the Iraqi Assyriologist Taha Baqir (1912–1984) published four papers in the journal *Sumer* with new texts excavated between 1945 and 1962 (Baqir 1950a, b, 1951, 1962). These were highly important for several reasons: They came from a region from which until then no mathematical texts were known; like the Susa texts their provenience was known, since they were regularly excavated; but unlike what had happened to the Susa texts, the excavations were carefully made, for which reason the texts can also be dated.[34] von Soden (1952) suggested a number of improved readings with implications for the interpretation,[35] and Bruins (1953) tried (as usually) to show that everything von Soden had said was absurd; but the impact of Baqir's papers on historians of mathematics was almost imperceptible—one joint article by the mathematician

[33]He returns to this link time and again in the numerous angry letters I have from his hand. I suppose he can be believed on this account, his general unreliability notwithstanding. According to the preface (TMS, xi), Rutten made the hand copies and collaborated with Bruins on the translation. However, already the translation of word signs into Akkadian contains so many blunders of a kind no competent Assyriologist would commit that Bruins can be clearly seen to have had the upper hand concerning everything apart from the hand copies.

[34]A further text covering three tablets was found on the ground, apparently left behind by illegal diggers as too damaged. It was published by Albrecht Goetze (1897–1971) in (1951).

[35]Until then, von Soden had never worked directly on mathematical questions himself; but he had always been interested in the topic, as can be seen from his careful and extensive reviews of Neugebauer 1935–1937 (1937) and TMB (1939). He also made a review of TMS in (1964), an indispensable companion piece to the edition itself.

Karl-Bernhard Gundlach (*1926) and Wolfram von Soden (1908–1995) from (1963) deals with one of Baqir's texts and a text from Susa.

Already in 1945, Goetze had contributed a chapter "The Akkadian Dialects of the Old–Babylonian Mathematical Texts" to (Neugebauer and Sachs 1945, 146–151). In contrast to the volume as a whole, this chapter falls outside what had been done in the 1930s.[36] In these pages, Goetze makes a careful classification of all Old Babylonian mathematical texts known by then that contained enough syllabic writing to allow orthographic analysis.

Occasionally, some Assyriological publication would touch at numero-metrological questions, but not very often.[37] We have to wait until the early 1970s before an Assyriologist took up systematically the kind of work which Thureau-Dangin and others had pursued in the 1920s. In (1971), Marvin Powell submitted his doctoral dissertation on *Sumerian Numeration and Metrology*, soon followed by a major paper on "Sumerian Area Measures and the Alleged Decimal Substratum" (1972a). Also in 1972 a short paper from his hand (1972b) on "The Origin of the Sexagesimal System: the Interaction of Language and Writing", followed, and in (1976) a longer one on "The Antecedents of Old Babylonian Place Notation and the Early History of Babylonian Mathematics", published in *Historia Mathematica*. The latter two articles took up topics which both Thureau-Dangin and Neugebauer had tried their teeth on around 1930, yet for lack of adequate sources from the third millennium without reaching solid results. In (1978) and (1979), Jöran Friberg, paradoxically a mathematician of merit and no Assyriologist but using approaches and methods that had been characteristic of the Assyriological tradition, made a break-through on the numerical and metrological notations of the fourth millennium; with minor corrections, his results were later confirmed by the Berlin Uruk project (Damerow and Englund 1987).

7 Historians of Mathematics

During the same decades, little original work on Mesopotamian mathematics was made by scholars who would primarily be classified as historians of mathematics. They can be seen to have regarded the analysis in (Neugebauer 1935–1937) and (Neugebauer and Sachs 1945) as exhaustive—as it actually was on most accounts,

[36]The outcome can be seen as an extension of a division of the corpus into a "northern" and a "southern" group which Neugebauer had suggested in (1932, 6f); but Neugebauer's arguments had been of a wholly different nature.

[37]In 1978–79, Carlo Zaccagnini thus published at least four papers on the metrologies of peripheral areas.I disregard publications in Russian, most noteworthy of which is (Vajman 1961)—my reading of Russian, which reached the level of "rudimentary" 25 years ago, has vanished completely since then for lack of practice. An exhaustive survey, often with discussion, of all at least minimally pertinent publications (also those in Russian) for the period 1945–1980 will be found in (Friberg 1982, 67–130).

as long as Neugebauer's and Sachs's approach *as understood by historians of mathematics* was taken for granted.

There are again a few exceptions. The most substantial of these is a sequence of proposed interpretations of the famous text Plimpton 322, originally published in (Neugebauer and Sachs 1945, 38–41) and considered there as an early instance of number theory. Most noteworthy during the early period is (Bruins 1957), where a derivation of its Pythagorean triples from pairs of reciprocals is proposed (an interpretation which has been confirmed with modifications and extra arguments since then by Friberg (1981) and Robson (2001)). It may be considered a manifestation of the new "modernizing" orientation of (Neugebauer and Sachs 1945) that this possibility had been overlooked, given that Neugebauer had believed in the 1930s that the whole second-degree "algebra" came from the place-value system (above, text around note 28).

Other exceptions are a publication of some merit by Gandz (1948), sent to the journal *Osiris* around 1938 but then delayed by the war; and a republication of one of Gandz's results in (1955) by Peter Huber, who had not noticed Gandz's work.

However, while little new research was done on Mesopotamian mathematics by historians of mathematics, "Babylonian mathematics" was close to becoming the standard introduction to histories of mathematics.

The way it was dealt with is well illustrated by Asger Aaboe's (1922–2007) *Episodes from the Early History of mathematics* (1964). Aaboe starts by observing that a modern schoolboy transposed to Babylonia or ancient Greece would find the "physics" of classical Antiquity utterly unfamiliar (p. 1). Mathematics, however, would

> look familiar to our schoolboy: he could solve quadratic equations with his Babylonian fellows and perform geometrical constructions with the Greeks. This is not to say that he would see no differences, but they would be in form only, and not in content; the Babylonian number system was not the same as ours, but the Babylonian formula for solving quadratic equations is still in use.

That is, firstly: mathematics is a topic outside history, changing "in form" only. Secondly, the "contents" of mathematics consists in "formulae". Aaboe himself may have believed to continue Neugebauer's approach, but in reality the programme of *Quellen und Studien* has been betrayed. The "seemingly so a-historical 'exact sciences' " have become, precisely, *a-historical*. The lack of information about the social context of Babylonian mathematics is no longer a deplorable fact, as for Neugebauer in (1934)—the absence of information about its creators is just taken note of, while institutional setting etc. constitute non-questions.[38]

[38] "Of the creators of Babylonian mathematics we know nothing whatsoever except the result of their work" (p. 6). That the texts are school texts is intimated by photos of presumed schoolrooms from Mari (which are actually store-rooms) and occasional references to a "schoolboy"—but schooling seems to be just as timeless as mathematics. In 1964, we may observe, more was known about the Old Babylonian scribe school than in 1934—cf. (Kramer 1949; Falkenstein 1953; Gadd 1956).

Turning to the contents, we see that the reader learns that the sexagesimal place-value system is *the* Babylonian number system. Aaboe ignores that it was used only for intermediate calculations; in school; and in (late Babylonian) mathematical astronomy. He is unaware that a different system was used in "real-life" juridical and economical documents[39]—he only knows about inconsistency and failing rationality.[40]

When going beyond place-value computation Aaboe deals with three more advanced topics. The first is treated through two "algebraic" problems about square areas and appurtenant sides from BM 13901, quoted in Neugebauer's translation but then immediately transformed into modern algebraic symbols; the second is YBC 7289, a tablet showing a square with diagonals and three inscribed numbers corresponding to the side, the diagonal and their approximate ratio, which allows immediate discussion in terms of $\sqrt{2}$; the third the calculation of a height in an isosceles trapezium. It is mentioned (p. 23) that the first two are Old Babylonian and the third Seleucid, but it is claimed (as does *not* correspond to the information that could be extracted from (Neugebauer 1935–1937), and as is in any case quite wrong) that all three could have been written in any period. The conclusion discusses "algebra" once again, and states that

> Quadratic equations are often given in the equivalent form of two equations with two unknowns, such as
>
> $$x+y=a, \quad xy=b,$$
>
> whence one finds immediately that x and y are the solutions of
>
> $$z^2-az+b=0$$

without mentioning that such problems deal with rectangular areas and sides, nor that the "one" who "finds immediately" is Aaboe himself or some other modern calculator, and that no corresponding step can be found in the original texts.

Aaboe's book was intended as supplementary high-school reading, and can thus be understood according to Toeplitz' "genetic method" (1927), the introduction of modern concepts through pedagogically motivating idealized quasi- or

[39]However, all of this is described in (Thureau-Dangin 1939), who distinguishes the "abstract" (namely place-value) system "intended only to serve as an instrument of calculation" (p. 117) from the ordinary sexagesimal but non-positional system.

[40]"It should be added that an entirely consistent use of the sexagesimal system is to be found only in the mathematical and astronomical texts, and even in astronomical texts one can find year numbers written as, e.g., 1–*me* 15 (meaning 1 hundred 15) instead of 155. In practical life the Babylonians showed the same profound disregard for rationality in their use of units for weight and measure as does the modern English-speaking world" (p. 20). The year number in question is written in precisely that number system which Hincks had deciphered in 1847, cf. note 13—the very first contribution to the study of Assyro-Babylonian mathematics.

pseudo-history.[41] However, the typical general histories of mathematics published during the period share the basic character of Aaboe's presentation—see my anatomies of (Hofmann 1953), (Boyer 1968) and (Kline 1972) in (Høyrup 2010). Only another book written for the high-school level (but here the German *Gymnasium*), Vogel's *Die Mathematik der Babylonier* from (1959) stands out—with its awareness that the place-value system was a scholarly system; because of its interest in metrologies and in computations dealing with everyday life; and with its discussions of ways of thought.[42] Vogel, indeed, had worked on the material himself already in the 1930s, and he had always been interested in ways of thought and in the mathematics of practical life, while Aaboe had only worked on Seleucid astronomical texts, and Joseph Ehrenfried Hofmann (1900–1973), Carl Boyer (1906–1976) and Morris Kline (1908–1992) at best on Neugebauer's translations—but apparently more often on his popularizations and his explanatory commentaries without distinguishing the latter from what was done in the sources.

8 After 1980

After c. 1945, the historiography of Mesopotamian mathematics had thus been an almost dead topic, little considered by Assyriologists and treated under the point of view of "historical mathematics" by those who otherwise wrote about the history of mathematics.[43]

Beginning with the above-mentioned works of Powell and Friberg this situation was going to change once again. But this is where my own work in the field started, first on the connection between mathematics, general socio-cultural context and educational situation, from 1982 onward on the concepts and operations of Old Babylonian mathematics, so here I shall stop—adding only that in recent years a number of younger scholars trained in mathematics as well as Assyriology have entered the field, adding new approaches and returning to the Assyriological questions of the earlier twentieth century with the luggage of a century of extra textual and archaeological discoveries, thus being able to integrate the mathematical

41"Nothing is farther from me than to teach a history of the infinitesimal calculus; I myself as a student ran away from a lecture of that kind. History is not at stake, but the genesis of problems, acts and demonstrations, and the decisive turning points in this genesis" (Toeplitz 1927, 94).

42Dirk Struik's (1894–2000) *Concise History of Mathematics* from (1948) deals with Mesopotamian mathematics too briefly to allow description in similar depth (pp. 23–32). Struik's layout, however, is similar: The analysis is embedded in general social history; non-positional as well as place-value system are described; but like Vogel, Struik has no possibility to go beyond Neugebauer.

43Boyer had written about the "*concepts* of the calculus" (1949), and Kline's title refers to "mathematical *thought*". Hofmann had written among other things about Ramon Lull's squaring of the circle in (1942), and had tried there to penetrate the thinking and motives of Lull (without which he would indeed have been unable to conclude anything of interest).

dimension with studies of social, political and economic history. The field remains alive—but mathematicians may not find it very interesting for their purpose.

References

Aaboe, A. (1964). *Episodes from the early history of mathematics*. New York: Random House.

Baqir, T. (1950a). An important mathematical problem text from Tell Harmal. *Sumer, 6*, 39–54.

Baqir, T. (1950b). Another important mathematical text from Tell Harmal. *Sumer, 6*, 130–148.

Baqir, T. (1951). Some more mathematical texts from Tell Harmal. *Sumer, 7*, 28–45.

Baqir, T. (1962). Tell Dhiba'i: New mathematical texts. *Sumer, 18*, 11–14, pl. 1–3.

Berriman, A. E. (1953). *Historical metrology: A new analysis of the archaeological and the historical evidence relating to weights and measures*. London: Dent, New York: Dutton.

Bezold, C. (1886). *Kurzgefasster Überblick über die babylonisch-assyrische Literatur*. Leipzig: Otto Schulze.

Biancani, G. S. J. (1615). *De mathematicarum natura dissertatio. Una cum clarorum mathematicorum chronologia*. Bologna: Bartholomeo Cocchi.

Boyer, C. B. (1949). *The concepts of the calculus: A critical and historical discussion of the derivative and the integral*. New York: Columbia University Press.

Boyer, C. B. (1968). *A history of mathematics*. New York: Wiley.

Brandis, J. (1866). *Das Münz-, Mass- und Gewichtswesen in Vorderasien bis auf Alexander den Großen*. Berlin: Wilhelm Herz.

Bruins, E. M., & Rutten, M. (1930). *Textes mathématiques de Suse*. TMS. Paris: Paul Geuthner.

Bruins, E. M. (1953). Revision of the mathematical texts from Tell Harmal. *Sumer, 9*, 241–253.

Bruins, E. M. (1957). Pythagorean triads in Babylonian mathematics. *The Mathematical Gazette, 41*, 25–28.

Bruins, E. M., & Rutten, M. (1961). *Textes mathématiques de Suse*. Paris: Paul Geuthner.

Cantor, M. (1863). *Mathematische Beiträge zum Kulturleben der Völker*. Halle: H. W. Schmidt.

Cantor, M. (1880). *Vorlesungen über Geschichte der Mathematik. Erster Band, von den ältesten Zeiten bis zum Jahre 1200 n. Chr*. Leipzig: Teubner.

Cantor, M. (1907). *Vorlesungen über Geschichte der Mathematik. Erster Band, von den ältesten Zeiten bis zum Jahre 1200 n. Chr. Dritte Auflage*. Leipzig: Teubner.

Cantor, M. (1908). Babylonische Quadratwurzeln und Kubikwurzeln. *Zeitschrift für Assyriologie und verwandte Gebiete, 21*, 110–114.

Cory, I. P. (1832). *Ancient fragments of the Phoenician, Chaldæan, Egyptian, Tyrian, Carthaginian, Indian, Persian, and other writers* (2nd ed.). London: William Pickering.

Damerow, P., & Englund, R. K. (1987). Die Zahlzeichensysteme der Archaischen Texte aus Uruk. In M. W. Green & H. J. Nissen (Eds.), *Zeichenliste der Archaischen Texte aus Uruk* (Band II, Kapitel 3). Berlin: Gebr. Mann.

Daniels, P. T. (1988). 'Shewing of hard sentences and dissolving of doubts': The first decipherment. *Journal of the American Oriental Society, 108*, 419–436.

Falkenstein, A. (1953). Die babylonische Schule. *Saeculum, 4*, 125–137.

Fossey, C. (1904). *Manuel d'assyriologie. Tôme premier. Explorations et fouilles, déchiffrement des cunéiformes, origine et histoire de l'écriture*. Paris: Leroux.

Frank, C. (1928). *Straßburger Keilschrifttexte in sumerischer und babylonischer Sprache*. Berlin & Leipzig: Walter de Gruyter.

Friberg, J. (1978). The third millennium roots of Babylonian mathematics. I. A method for the decipherment, through mathematical and metrological analysis, of proto-Sumerian and proto-Elamite semi-pictographic inscriptions. Department of Mathematics, Chalmers University of Technology and the University of Göteborg No. 1978-9.

Friberg, J. (1979). The early roots of Babylonian mathematics. II: Metrological relations in a group of semi-pictographic tablets of the Jemdet Nasr type, probably from Uruk-Warka. Department of Mathematics, Chalmers University of Technology and the University of Göteborg No. 1979-15.

Friberg, J. (1981). Methods and traditions of Babylonian mathematics. Plimpton 322, Pythagorean triples, and the Babylonian triangle parameter equations. *Historia Mathematica, 8*, 277–318.

Friberg, J. (1982). A survey of publications on Sumero-Akkadian mathematics, metrology and related matters (1854–1982). Department of Mathematics, Chalmers University of Technology and the University of Göteborg No. 1982-17.

Gadd, C. J. (1922). Forms and colours. *Revue d'Assyriologie, 19*, 149–159.

Gadd, C. J. (1956). *Teachers and students in the oldest schools. An inaugural lecture delivered on 6 March 1956*. London: School of Oriental and African Studies, University of London.

Gandz, S. (1948). Studies in Babylonian mathematics I. *Indeterminate Analysis in Babylonian Mathematics. Osiris, 8*, 12–40.

Grotefend, G. F. (1802). Bericht über "Praevia de cuneatis, quas vocant, inscriptionibus Persepolitanis legendis et explicandis relatio". Göttingische gelehrte Anzeigen, 18. September 1802, 1481–1487; 6. November 1802, 1769–1772; 14. April 1803, 593–595; 23. Juli 1803, 1161–1164.

Gundlach, K.-B., & von Soden, W. (1963). Einige altbabylonische Texte zur Lösung »quadratischer Gleichungen«. *Abhandlungen aus dem mathematischen Seminar der Universität Hamburg, 26*, 248–263.

Hankel, H. (1874). *Zur Geschichte der Mathematik in Alterthum und Mittelalter*. Leipzig: Teubner.

Heuzey, L. (1906). À la mémoire de Jules Oppert. *Revue d'Assyriologie, 6*(1904–07), 73–74.

Hicks, R. D. (Ed., Trans.). (1925). Diogenes Laertius, Lives of eminent philosophers (Vol. 2). London: Heinemann/New York: Putnam.

Hilprecht, H. V. (1906). *Mathematical, metrological and chronological tablets from the temple library of Nippur*. Philadelphia: Department of Archaeology, University of Pennsylvania.

Hincks, E. (1848). On the third Persepolitan writing, and on the mode of expressing numerals in cuneatic characters. *Transactions of the Royal Irish Academy. Polite Literature, 21*, 249–256.

Hincks, E. (1854a). Cuneiform inscriptions in the British Museum. *Journal of Sacred Literature*, New Series, *13*(October 1854), 231–234. Reprint after *The Literary Gazette, 38*, 707.

Hincks, E. (1854b). On the Asssyrian mythology. *Transactions of the Royal Irish Academy, 22*(6), 405–422.

Hofmann, J. E. (Ed.). (1942). Ramon Lulls Kreisquadratur; Raimundus Lullus, *De quadratura et triangulaturo circuli. Sitzungsberichte der Heidelberger Akademie der Wissenschaften: Philosophisch-historische Klasse* 1941/42 Nr. 4.

Hofmann, J. E. (1953–1957). *Geschichte der Mathematik* (Vol. 3). Berlin: Walter de Gruyter.

Hommel, F. (1885). *Geschichte Babyloniens und Assyriens*. Berlin: Grote'sche Verlagsbuchhandlung.

Høyrup, J. (2002). *Lengths, widths, surfaces: A portrait of Old Babylonian algebra and its kin*. New York: Springer.

Høyrup, J. (2010). Old Babylonian 'Algebra', and what it teaches us about possible kinds of mathematics. *Gaṇita Bhāratī 32*, (published 2012), 87–110.

Huber, P. (1955). Zu einem mathematischen Keilschrifttext (VAT 8512). *Isis, 46*, 104–106.

Jaritz, K. (1993). Ernst Weidner—Gelehrter und Mensch. In H. D. Galter (Ed.), *Grazer Morgenländische Studien. Band 3: Die Rolle der Astronomie in den Kulturen Mesopotamiens* (pp. 11–20). Graz: Universität Graz.

Kästner, A. G. (1796–1800). *Geschichte der Mathematik seit der Wiederherstellung der Wissenschaften bis an das Ende des achtzehnten Jahrhunderts* (Vol. 4). Göttingen: Johann Georg Rosenbusch.

Kline, M. (1972). *Mathematical thought from ancient to modern times*. New York: Oxford University Press.

Kramer, S. N. (1949). Schooldays: A sumerian composition relating to the education of a scribe. *Journal of the American Oriental Society, 69*, 199–215.

Kramer, S. N. (1963). *The Sumerians: Their history, culture, and character*. Chicago: Chicago University Press.

Kuhn, T. S. (1970). *The structure of scientific revolutions* (2nd ed.). Chicago: University of Chicago Press.

Lelgemann, D. (2004). Recovery of the ancient system of foot/cubit/stadion-length units. Contribution, Workshop—History of Surveying and Measurement, Athens, May 22–24, 2004. http://www.fig.net/pub/athens/papers/wshs2/WSHS2_1_Lelgemann.pdf. Accessed May 31, 2011.

Meyer, W. (1893). G. Fr. Grotefends erste Nachricht von seiner Entzifferung der Keilschrift. Nachrichten von der Königlichen Gesellschaft der Wissenschaften und der Georg-Augusts-Universität zu Göttingen, 13. September 1893, 573–616.

Montucla, J. F. (1758). *Histoire des mathematiques* (Vol. 2). Paris: Jombert.

Montucla, J. F. (1799). *Histoire des mathématiques* (Vol. 4) (III–IV achevés et publiés par Jérôme de la Lande). Paris: Henri Agasse, an VII—an X (1799–1802).

Münter, F. (1802). *Versuch über die keilformigen Inschriften zu Persepolis*. København: C. G. Proft, 1802.

Neugebauer, O. (1926). *Die Grundlagen der ägyptischen Bruchrechnung*. Berlin: Julius Springer.

Neugebauer, O. (1927). Zur Entstehung des Sexagesimalsystems. *Abhandlungen der Gesellschaft der Wissenschaften zu Göttingen. Mathematisch-Physikalische Klasse*, Neue Folge, *13*(1).

Neugebauer, O. (1929). Zur Geschichte der babylonischen Mathematik. *Quellen und Studien zur Geschichte der Mathematik, Astronomie und Physik. Abteilung B: Studien*, *1*(1929–31), 67–80.

Neugebauer, O. (1932). Studien zur Geschichte der antiken Algebra I. *Quellen und Studien zur Geschichte der Mathematik, Astronomie und Physik. Abteilung B: Studien*, *2*(1932–33), 1–27.

Neugebauer, O. (1934). *Vorlesungen über Geschichte der antiken mathematischen Wissenschaften. I: Vorgriechische Mathematik*. Berlin: Julius Springer.

Neugebauer, O. (1935–1937). *Mathematische Keilschrift-Texte* (Vol. 3). Berlin: Julius Springer.

Neugebauer, O. (1951). *The exact sciences in antiquity*. København: Munksgaard.

Neugebauer, O., & Sachs, A. (1945). *Mathematical cuneiform texts*. New Haven, Connecticut: American Oriental Society.

Neugebauer, O., Stenzel, J., & Toeplitz, O. (1929). Geleitwort. *Quellen und Studien zur Geschichte der Mathematik, Astronomie und Physik. Abteilung B: Studien*, *1*(1929–31), 12.

Neugebauer, O., & Struve, W. (1929). Über die Geometrie des Kreises in Babylonien. *Quellen und Studien zur Geschichte der Mathematik, Astronomie und Physik. Abteilung B: Studien*, *1* (1929–31), 81–92.

Niebuhr, C. (1774–1778). *Reisebeschreibung nach Arabien und andern umliegenden Ländern* (Vol. 2). København: Nicolaus Møller.

Norris, E. (1856). On the Assyrian and Babylonian weights. *Journal of the Royal Asiatic Society of Great Britain and Ireland, 16*, 215–226.

Oppert, J. (1856). Schreiben an den Präsidenten der Hamburger Orientalisten-Versammlung und an Prof. Brockhaus. *Zeitschrift der Deutschen Morgenländischen Gesellschaft, 10*, 288–292.

Oppert, J. (1859). *Expédition scientifique en Mésopotamie exécutée par ordre du Gouvernement de 1851 à 1854. II. Déchiffrement des inscriptions cunéiformes*. Paris: Imprimerie Impériale.

Oppert, J. (1872). L'étalon des mesures assyriennes fixé par les textes cunéiformes. *Journal asiatique,* sixième série, *20*(1872), 157–177; septième série, *4*(1874), 417–486.

Oppert, J. (1886). La notation des mesures de capacité dans les documents juridiques cunéiformes. *Zeitschrift für Assyriologie, 1*, 87–90.

Planck, M. (1950). *Scientific autobiography and other papers*. London: Williams & Norgate.

Powell, M. A. (1971). Sumerian numeration and metrology. *Dissertation*. University of Minnesota.

Powell, M. A. (1972a). Sumerian area measures and the alleged decimal substratum. *Zeitschrift für Assyriologie und Vorderasiatische Archäologie, 62*(1972–73), 165–221.

Powell, M. A. (1972b). The origin of the sexagesimal system: The interaction of language and writing. *Visible Language, 6*, 5–18.

Powell, M. A. (1976). The antecedents of Old Babylonian place notation and the early history of Babylonian mathematics. *Historia Mathematica, 3*, 417–439.
Ramus, P. (1569). *Scholarum mathematicarum libri unus et triginta*. Basel: Eusebius Episcopius.
Rawlinson, H. (1855). Notes on the early history of Babylonia. *Journal of the Royal Asiatic Society of Great Britain and Ireland, 15*, 215–259.
Robson, E. (2001). Neither Sherlock Holmes nor Babylon: A reassessment of Plimpton 322. *Historia Mathematica, 28*, 167–206.
Rottländer, R. C. A. (2006). *Vormetrische Längeneinheiten*. http://Vormetrische-laengeneinheiten.de. Accessed May 23, 2011.
Sayce, A. H. (1875). The astronomy of the Babylonians. *Nature, 12*, 489–491.
Sayce, A. H. (1887). Miscellaneous notes. *Zeitschrift für Assyriologie, 2*, 331–340.
Schuster, H. S. (1930). Quadratische Gleichungen der Seleukidenzeit aus Uruk. *Quellen und Studien zur Geschichte der Mathematik, Astronomie und Physik. Abteilung B: Studien, 1* (1929–31), 194–200.
Smith, G. (1872). On Assyrian weights and measures. *Zeitschrift für Ägyptische Sprache und Altertumskunde, 10*, 109–112.
Strassmaier, J. N., & Epping, J. (1881). Zur Entzifferung der astronomischen Tafeln der Chaldäer. *Stimmen aus Maria-Laach. Katholische Blätter, 21*, 277–292.
Struik, D. J. (1948). *A concise history of mathematics* (Vol. 2). New York: Dover.
Thureau-Dangin, F. (1897). Un cadastre chaldéen. *Revue d'Assyriologie, 4*(1897–98), 13–27.
Thureau-Dangin, F. (1931). Notes sur la terminologie des textes mathématiques. *Revue d'Assyriologie, 28*, 195–198.
Thureau-Dangin, F. (1932a). Notes assyriologiques. LXIV. Encore un mot sur la mesure du segment de cercle. LXV. BAL = «raison (arithmétique ou géométrique)». LXVI. Warâdu «abaisser un perpendiculaire»; elû «élever un perpendiculaire». LXVII. La mesure du volume d'un tronc de pyramide". *Revue d'Assyriologie, 29*, 77–88.
Thureau-Dangin, F. (1932b). La ville ennemi de Marduk. *Revue d'Assyriologie, 29*, 109–119.
Thureau-Dangin, F. (1932c). *Esquisse d'une histoire du système sexagésimal*. Paris: Geuthner.
Thureau-Dangin, F. (1934). Une nouvelle tablette mathématique de Warka. *Revue d'Assyriologie, 31*, 61–69.
Thureau-Dangin, F. (1936). L'Équation du deuxième degré dans la mathématique babylonienne d'après une tablette inédite du British Museum. *Revue d'Assyriologie, 33*, 27–48.
Thureau-Dangin, F. (1938a). La méthode de fausse position et l'origine de l'algèbre. *Revue d'Assyriologie, 35*, 71–77.
Thureau-Dangin, F. (1938b). *Textes mathématiques Babyloniens*. Leiden: Brill.
Thureau-Dangin, F. (1939). Sketch of a history of the sexagesimal system. *Osiris, 7*, 95–141.
Toeplitz, O. (1927). Das Problem der Universitätsvorlesungen über Infinitesimalrechnung und ihrer Abgrenzung gegenüber der Infinitesimalrechnung an den höheren Schulen. *Jahresbericht der deutschen Mathematiker-Vereinigung, 36*, 88–100.
Tropfke, J. (1902). *Geschichte der Elementar-Mathematik in systematischer Darstellung. I. Rechnen und Algebra. II. Geometrie. Logarithmen. Ebene Trigonometrie. [...]*. Leipzig: von Veit.
Ungnad, A. (1916). Zur babylonischen Mathematik. *Orientalistische Literaturzeitung, 19*, 363–368.
Ungnad, A. (1918). Lexikalisches. *Zeitschrift für Assyriologie, 31*(1917–18), 248–276.
Vajman, A. A. (1961). *Šumero-vavilonskaja matematika. III-I Tysjačeletija do n. e.* Moskva: Izdatel'stvo Vostonoj Literatury.
Vergilius, P. (1546). *De rerum inventoribus libri octo*. Lyon: Gryphius.
Vogel, K. (1936). Beiträge zur griechischen Logistik. Erster Theil. *Sitzungsberichte der mathematisch-naturwissenschaftlichen Abteilung der Bayerischen Akademie der Wissenschaften zu München* 1936, 357–472.
Vogel, K. (1959). *Vorgriechische Mathematik. II. Die Mathematik der Babylonier*. Hannover: Hermann Schroedel/Paderborn: Ferdinand Schöningh.

von Soden, W. (1937). [Review of Neugebauer 1935–1937]. *Zeitschrift der Deutschen Morgenländischen Gesellschaft, 91*, 185–203.

von Soden, W. (1939). [Review of Thureau-Dangin 1938b]. *Zeitschrift der Deutschen Morgenländischen Gesellschaft, 93*, 143–152.

von Soden, W. (1952). Zu den mathematischen Aufgabentexten vom Tell Harmal. *Sumer, 8*, 49–56.

von Soden, W. (1964). [Review of Bruins & Rutten 1961]. *Bibliotheca Orientalis, 21*, 44–50.

Vossius, G. I. (1650). *De universae mathesios natura et constitutione liber, cui subjungitur Chronologia mathematicorum*. Amsterdam: Ioannes Blaev.

Waschow, H. (1932a). Verbesserungen zu den babylonischen Dreiecksaufgaben S.K.T. 8. *Quellen und Studien zur Geschichte der Mathematik, Astronomie und Physik. Abteilung B: Studien, 2* (1931–32), 211–214.

Waschow, H. (1932b). Reihen in der babylonischen Mathematik. *Quellen und Studien zur Geschichte der Mathematik, Astronomie und Physik. Abteilung B: Studien, 2*(1932–33), 298–302.

Waschow, H. (1932c). Angewandte Mathematik im alten Babylonien (um 2000 v. Chr.). Studien zu den Texten CT IX, 8–15. *Archiv für Orientforschung, 8*(1932–33), 127–131, 215–220.

Waschow, H. (1936). Babylonische Briefe aus der Kassitenzeit. *Inaugural-Dissertation*. Gräfenhainichen: A. Heine.

Weißbach, F. H. (1907). Über die babylonischen, assyrischen und altpersischen Gewichte. *Zeitschrift der Deutschen Morgenländischen Gesellschaft, 61*, 379–402, 948–950.

Zeuthen, H. G. (1896). *Geschichte der Mathematik im Altertum und im Mittelalter*. Vorlesungen. København: Høst & Søn.

Zimmern, H. (1916). Zu den altakkadischen geometrischen Berechnungsaufgaben. *Orientalistische Literaturzeitung, 19*, 321–325.

Appropriating Role Models for the Mathematical Profession: Biographies in the *American Mathematical Monthly* Around 1900

Henrik Kragh Sørensen

Abstract

During the first decade of its existence, the *American Mathematical Monthly* regularly published short biographies of mathematicians. When read as appropriations of past lives, these biographies can be analysed to provide new insights into the images of mathematics, and of American mathematics in particular, held by groups of authors and welcomed by the readership of the *Monthly*. Thus, the approach in this paper is a "meta-biographical" one in which biographies are analysed not for their content about past mathematicians but for their appropriation and framing in the context where they were produced. This approach brings forward new insights into the professionalization of mathematics in the United States, the promotion of different disciplines, and the efforts of individuals to cast themselves as following in the footsteps of some of the greatest heroes of mathematics.

Keywords

Appropriation · Biography · Professionalization · US mathematics around 1900 · Halsted · American Mathematical Monthly

1 Introduction and Methodology

During the decades around 1900, American mathematics underwent monumental developments that took it from a provincial position at the fringes of the mathematical community into a position from which it could compare to other nations and at some point come to dominate global mathematics. At that time, institutions were

H.K. Sørensen (✉)
Department of Science Education, University of Copenhagen,
Øster Voldgade 3, 1350 København K, Denmark
e-mail: henrik.kragh@ind.ku.dk

V.R. Remmert et al. (eds.), *Historiography of Mathematics in the 19th and 20th Centuries*, Trends in the History of Science,
DOI 10.1007/978-3-319-39649-1_5

devised and shaped to promote mathematical research and instruction, and an intriguing group of individuals were moulded into the first generation of really professional American mathematicians. Among the important institutional developments was the creation of a number of American journals of mathematics, which targeted various audiences and fared with varied success. Some of these journals featured historical—including biographical—columns among their content; the *American Mathematical Monthly*, in particular, regularly included short biographies from its inception in 1894. These short biographical articles are, of course, not to be seen as truly historical scholarship—indeed, they are better analysed and understood, I claim, as activities and documents that helped to shape the professionalization of American mathematics.

Therefore, what I propose is to analyse the writing of history of mathematics—in this particular case the biographies of mathematicians—as situated in a particular discourse of professionalization in the local settings of American mathematics around 1900. In this respect, this chapter is based on an adaptation of the meta-biographical methodology employed by Nicolaas Rupke to bring forth the "many Humboldts" appropriated for different (German) socio-political contexts (Rupke 2008). Rupke's approach was, however, a longitudinal one devoted to appropriations of one individual at different times over a period of almost 200 years. Here, focus will be much narrower in geographical and temporal terms, yet the subjects of the biographies will be more varied.

The professionalization of mathematics research and teaching is, of course, by no means a uniquely or specifically American phenomenon around 1900; and neither is the deployment of history or biography for that purpose. Yet, something particular is at stake in the United States, where no historical record of local mathematics was available for legitimisation such as could be done in, for instance, Scandinavia (see e.g. Sørensen 2006). Whereas the writing of (intellectual) biographies could serve a variety of national identity-building purposes in e.g. Scandinavia (see Sørensen 2012), the present scope is much more limited: In concord with their medium, the biographies studied in this chapter were first and foremost concerned with bolstering an identity for mathematics within the United States, I argue, and only secondarily with carving out a room for American mathematics and culture on the international scene.[1]

This chapter thus seeks to analyse the interplay between the appropriation of past mathematicians and the shaping of a new profession of mathematics in a local context. It first provides a brief overview of the biographical genre in the nineteenth century as it pertains to mathematics in order to briefly address the relations between studying biographies of mathematicians and history of mathematics. It then paints a sketch of American mathematics around 1900 by drawing on

[1]For more on the internationalization of (American) mathematics, see also (Fenster 2002; Parshall 2009).

biographies from the *American Mathematical Monthly*. This leads into three topical case studies devoted to detailed analyses of biographies published in the *American Mathematical Monthly* between the inception of the journal in 1894 and 1904 when the editorial line changed to devote much less attention to biographies.

2 Biographies of Mathematicians in the Nineteenth Century

During the eighteenth and nineteenth century, biographies of mathematicians developed and proliferated in various media. Biographies in the form of obituaries and eulogies (éloges) were already established practices in the eighteenth century within the learned societies and academies. In that form, obituaries were circulated in the proceedings of the major academies in Paris, London, and Berlin. As such, they not only provided news of the profession but also helped to shape and profile the society and—perhaps more indirectly—the biographer on the international scene. Thus, in the eighteenth and nineteenth centuries, biographies were deployed as commemorative practices with a set of associated methodological consequences: The biographies were *life accounts* in that they started with the birth and ended with the death of the portrayed individual. They were largely *professional* in that they were written with a specific eye to the professional accomplishments of the protagonist—in our case their mathematical contributions—which they accounted for in ways that were to signal objectivity. And they were also largely *hagiographies* in the sense that the selection of emphasis and topics was made to underpin the greatness of the protagonist in the above-mentioned objective account of history that the biography presented. Thus, in particular, connections to other lines of thought or other more implicit external influences were seldom mentioned except where to establish priority of the protagonist against adversaries or explain major material changes in the protagonist's life.

However, during the nineteenth century, an additional purpose for biographies of mathematicians emerged as the genre developed from the authorised obituaries to include more partisan appropriations of the great lives. Upon Niels Henrik Abel's (1802–1829) death, for instance, his friends presented an opinionated commentary to the standard obituary when Christian Boeck (1798–1877) argued that the Norwegian state had not done enough for its young and brilliant mathematician (Holmboe 1829). Nevertheless, the biography and the obituary remain closely associated through their scope and their commemorative utility: In short, they simultaneously honour the life of the protagonist and express normative statements about values and virtues (see e.g. Lee 2009; Parke 2002).[2]

[2]As shown in (Rothman 1982), the appropriation of past mathematicians such as Évariste Galois (1811–1832) may even be distorting to the point of "fictionalization".

The tendency towards a more subjective author-stance in the biographies of mathematicians may not have been new in the broader context, but as the market for biographies of mathematicians grew and diversified, more profound appropriations for particular professional as well as non-mathematical purposes became clearer. By the last decades of the nineteenth century increasing numbers of biographies were produced—some at book length and some about long-dead individuals. For instance, the first major biography of Abel was produced in the early 1880s, some fifty years after his death, and the seminal book-length biographies of other nineteenth-century mathematicians such as Galois (Dupuy 1896) and Carl Gustav Jacob Jacobi (1804–1851) (Koenigsberger 1904) were being published around the turn of the century.[3] Complementing these more extensive works, the "objective" and factual biographies including short descriptions of scientific contributions were maintained in encyclopedias (such as *Encyclopedia Britannica*) and in the form of new "handbooks" of biography.[4]

These shifts imply that in order to read and understand biographies as sources for the context in which they were produced, careful attention must be paid to the particularity of those contexts: Biographies—on this account—should not be read as objective reports on the lives of great mathematicians. Instead, they should be read as local appropriations in particular contexts, for particular audiences, and for particular purposes. In order to utilize biographies as meta-biographical sources, they should thus be analysed as primary sources about the context in which they were written. This presents some particular methodological issues in considering the relation between the biographer and the biographee and the role and styling of the biographer. In particular, attention is directed to the biographer's way of characterising and describing the life and personality, including the virtues and the professional conditions, of the biographee. For the case of most biographies of mathematicians from the nineteenth century—and all of those considered here—the biographer is a mathematician and not a historian by training. Often connected by personal acquaintance or even stronger ties, the biographer would stand to gain from certain choices made in writing the biography. And reading the biography with this in mind allows the historian to discern some of these (often) hidden assumptions and agendas. Those areas where the appropriation becomes most visible are the subjective descriptions and valuations of the biographee's character,

[3]Leo Königsberger (1837–1921), who was a student of Jacobi, had also published his historical account of the theory of elliptic functions (Koenigsberger 1879), which contributed to Carl Anton Bjerknes' (1825–1903) biographical interest in Abel.

[4]Some of the most influential such handbooks are the *Nouvelle biographie générale* compiled under the direction of Jean-Chrétien-Ferdinand Hoefer (1811–1878) and published in 46 volumes from 1853 to 1866, the *Biographie universelle ancienne et moderne* edited by Louis-Gabriel Michaud (1773–1858) and first printed in 85 volumes in 1811 (second edition in 45 volumes from 1843), and Johann Christian Poggendorff's (1796–1877) famous handbook of biographies of scientists (Poggendorff 1853; see also Poggendorff 1863). In the United States, mathematicians were, of course, also included in later works such as the *American Dictionary of National Biography* which appeared from from 1926 and was then briefly mentioned in the *American Mathematical Monthly* "Mathematics in the American Dictionary of National Biography" 1926.

his personality, and his set of virtues. Initially pertaining only to the individual, it is a feature of the genre going back to its hagiographic origins, that the biography projects these virtues onto its audience. Thus, the virtues and characteristics emphasised in biographies not only say something about the biographee, or the biographer, but also about the set of norms that are envisioned for the community of readers of the biography, in this case about the mathematical community in the United States. When considering the appropriation of biographies of mathematicians aimed at a mathematical audience, we need to broaden our focus from the personal virtues towards the more professional ones. These are often expressed in the way the biographee's mathematical life and contributions are described. Paying attention to the mathematics posed challenges for the biographers of the nineteenth century (as it does today) in that although the readers were supposed to be mathematicians, they would not expect from a biography a very technical exposition. Yet, in analysing the descriptions of the mathematics of the biographee and comparing them to what is now known from historical research, we may get interesting (and otherwise often inaccessible) insights into the role of biographies in professionalizing mathematical communities such as the American one.

With these more general methodological concerns in mind, we turn our attention to the biographies published in the *American Mathematical Monthly*. In the following section, we present some background information and some general observations on the local context of the biographies; and in the subsequent three case studies, we zoom in on particular analyses of the appropriation of values from The Old World and on the role of biographies in fashioning individual and professional identities.

3 Biographies in the *American Mathematical Monthly* in Their Local Context

By the late nineteenth century, research mathematics was gradually institutionalising and maturing in the United States (see e.g. Parshall and Rowe 1989, 1994; Parshall 2000). The relevant components of the professionalization of American mathematics include important institutions, social factors, and key developments in mathematical research and education. Many of these dimensions have been researched intensively (see bibliography, including the chapters in Duren 1988–1989) in studies that have documented the poor state of research mathematics in the United States prior to the arrival of James Joseph Sylvester (1814–1897) in the 1870s and have often focused on the emergence of great, international centres such as Johns Hopkins University in Baltimore from the 1870s and University of Chicago around 1900. Similarly, the establishment of the *American Journal of Mathematics* or the *American Mathematical Society* have received considerable scholarly interest. So in the present context, these findings will be integrated with a portrayal of American mathematics as seen through the lens constituted by the biographies of Americans in the *American Mathematical Monthly*.

An initial bold move to bring Sylvester to Virginia in 1841 had resulted in a brief flurry of mathematical activity—but when Sylvester left the United States again in 1843, the field stagnated for decades. However, in 1877 Sylvester was hired as the first professor of mathematics at the newly founded Johns Hopkins University which position he held until 1883, and this coincided with (and largely caused) the revival of mathematical research in the United States. By 1890, a new set of clusters were forming—some in Baltimore, some in and around New York, and some around highly inspirational leader figures in Chicago (see e.g. Parshall 2003, 2004, 2006, 2013; Fenster 2002, 2003). Part of that institutional flourish had to do with the establishing of a university-based academic research tradition along European norms in the United States. Yet, other aspects involved the cultivation of better means of communication among the geographically dispersed, mathematically interested amateurs and professional mathematicians.[5] As is well known, the local club of mathematicians that used to meet as the *New York Mathematical Society* was transformed into the *American Mathematical Society* in 1894 (see Archibald 1938).

Next to the possibility of meeting in person, the second best thing would be to have proper avenues for mathematical publishing on the domestic scale, and different entrepreneurs vied for a career publishing mathematics. Thus, numerous mathematical journals were founded in the middle of the nineteenth century, but most of them quickly folded again as it was difficult and expensive to run a mathematical journal for a small community with dispersed interests while alleviating the relatively high costs of publishing in good quality. The main audience of these journals could not be professional research mathematicians, exclusively, and instead the first journals addressed themselves mainly to mathematics teachers (in particular at the collegiate level) and even students. Yet, the aim was often to provide a research journal, and for that the editor would have to rely on a steady stream of interesting contributions from the academic milieu which was thus also to be accommodated in the scope of the journal.

Some popular illustrated magazines also featured biographies of important and famous scientists, and some professional journals and magazines developed a taste for the biographical genre. In the United States, biographies of mathematicians could be found in general scientific journals such as *Science* or in national magazines such as *The Monist*, *The Nation* or *Century* (Merzbach 1989, pp. 642–643). Two mathematical journals made an explicit point of featuring biographies; these were the *American Mathematical Monthly* directed primarily at collegiate mathematics teachers in the United States and the *Bulletin of the American Mathematical Society* (until 1894 named *Bulletin of the New York Mathematical Society*).

[5]For the history of the mathematical publication community in Britain and the United States, see e.g. (Despeaux 2002; Timmons 2004).

From its inception in 1891 until 1931, the *Bulletin of the American Mathematical Society* proclaimed on its masthead that it provided "a historical and critical review of mathematical science" (ibid., pp. 642–643, 653–654), and it included a number of obituaries of prominent mathematicians. Similarly, during the first decade after the *American Mathematical Monthly* was founded in 1894, it also regularly featured short biographical notices, typically approximately three to five pages and often set at the front page of the issue and accompanied by a portrait.[6] Yet, whereas the biographies in the *Bulletin* were mainly obituaries, the scope of the *American Mathematical Monthly* was broader. Indeed, biographies had even been included in the purpose of the new journal (Fig. 1) as expressed by its editors in the first volume (Finkel and Colaw 1894, p. 1). The journal's editor Benjamin Finkel (1865–1947) wrote many of the biographies; others were penned by George Bruce Halsted (1853–1922) and Franklin P. Matz (1856–1908).[7]

During its first decade, the *American Mathematical Monthly* published 51 biographies of different mathematicians (a list of these biographies can be found in Fig. 2). The biographees formed a varied group—some were taken from the echelon of European mathematics of the past, some were among the most prominent contemporary research mathematicians in Europe, and others were among the main educators and contributors to the American journals of mathematics in the second half of the nineteenth century. Needless to say, the biographies written about such a heterogeneous population also varied considerably, not least because the authors of the biographies would add their individual perspectives to the stories they told.

About half of the biographies printed in the *American Mathematical Monthly* during its first decade of existence were devoted to American mathematicians. This fact reflected the scope of the journal, and many of those biographies were written by key contributors such as Finkel (7 in total), John M. Colaw (1860–1939) (3), and Matz (8). These biographies are in central respects comparable to the obituaries of the *Bulletin* in providing news of the profession in the United States, but they differ in that in the *American Mathematical Monthly* focus was on mathematics teachers. The other half of the biographies was devoted to prominent European mathematicians. Among those, Finkel wrote a few biographies of mathematicians from the seventeenth to the early nineteenth century (see Sect. 4), but the bulk of the biographies of Europeans were written by Halsted (20 in total) and concerned with contributors to non-Euclidean geometry in the nineteenth century. Halsted's use of

[6]For the early history of the *American Mathematical Monthly* and its first editors, see also (Finkel 1931; Maxwell 1993).

[7]Whereas most of the authors dealt with here are known today, my access to information about Matz is more sparse: He had studied at various American institutions before he received his PhD from Heidelberg and returned to take a post-graduate course at Johns Hopkins University. When he became professor of mathematics at the Military and Scientific School at King's Mountain, North Carolina, he also contributed problems and solutions to various American mathematics journals and edited the Department of Mathematics in the *New England Journal of Education*. There is an obituary in *Reading Times*, 30 September 1908, page 3, and his set of contributions to *The Mathematical Visitor* can be found in Rabinowitz (1996).

THE AMERICAN MATHEMATICAL MONTHLY.

DEVOTED TO THE
SOLUTION OF PROBLEMS IN PURE AND APPLIED MATHEMATICS,
PAPERS ON MATHEMATICAL SUBJECTS, BIOGRAPHIES
OF NOTED MATHEMATICIANS, ETC.

EDITED BY

B. F. FINKEL, A. M.,

AUTHOR OF FINKEL'S MATHEMATICAL SOLUTION BOOK, MEMBER OF THE NEW YORK MATHEMATICAL SOCIETY, AND PROFESSOR OF MATHEMATICS IN KIDDER INSTITUTE, KIDDER, MISSOURI.

J. M. COLAW, A. M.,

MEMBER OF THE NEW YORK MATHEMATICAL SOCIETY, AND PRINCIPAL OF HIGH SCHOOL, MONTEREY, VIRGINIA.

CHUBBUCK BROS., Publishers.
KIDDER, MISSOURI.

Fig. 1 Front page of the first issue of the *American Mathematical Monthly* 1894. As stated in the description, the *American Mathematical Monthly* was devoted in part to "biographies of noted mathematicians". ©Mathematical Association of America

Year	Author	Title
1894		"Biography: James W. Nicholson"
		"Biography: William Hoover"
	Aley	"Biography: Daniel Kirkwood"
	Burleson	"Biography: Samuel Gardner Cagwin"
	Colaw	"Biography: Simon Newcomb"
	Colaw	"Biography: Joel E. Hendricks"
	Dickson	"Biography: Dr. George Bruce Halsted"
	Finkel	"Biography: E. B. Seitz"
	Finkel	"Biography: Elisha Scoot Loomis"
	Finkel	"Biography: Artemas Martin"
	Halsted	"Biography: Felix Klein"
	Halsted	"Biography: James Joseph Sylvester"
	Matz	"Biography: James Matteson"
1895	Colaw	"Biography: Alexander Macfarlane"
	Halsted	"Biography: Paenutij Lvovitsch Tchebychev"
	Halsted	"Biography: Arthur Cayley"
	Halsted	"Biography: Lobachevsky"
	Halsted	"John Henry Lambert"
	Macfarlane	"Biography: Arthur Cayley"
	Matz	"Biography: Benjamin Peirce"
	Matz	"Hudson A. Wood"
	Matz	"Ormond Stone"
	Matz	"Biography: De Volson Wood"
	Matz	"Biography: William Chauvenet"
1896	Finkel	"W. J. C. Miller"
	Halsted	"Biography: Bolyai Farkas"
	Matz	"John Newton Lyle"
	Miller	"Lie's views on several important points in modern mathematics"
	Smith	"Emile-Michel-Hyacinthe Lemoine"
1897		"De Volson Wood"
	Finkel	"Biography: Leonhard Euler"
	Halsted	"Biography: Quillaume Jules Hoüel"
	Halsted	"Biography: De Morgan"
	Halsted	"Biography: Vasiliev"
	Halsted	"Biography: James Joseph Sylvester"
	Philips	"Hubert Anson Newton"
1898	Finkel	"Biography: René Descartes"
	Halsted	"Biography: Tchébychev"
	Halsted	"Biography: Bolyai János"
1899	Darboux	"Sophus Lie"
	Halsted	"Biography: Sophus Lie"
	Halsted	"Biography: Percival Frost"
1900	Halsted	"Robert Tucker"
1901	Craig	"Biography: Isaac Newton"
	Finkel	"Biography: Karl Frederich Gauss"
	Halsted	"Biography: Charles Hermite"
	Halsted	"Franz Schmidt"
	Matz	"Biography: Thomas Craig"
1902	Halsted	"Biography: Cristoforo Alasia"
	Halsted	"Eugenio Beltrami"
1903	Tyler	"Biography: John Daniel Runkle"

Fig. 2 Overview of the 51 biographies in the *American Mathematical Monthly* between its foundation in 1894 and 1904

the biographies for his quest to promote non-Euclidean geometry will be taken up separately in Sect. 5. A separate, smaller cluster of biographies of British mathematicians can also be identified as a special interest in appropriating the legacy of Sylvester (see Sect. 6). Here, again, Halsted was the major contributor, but on the few occasions where multiple biographies of the same person were presented, marked debate is discernible among the authors writing for the *American Mathematical Monthly*.[8]

In accordance with the general characteristics of the biographical genre, many of the biographies of mathematicians in the *American Mathematical Monthly* would start out by a short description of the protagonist's background before typically treating his early education with special emphasis on his first encounters with mathematics. For many of the American mathematicians biographed, the family background in Europe would also be mentioned along with when and how the mathematician's ancestors arrived in the New World. The biography would then typically outline the protagonist's mathematical contributions and include mention of relevant (often international) accolades. A large proportion of the biography could be devoted to describing the teaching role of the biographee (for Americans) or his other relevance in the American context as a mentor or correspondent. The theme of catching up to Europe was sometimes explicitly addressed, for instance by the young George Abram Miller (1863–1951) who—describing how "[i]t is generally admitted that America has contributed comparatively little towards the advancement of the science of mathematics", it was still the case that "our relative position is not improving as rapidly as might be desired" (Miller 1896, p. 295). Thus, biographies about European mathematicians could also be used to comment upon contemporary domestic affairs. Finally, an assessment of character traits is an integral part of any biography, and the characterisation of mathematical genius sometimes formed part of it. In general, external circumstances are described as hindrances to be overcome which projects an image of mathematics as a solitary endeavour of creativity. Thus, among the common themes of the set of biographies, the construction of mathematical genius and of national American identity can be identified in accordance with the typical traits of the genre.

Thus largely conforming to the genre, some of the biographies included an additional element of reflexivity in which statements were explicitly made about the biographical genre. Many of the biographies were explicitly described as "imperfect sketches" pointing to their fragmentary nature (see e.g. Halsted 1895a, p. 106). The presentations were—as was expected—almost always positive in their attitudes towards the biographee, with only little or indirect criticism. In describing character traits, some biographers explicitly noticed the congruence of the particular individual and the ideal (see e.g. Aley 1894, p. 144). Yet, while many of the biographies were more or less explicitly hagiographic, some took objection to this trend. For instance, implicitly reacting to the biographies by Halsted, Miller sought to

[8]On a few occasions, biographers even became biographees in the *American Mathematical Monthly* (this is true of Alexander Macfarlane (1851–1913) and Halsted during the period studied here).

bring a more nuanced picture of Sophus Lie (1842–1899), arguing that "Lie would have liked his faults to go with his merits" (Miller 1899, p. 193).

In most of the biographies of contemporary mathematicians, a special relationship between the biographer and the biographee can be discerned; often, it was explicitly addressed in the biography as a means of situating and fashioning the biographer. In many instances, particularly those biographies written by Halsted, privileged access to sources—either written or by personal acquaintance—would form part of the biography. Sometimes, the biographer would have access to letters to or from the biographee, sometimes he could base his description on first-hand or second-hand testimony, as when Halsted could point to his ability to add to the biography of Augustus De Morgan (1806–1871) which had been adopted from the *Encyclopedia Britannica*: "If I am able to add anything specifically new to this part of the paper, it is from a life-long study of his works, and many conversations about him with the great Sylvester, who knew him very intimately" (Halsted 1897a, p. 1).[9] Thereby, Halsted would position himself doubly as a first-hand reader of De Morgan's work and as a confidant of Sylvester. And as will be further analysed in Sect. 6, such privileged access could even amount to fashioning an intellectual genealogy (see also Sørensen 2012). In this way, the biographer's voice could become quite explicit and quite personal. For instance, Matz would write of Thomas Craig (1855–1900) (see also further below): "Speaking from a personal acquaintance extending over many years, the writer always found Dr. Craig to be an efficient workman, a pleasant companion, a warm friend, and a good man" (Matz 1901, p. 184) and seek to back this assessment by including laudations by others.

A special ambition for the biographical genre is to present a life which could be inspiring or even worthy of emulation. And in the biographies of mathematicians, the connections between life and mathematics could be quite intricate. Thus, Halsted argued in his biography of János Bolyai (1802–1860) that since the history of non-Euclidean geometry is so important in many branches of human knowledge, the life of Bolyai "possesses imperishable interest for mankind" (Halsted 1898a, p. 35). And concerning Nikolai Lobachevsky (1792–1856), Halsted argued that the biography "will arouse a deeper enthusiasm for scientific advancement and widen the horizon of every reader" (Halsted 1895b, p. 139). Clearly, such statements not only serve to legitimise the biographies but also state as obvious that non-Euclidean geometry is of the greatest importance. And as discussed in Sect. 5, this claim formed a core argument for Halsted's biographies in the *American Mathematical Monthly*.

Common to many of the biographies were descriptions of the character of the biographee which expressed intellectual generosity: Mentoring of younger scholars was a particular concern, but it also took other forms. Thus, generosity was a trope frequently connected with the truly great mathematical geniuses who could be aloof of petty priority concerns. Yet, for others such as Carl Friedrich Gauss (1777–1855), the verdict could not be so unequivocally positive: In his biography of

[9]Some of the biographies such as (Colaw 1894b; "De Volson Wood" 1897) were adopted from other media, in particular from the *Encyclopedia Britannica*, or translated from foreign languages.

Gauss, Finkel invoked the typically positive trait of an independent thinker only to add a more critical spin: "Gauss had a strong will, and his character showed a curious mixture of self-conscious dignity and child-like simplicity. He was little communicative, and at times morose" (Finkel 1901, p. 30). By pointing out that Caspar Wessel (1745–1818) held priority for the geometrical representation of complex numbers, Finkel stressed that Gauss "needs no undue credit to make him famous" (ibid., p. 31), thereby siding with a more critical approach to biography.

In describing the personality of the biographee, the biographer would often invoke childhood and social background to account for either the biographee's hard-working nature (in the case of American farmers) or his erudition (in the case of more academic backgrounds). Special emphasis would often be given to describe the personal relation to mathematics in such terms as "an inordinate love for mathematics" (Burleson 1894, p. 375), and sometimes the talent for mathematics could be connected to an aptitude for languages as a "very general characteristic of eminent mathematicians" (Finkel 1901, p. 25). And sometimes, mathematics, erudition, and the necessity to support oneself could be linked as when the Scottish mathematician Macfarlane was described as paying for his own university education by winning prizes and scholarships in mathematics and having to choose between studying classics and mathematics (Colaw 1895, p. 1).

In an age when pivotal mathematicians such as Gösta Mittag-Leffler (1846–1927) were collecting and distributing autographs and photographs of each other, it is no surprise that portraits accompany many of the biographies. Yet, the portraits and physical appearance of the biographee are rarely explicitly taken into consideration in the main text of the biography, although it was a theme sometimes drawn upon in late-nineteenth century biographies. Yet, on some occasions such as the biography of Arthur Cayley (1821–1895) whose father was Russian and whose "features had a Russian cast", speculative arguments could be made concerning the relative merits for mathematics by drawing on the equal weights of the brains of Russian men and women (Halsted 1895a, p. 103; see also Halsted 1898b).

Due to their short format and the rather general nature of the journal, the biographies in the *American Mathematical Monthly* were typically not very technical. Some included a bibliography of works, and some featured descriptions of the biographee's mathematics to varying levels of detail. For some, such as Daniel Kirkwood (1814–1895), a more thorough description would be included to argue for priority of discoveries; yet, for the most part, the description of mathematics would be rather superficial. Some arguments were included concerning certain evaluations in mathematics, for instance when Felix Klein's (1849–1925) doctrine about recasting mathematics as the study of groups was implicitly debated between Halsted who ascribed it to Lie (Halsted 1899b, p. 97) and Miller who described Lie as not being very serious about the claim that "everything could be done by means of groups" although he frequently made that argument (Miller 1899, p. 193). Thus, such discussions were not only about the proper biographical depiction of Lie but rather about the proper interpretation of more general statements about mathematics.

The first ten and a number of subsequent biographies published in the *American Mathematical Monthly* deal with American mathematicians. These raise an interesting set of questions that revolve around the locality and context in which the new journal was establishing itself. Resonating with images of the American Frontier, locality is seen to play into the biographies—and therefore, indirectly, into the development of mathematics in the United States—in an interesting variety of ways. In a sense, the biographer could appropriate locality for different purposes to signal aspects of the professionalization of mathematics. At the same time, the biographies also served to construct a *national* story of mathematics that bring to mind the trope of the American Dream through interesting appropriations of common biographical themes such as the intellectual mobility of the mathematical genius. They thus point towards the United States carving out a role for itself on the global mathematical scene.

These tropes can be seen, for instance, in the biography of Kirkwood, who was among the American mathematicians of international renown. This biography appeared in the *American Mathematical Monthly* in 1894 when Kirkwood was still alive in his 80th year, and it was written by Robert J. Aley (1863–1935) of Indiana University, where Kirkwood had been professor from 1856 until 1886 when health problems forced him to retire. As Aley was writing his biography, Indiana University was in the process of naming a new building after Kirkwood such was the veneration for and importance of Kirkwood for academia in Indiana. Kirkwood's specialisation was in astronomy, as was the case for many of the other American mathematicians whose biographies appeared in the *American Mathematical Monthly* (noticeably Simon Newcomb (1835–1909) and Samuel Gardner Cagwin). Yet, Kirkwood's situation in Indiana without an astronomical observatory was given a twist in the biography: The "lack of an observatory has been a real benefit to Astronomy" (Aley 1894, p. 142), Aley explained, since it allowed (or forced) Kirkwood to take up the theoretical study of observations made by others from which "[a]n observatory might have turned him aside". Instead, based on his "natural bent for mathematics", Kirkwood had—in the words of Richard A. Proctor (1837–1888)—become the "Kepler of America" (ibid., p. 142).[10] The substantial bibliography that accompanied this biography shows that Kirkwood's many contributions were made in American journals, and as a partial consequence, the claim for Kirkwood's priority for what is known as "Kirkwood's Law" was also described and insisted upon in the biography (Aley 1894, pp. 142–143). Thus, locality played into the biography of Kirkwood twice: The biography of the internationally recognised astronomer was connected to Indiana University, and the apparently adverse conditions for an astronomer without an observatory was made into an asset

[10]No explicit reference is given by Aley, and the exact origin of the phrase is not easily identified. Yet, it made it into a number of the American commemorative publications such as biographies and obituaries of Kirkwood or lectures at the dedication of the Kirkwood Observatory (see e.g. Swain 1901, p. 144; Jordan 1922, p. 188) and now seems to be part of common folklore about Kirkwood, in particular in Indiana. Indeed, the connection between Johannes Kepler (1571–1630) and Kirkwood was a recurring feature during Kirkwood's career, at least in the American context (see Numbers 1973).

by the biographer. Moreover, locality was also very important as it linked the biographer and the biographee. Aley had personal acquaintance with Kirkwood and could draw upon it for anecdotes and character assessment for his biography. The role as teacher and—we would say—mentor of young students was prominent in many of the biographies, making the explicit claim that the older generation served in preparing the ground for and facilitating the mathematicians now contributing to American mathematics. Aley's biography also featured a reflective stance as mentioned above, when he wrote of Kirkwood: "To write the plain truth about the personal character of Daniel Kirkwood is to write such an eulogy as most men give to their ideal hero" (ibid., p. 144). Projecting important cultural values, that ideal character would often encompass such diverse traits of the American intellectual as hard-working, energetic, friendly, God-revering, and austere.

An additional set of locality dimensions appears in some of the other biographies of American mathematicians such as Artemas Martin (1835–1918) and Craig. These two mathematicians were centrally involved in another aspect of the cultivation and professionalization of mathematical culture in the United States during the last decades of the nineteenth century, namely the aspect of communication. The biography of Martin was written by Finkel and appeared in the first volume of the *American Mathematical Monthly* (Finkel 1894a). Thereby, it served to introduce the readers of the new journal to the editor of two journals that would potentially compete with Finkel's new endeavour for audience and contributors.[11] By 1894, Martin was editing the *Mathematical Magazine* and the *Mathematical Visitor* which Finkel described as "two of the best mathematical periodicals published in America" (Finkel 1894a, p. 111). The two journals were related in the way that the latter was devoted to "higher mathematics" whereas the former included "solutions of problems of a more elementary nature", and "[t]he best mathematicians from all over the world contribute to these two journals" (ibid., p. 111). Finkel did not say much about neither the mathematical contents of the two journals nor about the mathematical merits of Martin, but from the voice of one editor, his laudation of Martin's "handsomely arranged and profusely illustrated" journals in which the editor was so deeply involved to the point of "doing the typesetting with his own hand" speaks as much about the ambitions, aspirations, and adversities of the newly founded *American Mathematical Monthly* as about Martin (ibid., p. 111). In the biography, Finkel also described how Martin had not received any regular education as a child and had learnt to read and write at home, only coming into contact with arithmetic when he was 14 years old (ibid., p. 108). But by "self-application and indefatigable energy", Martin as many a great man worked his way into teaching mathematics in the winter while working at farming and gardening in the summers (ibid., p. 108). This combination of humble backgrounds, an early affinity for mathematics, and an amateur career in mathematics is shared by a

[11]Other biographies in the *American Mathematical Monthly* also pointed to earlier American journals of mathematics such as the *Mathematical Monthly* founded by John Daniel Runkle (1822–1902) in 1858 and folded soon thereafter (Tyler 1903) and *The Analyst* run by Joel E. Hendricks (1818–1893) between 1873 and 1883.

number of the biographies of American mathematicians in the *American Mathematical Monthly* (see e.g. Finkel 1894b, c; Colaw 1894a; Matz 1894, 1895b). These topics—which are also found in the European context—played particularly into the construction of the specifically American mathematician as an austere, hard-working mathematician for whom the realisation of mathematical talents could also reflect social mobility.

The biography of Craig was written by Matz and appeared in the *American Mathematical Monthly* in 1901, i.e. towards the end of the period when biographies were such a prominent part of the journal (Matz 1901). Just like Martin, Craig was involved with a journal potentially competing with the *American Mathematical Monthly*, namely the *American Journal of Mathematics* which he had edited from 1894 to 1899 (ibid., p. 184).[12] But whereas Finkel's treatment of Martin's role as an editor had focused mainly on the daily operation of the journal and had only noted the international contributions in passing, Matz made a virtue out of Craig's ability to keep "pace with the most rapid advances in mathematical learning" and actively engaging with these "for the benefit of the readers of the *American Journal of Mathematics*" (Matz 1901, p. 184). In particular, Craig was "eminently successful in securing the contributions of distinguished mathematicians of England and the Continent" (ibid., p. 184). And during his time as editor of the *American Journal of Mathematics*, Craig had undertaken two trips to Europe during which "the principal objective […] was to interest European geometers in" the journal (ibid., p. 186). Although Matz devoted more energy to Craig's mathematical contributions than Finkel had done to Martin's, we are still faced with a "sketch" of a mathematician who possessed "extraordinary powers of acquisition" (ibid., p. 185), but whose "productive faculty was developed more slowly" (ibid., p. 186). Thus, Matz's biography of Craig is centred on Craig's role as editor of the *American Journal of Mathematics*, and the description of his editorship is directed towards the internationalization of research and journal publication in the United States. These two streams converge in that Craig was doubly centrally positioned as a student of Sylvester and as connected with the Johns Hopkins University from its foundation (see also Sect. 6).

As these general impressions have already shown, biographies can be used as contextual sources for a number of different research questions. In the following, three topical cases serve to direct our focus to two specific ways in which the biographies can be seen as appropriations that contribute to shaping the American research community. The first of these cases is concerned with the presentation of excellent researchers to the American readership and exemplified by Finkel's biography of Leonhard Euler (1707–1783). The second case analyses the biographies written by Halsted on prominent geometers as part of Halsted's mission to promote the study of non-Euclidean geometry. Then, the third and last case continues these themes by focusing on the biographical representation of the merits of

[12]Previously, Newcomb had been the editor of the *American Journal of Mathematics* until he had "recently severed his immediate active connection with the Johns Hopkins University for the next two or three years" (Colaw, 1894b, p. 254).

Sylvester in building up mathematics at Johns Hopkins University as portrayed by his friends and students. Taken together, the three cases illustrate how the biographical genre served in the efforts to fashion both personal and professional identities in American mathematics around 1900.

4 Euler and the Appropriation of European Virtues

For the first 10 volumes of the *American Mathematical Monthly*, its main editor Finkel produced biographies of such luminous European mathematicians as Gauss (Finkel 1901), René Descartes (1596–1650) (Finkel 1898) and Euler (Finkel 1897). The first of these appeared in the December issue of the *American Mathematical Monthly* 1897, where Finkel published a biography of Euler spanning the first six pages of the issue (ibid.). Finkel's biography was based on the histories of mathematics published by Florian Cajori (1859–1930) and W.W. Rouse Ball (1850–1925) as well as the entry on Euler in the *Encyclopedia Britannica*. Apart from these two standard references of the day on history of mathematics, Finkel included references to modern works treating a few of Euler's results.

In accordance with the standard structure of biographical writing, Finkel's biography started with Euler's birth, his youth, his career, old age, and eventual death. Following the narrated life, Finkel included two pages reviewing some of Euler's mathematical contributions. In a characteristic use of the available sources, Finkel's biography attempted to resolve a question about Euler's congregational affiliations, which had been contradictorily presented by Cajori, Ball, and the *Encyclopedia Britannica*. Itself perhaps an issue of minor importance today, this was apparently considered of some relevance in the late nineteenth century and Finkel's resolution of the question is also interesting: Noticing in a footnote that Ball had described Euler's father as a "Lutheran minister" whereas the *Encyclopedia Britannica* had said he was a "Calvinistic minister", Finkel analysed a quotation from Euler (with no reference given) to reach the conclusion, that Euler had been a Calvinist in doctrine, thereby also tacitly agreeing with the *Encyclopedia Britannica* (ibid., p. 297). This footnote constitutes one of the instances where Finkel contributed critically to the biography that he was otherwise distilling from existing sources. Finkel's biography proceeds to enlist now-known themes of Euler's biography, including his initiation to mathematics, his life-long friendship with the Bernoulli family, his patronage in St. Petersburg, and his "command" to Berlin (ibid., p. 298). Hardly any mention of family life is made except when we accidentally learn that Euler died from apoplexy in the presence of one of his grand-children and that his wives were half-sisters and provided him with a family of which many attained distinction. Special attention is, instead, given to two aspects worth elaborating on: (1) Euler's health and personality, and (2) characterising his work.

Finkel related how Euler's eyesight was compromised by hard scientific work, how he was temporarily cured but strained himself again and lost sight completely, so that his massive mental calculations had to be committed to paper by an assistant. Yet, in Finkel's words his "constitution was uncommonly vigorous and his general health was always good" so that he "ceased to calculate and to breathe at nearly the same moment" (ibid., p. 300).

On the character of Euler, Finkel further noted how Euler's initial change of study was undertaken with the "consent of his father", how Euler was "a very timid man" who devoted "all his time to science", how "this wonderful genius" whose "knowledge was more general than might be expected" was a "simple, upright, affectionate" man who "had a strong religious faith" and possessed a "single and unselfish devotion to the truth" and found "joy at the discoveries of science whether made by himself or others" (ibid., pp. 298–300). Finkel thus highlighted typical character traits found also in the biographies of American mathematicians, and in his appropriation, pecuniary problems, faith and morality became particularly active parts of the biography. Such statements about Euler's sentiments and motives and the emphasis they are given are—since they are not directly backed by evidence—to be seen in the context in which they were written: They are examples of the normative character of nineteenth-century biography as they express values that should to be seen as virtues, communicated by Finkel for inspiration and imitation.

Finkel noted with further awe Euler's immense production in mathematics. He emphasised the "admirably clear exposition of the principal facts of mechanics, optics, acoustics and physical astronomy" (ibid., p. 299) and Euler's work on the motion of the moon. Concerning these, however, he noted that "Theory, however, is frequently unsoundly applied in it, and it is to be observed generally that Euler's strength lay rather in pure than in applied mathematics" (ibid., p. 299). Euler's work in pure mathematics was characterised as a "revolution" because it presented analysis in "so general and systematic a manner" as had never been done before (ibid., p. 301). Finkel emphasised that Euler would explicitly point out that "an infinite series cannot be safely employed in mathematical investigations unless it is convergent" (ibid., p. 301). Yet, his "exposition of the principles of the subject is often prolix and obscure, and sometimes not quite accurate" and his treatment of "elliptic integrals is superficial" (ibid., p. 302). Finkel then noted that through Joseph-Louis Lagrange's (1736–1813) reworking of Euler's analysis, it has made its way into modern textbooks. Many of these statements are debatable to modern-day historians of mathematics, but Finkel's biography shows that they were appropriated for the local context: Finkel's biography of Euler can be seen to follow the general composition of a biography of a mathematician with the inclusion of an assessment of his work. It projects virtues of character and work for the inspiration of the reader. Thus, we can learn both about the expected readership of the *Monthly* and of the context of professionalization from it. Introducing the readers of the *Monthly* to Euler was no accident—through Finkel's assessments of Euler, Americans could be shown the values and virtues of research, erudition, genius, and moral integrity.

5 Halsted's Promotion of Non-Euclidean Geometry

One of the most prolific contributors to the biographical column of the *American Mathematical Monthly* during its first decade of existence was the American geometer Halsted, and his voice as a biographer reveals some interesting perspectives that are otherwise more difficult to discern. He wrote biographies of many important research mathematicians of the nineteenth century, in particular some of the celebrated heroes of his own discipline, non-Euclidean geometry, a topic on the history of which he also wrote papers. Moreover, Halsted was an early financial contributor to the *Monthly* and an advocate of its relevance on the national scale as well as a liaison towards the *American Mathematical Society* and the international mathematical community. Halsted's biographical stance was often that of the hagiographer; for instance, in his short mathematical obituary of Lie published soon after Lie's death, the opening sentence reads: "On the eighteenth of February, 1899, the greatest mathematician in the world, Sophus Lie, died at Christiania in Norway" (Halsted 1899b, p. 97). As should be clear, Halsted was not afraid to express evaluations—even controversial ones—in his biographies. In the case of Lie, the more nuanced and factual biography was then provided by a translation of Jean-Gaston Darboux's (1842–1917) French biography printed in the *Comptes rendus* (Darboux 1899).

In 1878–1879, just around the time when Halsted obtained his PhD from Johns Hopkins University, he published three papers in the first volumes of the *American Journal of Mathematics* containing "his extraordinarily influential Bibliography of Hyper-Space and Non-Euclidean Geometry, since continually referred to in every learned country in the world" (Dickson 1894, p. 339; see also Halsted 1878a, b, 1879). This bibliography was a pride both for Halsted, who mentioned it and its translation into Russian in some of his biographies (Halsted 1895b, p. 139), and for his American colleagues such as the young Leonard Eugene Dickson (1874–1954). Having established himself as an expert—*the* American expert—on non-Euclidean geometry, Halsted then used the pages of the newly founded *American Mathematical Monthly* to promote his interest in the topic. By far the most productive biographer in the *American Mathematical Monthly*, Halsted was not only a very central supporter of the journal (as is evidenced by the editorials) but also contributed a large number of other papers. In particular, he contributed a series of 37 papers on "Non-Euclidean Geometry: Historical and Expository" which were published sequentially over many issues during 1894–1898. Each of them short, they combined into a translation and exposition of Giovanni Girolamo Saccheri's (1667–1733) *Euclides Vindicatus* (Cajori 1922) spanning 97 pages. Thus, Halsted established himself also as the master of the original literature in the field, and he continued to translate other classical works on non-Euclidean geometry into English (ibid.).

More interesting, in the present context, is Halsted's special agenda in setting up the biographies of mathematicians who contributed to the development of non-Euclidean geometry. There, a dual fashioning was involved: First, Halsted sought to promote the field by promoting his own heroes, and second, he also fashioned himself as an expert contributor to the field in the American context although his research publications in the field are actually neither many nor deep despite his two reports on contemporary developments in non-Euclidean geometry (Halsted 1899c, 1901c). He was—and took upon himself to be—a *populariser* of non-Euclidean geometry in the English language (see e.g. Halsted 1901d, 1902a). For his true heroes—Lobachevsky and, in particular, Bolyai—he went to considerable length to establish their remarkable, "wonderful genius" (Halsted 1895b, p. 139) for the audience to gaze at. In his biography of Lobachevsky (ibid.), Halsted combined an interpretative biography of Lobachevsky with a number of interesting side issues. For instance, he took the opportunity to counter a review by Charles Sanders Peirce (1839–1914) of Halsted's own translation of an address by Alexander V. Vasiliev (1853–1929) on Lobachevsky (ibid., p. 137). Thus, not only did he manage to assert his own authority on Lobachevsky and non-Euclidean geometry by pointing to his connection to Vasiliev, he also took the opportunity to correct one of the dominant American mathematicians on a historical interpretation. Later in the same biography of Lobachevsky, Halsted managed to include the information that he was presently translating "important documents recently obtained in regard to the two Bolyais" into English (ibid., p. 138). All this was done in order to establish Halsted's own position while assessing Lobachevsky's contributions. This promised translation of documents into English was raised again in Halsted's biography of Farkas Bolyai (1775–1856) published the following year (Halsted 1896). There, Halsted provided a rather staccato biography based on translations of materials recently obtained from Hungary. Yet, again, it also served as an advertisement for a planned, book-sized work on the life of the Bolyais. Although this never materialised, Halsted did produce a biography of Bolyai for the *American Mathematical Monthly* (Halsted 1898a) which says almost as much about Halsted as about Bolyai. It repeatedly invokes the theme of the mathematical genius —the word "genius" appears five times over the four pages. And Halsted connected it to many of its nineteenth-century connotations such as the child prodigy, youthful creativity, the misunderstood and neglected romantic hero who was far ahead of his times (see both Halsted 1895b, 1898a).[13] Referring to an autobiographical note which Halsted had found, Bolyai was described as the "the phoenix of Euclid" (Halsted 1898a, p. 35) which also both establishes Halsted's privileged access to sources and the romantic genius of Bolyai.

This theme of privileged access to sources in the form of letters and personal contact and correspondence was even more important for Halsted's fashioning of himself in the case of the biographies that he wrote about more contemporary mathematicians such as Vasiliev (Halsted 1897d), Pafnuty Chebyshev (1821–1894) (Halsted 1895c, 1898b), and Jules Hoüel (1823–1886) (Halsted 1897c). There, privileged access served explicitly to showcase Halsted's international network to

[13]The construction of the trope of the mathematical genius is an interesting nineteenth-century development which cannot, however, be pursued further in the present context, although this biography would certainly be an important source for its transformation in the English language.

the readers of the *American Mathematical Monthly*, something that he was explicitly keen on when he acted as the American translator and intermediary for the efforts of Vasiliev and others to found an international mathematical congress (Halsted 1895e). It thus seems fair to describe Halsted's biographies as two-fold partisan: First, he wanted to argue for the relevance of non-Euclidean geometry, describing it as the greatest development of nineteenth-century mathematics with which all the great mathematicians of the past century had been engaged (Halsted 1895b, p. 139). Second, he wanted to argue for the importance of the original contributions by Bolyai and Lobachevsky while not neglecting the work done later in the century.

This second argument took a particular form as an argument *against* ascribing any positive relevance to Gauss in the discovery of non-Euclidean geometry, which was an occasion where Halsted's different streams—biographical, historical, and mathematical—came together. Before all the relevant sources were made available, Gauss's role in the development of non-Euclidean geometry was quite controversial. In Finkel's biography of Gauss, non-Euclidean geometry is barely mentioned when the discovery was ascribed to Bolyai who belonged to Gauss' "small circle of intimate friends" (Finkel 1901, p. 26). Halsted on the other hand, had a more definite and controversial position on this question: To him, Gauss had been detrimental to the development of non-Euclidean geometry. This position shines through in Halsted's biographies of Lobachevsky and the Bolyais (Halsted 1895b, 1896, 1898a), but it is made explicit on the occasion of the long-awaited publication of the seventh volume of Gauss' collected works in 1900. In that supplementary volume containing unpublished manuscripts and letters by Gauss, Halsted found the evidence which he had sought (Halsted 1900b, p. 247). In this volume, Halsted found a warrant for arguing that neither Bolyai nor Lobachevsky had been "prompted, helped, or incited by Gauss, or by any suggestion emanating from Gauss", adding "quite the contrary" (ibid., p. 247). The "warrant for saying this with final and overwhelming authority", Halsted explained, was that the material in the newly published volume did not include any letters from Gauss in response to or recognition of the ground-breaking works by these mathematicians (ibid., pp. 247–248). Thus, Halsted was reinforced in his very critical attitude towards Gauss who he argued by 1805 "was still a baby on this subject" of non-Euclidean geometry (ibid., p. 249). The fact that nothing can really be claimed from an absence of sources or that Gauss might have thought about non-Euclidean geometry without communicating it, did not appear to Halsted as worthy of consideration. Thus, according to Halsted's own biographer, "[a]t his hands Gauss, for instance, received scant justice" (Cajori 1922, p. 339). The origin of the 'misconceived' importance of Gauss was not explicitly given in Halsted's criticism in 1900. When, in 1903, Jacob William Albert Young (1865–1948) sought to outline the necessity for increased bibliographic surveys in mathematics, the issue emerged in the *American Mathematical Monthly* again. Young did not forget to include Halsted's bibliography of non-Euclidean geometry as a very specialised case; yet, a different section of Young's argument caught Halsted's attention. Young argued that mathematics is a cumulative science void of revolutions, and as an example he discussed non-Euclidean geometry. To Halsted's dismay, Young mentioned in a footnote that in the inaugural volume of the *Jahresbericht der DMV*, Max Simon (1844–1918) had made the statements that Gauss knew that the parallel postulate was not a logical necessity by 1792 and that this had influenced Bolyai and Lobachevsky (Young 1903, p. 186; see also Simon 1890, p. 39). This provoked Halsted to repeat almost verbatim his earlier repost, now with a clearer target (Halsted 1904). At a time when history of mathematics was also sparsely professionalized, Halsted's numerous swings at historians of mathematics (see also e.g. Halsted 1894c) and his attempts to *set the record straight* point to an interesting confluence of biography, history and mathematics at his hands: To Halsted, these three were deeply intertwined with each other and with his own, quite elaborate self-fashioning.

6 Sylvester's Legacy and Johns Hopkins University

Two biographies in the very first volume of the *American Mathematical Monthly* deal directly and explicitly with the professionalization of American mathematics in the late nineteenth century; these are the biographies of Sylvester and Klein, both written by Halsted (Halsted 1894a, b). These two towering individuals, Sylvester and Klein, would succeed each other in bringing about the 'emergence of the American mathematical research community' (Parshall and Rowe 1989, 1994). Following his visit to the United States in 1893 and his lecture tour (Klein 1894), Klein and Göttingen developed into a hub of Americans going abroad to pursue studies. In the *Monthly*, Halsted provided a biography of Klein (Halsted 1894a) which, coming from his hand, was untypical in its factual style. However, concerning Sylvester, the interpretation and appropriation was much more explicit.

In 1894, when Halsted wrote his first biography of Sylvester for the *American Mathematical Monthly* (Halsted 1894b), Sylvester was still alive but back in Europe for a decade. There was therefore no direct commemorative motive for the biography which, instead, was more focused on a local and contextual purpose. It was, however, succeeded three years later, after Sylvester's death, by another one (Halsted 1897b); and that same year, another biography of Sylvester was published in the *Bulletin of the American Mathematical Society* (Franklin 1897).

Halsted's 1894-biography opened with some remarkable observations on the biographical genre:

> An adequate life of James Joseph Sylvester has never been written, and probably never will be while he lives. The present biography will aim neither at completeness nor even symmetry, since so brief a sketch of so great a man can be of permanent value only by giving what the writer alone knows, or for some particular reason happens to know better than others. (Halsted 1894b, p. 294)

Thus, as already emphasised in the previous section, Halsted had and utilised his privileged access, taking the opportunity to correct other historians, notably Cajori. Both of Halsted's two biographies of Sylvester were written in a lively, engaging, and valued language and made frequent use of metaphors as well as anecdotes from the author's personal acquaintance with the biographee and his possession of certain documents. And they convey roughly the same information, except for some anecdotes that were only included in the 1894 version where Halsted would, for instance, describe how classes had been held late at night or on excursions to Washington (ibid., p. 297).

Featuring importantly in the biographies is the personal acquaintance and apprenticeship of Halsted under Sylvester. We learn indirectly how Halsted had won a prize in the Intercollegiate Mathematical Contests that had meant that he was appointed as one of the first two fellows of Johns Hopkins University and desiring to learn from Sylvester "became alone his first class" (ibid., p. 297):

> I was his first pupil, his first class, and he always insisted that it was I who brought him back to the Theory of Invariative Forms. (Halsted 1897b, p. 165)

This led to a form of teaching by "mutual stimulus of student and professor" and left lasting traces in some of Sylvester's major mathematical publications. Although others had characterised Sylvester as a bad teacher, Halsted took a different stance explaining Sylvester's view that: "no teaching for a real university can be ranked high which is not vitalized by abundant original creative work" and that "if a mature man who had produced little or no original work applied for a professorship in a university, he must be either a contemptible ignoramus or a selfish hypocrite and an enemy to mankind" (Halsted 1894b, p. 298). Sylvester was portrayed as "the most extraordinary personage for half a century in the mathematical world" (Halsted 1897b, p. 159) and his resentment of religious injustice was cited as a major reason for his strange career and a certain bitterness. Halsted used many metaphors to explain how "Sylvester created a whole new continent, a new world in the universe of mathematics" while casting himself as "the Mathematical Adam" developing a "new language" (ibid., pp. 161–162).

Sylvester's involvement with America was strange at first as Halsted explained in this quote:

> Something of this irksomeness of the outside world, the world of matter, may have made him accept, in 1841, the professorship offered him in the University of Virginia. (ibid., p. 160)

And Halsted continued to stress that Sylvester's genius and his entire outlook on humanity and mathematics in particular was so foreign to the (implicitly provincial) young culture of the United States:

> The Virginians so utterly failed to understand Sylvester, his character, his aspirations, his powers, that the Rev. Dr. Dabney, of Virginia, has seriously assured me that Sylvester was actually deficient in intellect. (ibid., p. 160)

Compared to Halsted's rendering of Sylvester's time in Virginia, Sylvester's affiliation with Johns Hopkins University in Baltimore proved extremely important for American mathematics, which would be emphasized even more in the other biography written by Fabian Franklin (1853–1939) in 1897, then still at Johns Hopkins University. Franklin had also been a pupil of Sylvester, but his biography was more objective in its scope and style, and it even included references to other biographies by Cayley and Halsted. Yet, Franklin and Halsted can both be seen as vying for the position as Sylvester's intellectual heir in the United States. In Franklin's biography, the "retrograde condition of English mathematics" was noted (Franklin 1897, p. 300), and Sylvester together with William Rowan Hamilton (1805–1865) and Cayley were credited with changing that picture. Franklin included more areas of Sylvester's mathematics in his treatment and described his working method as revelling in new methods and working on crucial problems:

> His temperament was essentially poetic, and it would have been as impossible for him to concentrate the powers of his mind on one subject when the current of his thought was setting toward another. (ibid., p. 302)

An explicit ambition for Franklin was to assess Sylvester's time in Baltimore. This was portrayed as a mutual benefit; for Sylvester "at the age of 62, in Baltimore, he felt himself, for the first time, among a band of enthusiastic young workers pursuing pure mathematics for its own sake" (ibid., pp. 303–304). Elaborating on this line of argument, Sylvester's role for Johns Hopkins University and the *American Journal of Mathematics* was emphasised in a way that could easily extend to include American mathematics more generally:

> Of Sylvester's influence upon this university, not only through his teaching, through the foundation of the American Journal of Mathematics, and through the constant stimulation of mathematical interest here by his incessant productiveness, but also through the infection of his enthusiasm, which was felt in every department of the university, it would be impossible to speak too strongly. His aggressive and singular personality seemed to act the part of a ferment which spread itself through the entire body of the university. (ibid., p. 308)

Thus, Franklin's biography is a more balanced and less polemic and self-fashioning portrait of the complex personality and creative genius of Sylvester. Yet, it too was deeply grounded in the context from which it was produced.

Thus, we have seen in the biographies of Sylvester how the historical and biographical genre can be explicitly used to promote certain views about mathematics, its cultivation and the merits of individuals and institutions in the local context. These are, again, general features of the genre and they are here seen to be at work in shaping the outlook of the newly founded journal, to promote a specific research ideal under incorporation from The Old World, and a set of virtues about mathematical practice aimed at promoting creativity and high standards of research training in the postgraduate seminar.

7 The Professionalization of the *American Mathematical Monthly* and of American Mathematics

These general discussions and detailed case-studies have all pointed to a number of issues involved in the professionalization of mathematics in the United States and the role that the biographical genre played in that process.

During the formative decade for American mathematics around 1900, the *American Mathematical Monthly* developed and professionalized considerably. On the most technical level, production quality increased so as to eliminate many of the typographical errors that marred the first volumes. On the side of content, the *American Mathematical Monthly* continuously featured a column of problems and solutions which was also a common feature of most competing journals in the United States and elsewhere. However, a changing trend in the attitude towards such problems may be discernible over the first decade, and the biographies may offer a cue that we can analyse to extend the previous discussions: In 1894, Finkel had described the recently deceased Enoch Beery Seitz (1846–1883) as "the most distinguished mathematician of his day" and described him as a Napoleon or a Shakespeare of mathematics (Finkel 1894b, p. 3). Even taking into consideration the occasion for the biography, this seems an inflated evaluation: Seitz' only claim to mathematical fame was his interest in the field known as "averages and probabilities" which also attracted the attention of Martin (ibid., pp. 3–4). Obviously connected to general probability theory, the Americans seem to have approached the field mainly through problems such as 'Find the average area of a spherical triangle'. Yet, at least partly due to the vast biographical production of Halsted and the instillation of European research ideals, such problems were not valued nearly as highly after the first dozen volumes of the *American Mathematical Monthly*. Thus, it may be that the biographical genre—with its contingencies, biographer's voices and interests—contributed to reinforcing developments towards more abstract, theory-driven mathematics, even in the *American Mathematical Monthly*.

The conditions for existence of mathematical journals had not been good enough when Hendricks started *The Analyst* and other similar efforts were made in the mid-nineteenth century. By the turn of the century, though, these external circumstances were changing with the foundation of the *American Mathematical Society* which set up its own journal. By 1915, the *Mathematical Association of America* was consolidated and took over the *American Mathematical Monthly*, giving it a lasting backing while also turning it even more in the direction of mathematics teachers. Yet, for the first decade of its existence, the *American Mathematical Monthly* had been competing quite openly to represent the voice and identity of American mathematics. This double identity as catering to mathematicians involved both in research and teaching can, of course, also be seen from the selection of biographees in the first volumes of the *American Mathematical Monthly*. The earliest biographies in the *American Mathematical Monthly* thus sought to establish a shared identity among American mathematicians by pointing out and introducing key figures in the community. These individuals were presented through the biographical genre, and their biographies were appropriated to reflect virtues that the newly forming and solidifying group were to share. Those virtues valued hard work, efforts in teaching and mentoring, and certain kinds of mathematical work such as problem solving and astronomy. As the ultimate objective, American mathematicians would help build up intellectual strength so as to eventually contribute genuinely to the global culture of mathematics.

This line of argument was supplemented by the biographies of great European mathematicians from Euler to Bolyai, whose 'genius' was cast to represent quite different things. For Finkel's Euler, genius consisted in hard work and great skill at mathematical manipulation; for Halsted's Bolyai, it consisted in the ability to create entire new theories out of nothing. Such a partial opposition about the meaning of *doing* mathematics at the frontier of research was, of course, wide-spread throughout the nineteenth century; yet, in the American context, it gained a specific role through the biographies: New mathematicians were told how a strong mathematical mind bordering on the stubborn or absent-minded could overcome external adversary. Mathematics was imported from Europe as a vocation—and, in the case of Sylvester, most clearly as a research profession. Thus, in the appropriation of Sylvester's biography, the biographers were also debating his impact at Johns Hopkins University in setting up what has been called the first research institution in mathematics in the United States. This new emphasis on *research* as the objective of mathematics—over, for instance, more recreational problem solving—was also part of the professionalization of mathematics in Europe; but through the biographies of great mathematicians, these virtues were incorporated into American mathematical culture beyond the established research centres.

With the new, more international, outlook came an aspiration for American mathematics to measure up to the Old World. In the biographies, some American mathematicians were prematurely hyped into 'the American Kepler' or 'the greatest mathematician of his day'. Yet, the role as the leader of American mathematics was up for grabs around 1900, and Halsted seized upon the *American Mathematical Monthly* and, in particular, its biographical column as a means to establish his own authority. Thus, with a pointed pen, Halsted would fashion himself alternatingly as the American expert on (the history of) non-Euclidean geometry and the proper heir of Sylvester's legacy at Johns Hopkins University. And perpetually the outsider, Halsted's claims for both roles were not unequivocally successful. Again, such themes are also discernible in Europe, in particular in Mittag-Leffler's contemporary efforts to position himself and Stockholm in the dual legacy of Abel and Karl Weierstrass (1815–1897) (see Sørensen 2006, 2012). However, perhaps because power structures *were* more established in Europe, Halsted's dual quest to promote both himself and his field of interest does seem rather unusual.

In reviewing these appropriations—uses and adaptations—of the biographies of past mathematicians for a contemporary audience, we have utilised a source of insights into the professionalization that goes beyond the manifest institutional and personal developments. Instead, we have analysed how biographies carried with them virtues and norms of the mathematical community, how their selection and framing could serve a variety of purposes, and how among those purposes, the roles of both the community and the biographer need to be considered. As such, we have *used* the biographical genre as a means for historical analysis that transcends the classical view of biographies as presenting lives worthy of universal emulation. Instead, we have viewed the biographies as representations of people from the past to say something about relations, actions and development. Thus, while Halsted was not considered a historian of mathematics worthy of mention in Dauben and

Scriba (2002) except for a single footnote, we have here seen how much can be analysed from even a first treatment of Halsted's biographical and historical writings in the *American Mathematical Monthly*. Although grappling with the historical genre and with his biographical writing, Halsted's historical viewpoint was subjugated to his mathematical, personal and professional ambitions. Arguably, this makes sense given that the vast professionalization of the field of history of mathematics only occurred from the first decades of the nineteenth century. Yet, this way of analysing even 'poor' history of the past is, I claim, a more general route into researching past historiography by analysing its products in light of the context and agendas of our precursors as historians.

Biographies from the *American Mathematical Monthly*, 1894–1904

Aley, R. J. (1894). Biography: Daniel Kirkwood. *The American Mathematical Monthly, 1*(5), 140–149.

"Biography: James W. Nicholson" (1894). *The American Mathematical Monthly, 1*(6), 182–187.

"Biography: William Hoover" (1894). *The American Mathematical Monthly, 1*(2), 34–37.

Burleson, B. F. (1894). Biography: Samuel Gardner Cagwin. *The American Mathematical Monthly, 1*(11), 374–375.

Colaw, J. M. (1894a). Biography: Joel E. Hendricks. *The American Mathematical Monthly, 1*(3), 64–67.

Colaw, J. M. (1894b). Biography: Simon Newcomb. *The American Mathematical Monthly, 1*(8), 252–256.

Colaw, J. M. (1895). Biography: Alexander Macfarlane. *The American Mathematical Monthly, 2*(1), 1–4.

Craig, V. J. (1901). Biography: Isaac Newton. *The American Mathematical Monthly, 8*(8/9), 157–161.

Darboux, Gaston. (1899). Sophus Lie. *The American Mathematical Monthly, 6*(4), 99–101.

"De Volson Wood" (1897). *The American Mathematical Monthly, 4*(8–9), 197–199.

Dickson, L. E. (1894). Biography: Dr. George Bruce Halsted. *The American Mathematical Monthly, 1*(10), 336–340.

Finkel, B. F. (1894a). Biography: Artemas Martin. *The American Mathematical Monthly, 1*(4), 108–111.

Finkel, B. F. (1894b). Biography: E. B. Seitz. *The American Mathematical Monthly, 1*(1), 3–6.

Finkel, B. F. (1894c). Biography: Elisha Scoot Loomis. *The American Mathematical Monthly, 1*(7), 218–222.

Finkel, B. F. (1896). W. J. C. Miller. *The American Mathematical Monthly, 3*(6–7), 159–163.

Finkel, B. F. (1897). Biography: Leonhard Euler. *The American Mathematical Monthly, 4*(12), 297–302.

Finkel, B. F. (1898). Biography: René Descartes. *The American Mathematical Monthly, 5*(8/9), 191–195.

Finkel, B. F. (1901). Biography: Karl Frederich Gauss. *The American Mathematical Monthly, 8*(2), 25–31.

Halsted, G. B. (1894a). Biography: Felix Klein. *The American Mathematical Monthly, 1*(12), 416–420.

Halsted, G. B. (1894b). Biography: James Joseph Sylvester. *The American Mathematical Monthly, 1*(9), 294–298.

Halsted, G. B. (1895a). Biography: Arthur Cayley. *The American Mathematical Monthly, 2*(4), 102–106.
Halsted, G. B. (1895b). Biography: Lobachevsky. *The American Mathematical Monthly, 2*(5), 137–139.
Halsted, G. B. (1895c). Biography: Paenutij Lvovitsch Tchebychev. *The American Mathematical Monthly, 2*(3), 61–63.
Halsted, G. B. (1895d). John Henry Lambert. *The American Mathematical Monthly, 2*(7–8), 208–211.
Halsted, G. B. (1896). Biography: Bolyai Farkas. *The American Mathematical Monthly, 3*(1), 1–5.
Halsted, G. B. (1897a). Biography: De Morgan. *The American Mathematical Monthly, 4*(1), 1–5.
Halsted, G. B. (1897b). Biography: James Joseph Sylvester. *The American Mathematical Monthly, 4*(6–7), 159–168.
Halsted, G. B. (1897c). Biography: Quillaume Jules Hoüel. *The American Mathematical Monthly, 4*(4), 99–101.
Halsted, G. B. (1897d). Biography: Vasiliev. *The American Mathematical Monthly, 4*(11), 265–267.
Halsted, G. B. (1898a). Biography: Bolyai János. *The American Mathematical Monthly, 5*(2), 35–38.
Halsted, G. B. (1898b). Biography: Tchébychev. *The American Mathematical Monthly, 5*(12), 285–288.
Halsted, G. B. (1899a). Biography: Percival Frost. *The American Mathematical Monthly, 6*(8–9), 189–191.
Halsted, G. B. (1899b). Biography: Sophus Lie. *The American Mathematical Monthly, 6*(4), 97–98.
Halsted, G. B. (1900a). Robert Tucker. *The American Mathematical Monthly, 7*(11), 237–239.
Halsted, G. B. (1901a). Biography: Charles Hermite. *The American Mathematical Monthly, 8*(6/7), 131–133.
Halsted, G. B. (1901b). Franz Schmidt. *The American Mathematical Monthly*, *8*(5), 107–110.
Halsted, G. B. (1902b). Biography: Cristoforo Alasia. *The American Mathematical Monthly*, *9* (8/9), 183–185.
Halsted, G. B. (1902c). Eugenio Beltrami. *The American Mathematical Monthly*, *9*(3), 59–63.
Macfarlane, A. (1895). Biography: Arthur Cayley. *The American Mathematical Monthly, 2*(4), 99–102.
Matz, F. P. (1894). Biography: James Matteson. *The American Mathematical Monthly, 1*(11), 372–374.
Matz, F. P. (1895a). Biography: Benjamin Peirce. *The American Mathematical Monthly, 2*(6), 173–179.
Matz, F. P. (1895b). Biography: De Volson Wood. *The American Mathematical Monthly, 2*(9–10), 253–256.
Matz, F. P. (1895c). Biography: William Chauvenet. *The American Mathematical Monthly, 2*(2), 33–37.
Matz, F. P. (1895d). Hudson A. Wood. *The American Mathematical Monthly, 2*(12), 343–345.
Matz, F. P. (1895e). Ormond Stone. *The American Mathematical Monthly, 2*(11), 299–301.
Matz, F. P. (1896). John Newton Lyle. *The American Mathematical Monthly, 3*(4), 95–100.
Matz, F. P. (1901). Biography: Thomas Craig. *The American Mathematical Monthly, 8*(10), 183–187.
Miller, G. A. (1896). Lie's views on several important points in modern mathematics. *The American Mathematical Monthly, 3*(12), 295–296.
Philips, A. W. (1897). Hubert Anson Newton. *The American Mathematical Monthly, 4*(3), 67–71.
Smith, D. E. (1896). Emile-Michel-Hyacinthe Lemoine. *The American Mathematical Monthly, 3* (2), 29–33.
Tyler, H. W. (1903). Biography: John Daniel Runkle. *The American Mathematical Monthly, 10*(8–9), 183–185.

Other References

Archibald, R. C. (1938). *A semicentennial history of the American Mathematical Society, 1888–1938, with biographies and bibliographies of the past Presidents*. New York: American Mathematical Society.

Cajori, F. (1922). George Bruce Halsted. *The American Mathematical Monthly, 29*(9), 338–340.

Dauben, J. W., & Scriba, C. J. (Eds.). (2002). *Writing the history of mathematics. Its historical development*. Science networks—Historical studies (Vol. 27). Basel, Boston, and Berlin: Birkhäuser Verlag.

Despeaux, S. E. (2002). *The development of a publication community: Nineteenth-century mathematics in British scientific journals*. PhD thesis. Department of Mathematics, University of Virginia.

Dupuy, P. (1896). La vie d'Évariste Galois. *Annales scientifiques de l'É.N.S. 3e série, 13*, 197–266.

Duren, P., et al. (Eds.). (1988–1989). *A century of mathematics in America* (3 Volumes). Providence (RI): American Mathematical Society.

Fenster, D. D. (2002). American initiatives toward internationalization: The case of Leonard Dickson. In K. H. Parshall & A. C. Rice (Eds.), *Mathematics unbound: The evolution of an international mathematical research community, 1800–1945*. History of mathematics (Vol. 23). American Mathematical Society and London Mathematical Society, Chap. 15, pp. 311–333.

Fenster, D. D. (2003). Funds for mathematics: Carnegie Institution of Washington support for mathematics from 1902 to 1921. *Historia Mathematica, 30*, 195–216.

Finkel, B. F. (1931). The human aspect in the early history of the American mathematical monthly. *The American Mathematical Monthly, 38*(6), 305–320.

Finkel, B. F., & Colaw, J. M. (1894). Introduction. *The American Mathematical Monthly, 1*(1), 1–2.

Franklin, F. (1897). James Joseph Sylvester. *Bulletin of the American Mathematical Society, 3*(9), 299–309.

Halsted, G. B. (1878a). Addenda to Mr. Halsted's paper on the bibliography of hyper-space and non-euclidean geometry. *American Journal of Mathematics, 1*(4), 384–385.

Halsted, G. B. (1878b). Bibliography of hyper-space and non-euclidean geometry. *American Journal of Mathematics, 1*(3), 261–276.

Halsted, G. B. (1879). Addenda to bibliography of hyper-space and non-euclidean geometry. *American Journal of Mathematics, 2*(1), 65–70.

Halsted, G. B. (1894c). Note on Smith's review of Cajori. *Bulletin of the American Mathematical Society*, 248–251.

Halsted, G. B. (1895e). The international mathematical congress. *Science*. New Series, *1*(18), 486–487.

Halsted, G. B. (1899c). Report on progress in non-euclidean geometry. *The American Mathematical Monthly, 6*(10), 219–233.

Halsted, G. B. (1900b). Gauss and the Non-Euclidean Geometry. *The American Mathematical Monthly, 7*(11), 247–252.

Halsted, G. B. (1901c). Supplementary report on non-euclidean geometry. *The American Mathematical Monthly, 8*(11), 216–230.

Halsted, G. B. (1901d). The popularization of non-euclidean geometry. *The American Mathematical Monthly, 8*(2), 31–35.

Halsted, G. B. (1902a). A non-euclidean gem. *The American Mathematical Monthly, 9*(6/7), 153–159.

Halsted, G. B. (1904). Simon's claim for Gauss in non-euclidean geometry. *The American Mathematical Monthly, 11*(4), 85–86.

Holmboe, B. (1829). Necrolog. Kort Fremstilling af Niels Henrik Abels Liv og videnskabelige Virksomhed. *Magazin for Naturvidenskaberne*, 334–354.

Jordan, D. S. (1922). *The days of a man. Being memoirs of a naturalist, teacher and minor prophet of democracy* (Vol. 1). Yonkers-on-Hudson, New York: World Book Company.

Klein, F. (1894). *Lectures on mathematics delivered from Aug. 28 to Sept. 9, 1893 before members of the Congress of Mathematics held in connection with the Worlds Fair in Chicago at Northwester University*. New York and London: MacMillan and Co.

Koenigsberger, L. (1879). *Zur Geschichte der Theorie der elliptischen Transcendenten in den Jahren 1826–29*. Leipzig: B. G. Teubner.

Koenigsberger, L. (1904). *Carl Gustav Jacob Jacobi. Festschrift zur Feier der hundertsten Wiederkehr seines Geburtstages*. Leipzig: B. G. Teubner.

Lee, H. (2009). *Biography. A very short introduction*. Oxford: Oxford University Press.

Mathematics in the American Dictionary of National Biography (1926). *The American Mathematical Monthly, 33*(6), 318–319.

Maxwell, J. D. (1993). John Marvin Colaw: And the American Mathematical Monthly. *The American Mathematical Monthly, 100*(2), 117–118.

Merzbach, U. C. (1989). The study of the history of mathematics in America. A centennial sketch. In P. Duren et al. (Ed.), *A Century of Mathematics in America* (vol. 3, pp. 639–666), Providence (RI): American Mathematical Society.

Miller, G. A. (1899). Some reminiscences in regard to Sophus Lie. *The American Mathematical Monthly, 6*(8–9), 191–193.

Numbers, R. L. (1973). The American Kepler: Daniel Kirkwood and his analogy. *Journal for the History of Astronomy, 4*, 13–21.

Parke, C. N. (2002). *Biography. Writing lives*. Genres in context. New York and London: Routledge.

Parshall, K. H. (2000). Perspectives on American Mathematics. *Bulletin of the American Mathematical Society, 37*(4), 381–405.

Parshall, K. H. (2003). Entering the international arena: E. H. Moore, the University of Chicago, and Hilbert's *Grundlagen der Geometrie*". In R. Franci, P. Pagli, & A. Simi (Eds.), *Il sogno di Galois — scritti di storia della matematica dedicati a Laura Toti Rigatelli per il suo 60o compleanno* (pp. 199–227). Siena: Centro Studi della Matematica Medioevale, Università di Siena.

Parshall, K. H. (2004). Defining a mathematical research school: The case of algebra at the University of Chicago, 1892–1945. *Historia Mathematica, 31*(3), 263–278.

Parshall, K. H. (2006). *James Joseph Sylvester: Jewish Mathematician in a Victorian World*. Baltimore: The Johns Hopkins University Press.

Parshall, K. H. (2009). The internationalization of mathematics in a world of nations, 1800–1960. In E. Robson & J. Stedall (Eds.), *The Oxford handbook of the history of mathematics* (Chap. 1.4, pp. 85–104). Oxford: Oxford University Press.

Parshall, K. H. (2013). *James Joseph Sylvester: Life and work in letters*. Oxford: Oxford University Press.

Parshall, K. H., & Rowe D. E. (1989). American mathematics comes of age, 1875–1900. In P. Duren et al. (Eds.), *A century of mathematics in America* (Vol. 3, pp. 3–28). Providence (RI): American Mathematical Society.

Parshall, K. H., & Rowe D. E. (1994). *The emergence of the American mathematical research community, 1876–1900: J. J. Sylvester, Felix Klein, and E. H. Moore*. History of mathematics (Vol. 8). Providence (RI): American Mathematical Society and London Mathematical Society.

Poggendorff, J. C. (1853). *Lebenslinien zur Geschichte der exacten Wissenschaften seit Wiederherstellung derselben*. Berlin: Alexander Duncker.

Poggendorff, J. C. (1863). *Biographisch-Literarisches Handwörterbuch der exakten Naturwissenschaften* (2 volumes). Leipzig: Johann Ambrosius Barth.

Rabinowitz, Stanley (Ed.). (1996). *Problems and solutions from the mathematical visitor, 1877–1896*. Westford (MA): MathPro Press.

Rothman, T. (1982). Genius and biographers: The fictionalization of Evariste Galois. *The American Mathematical Monthly, 89*, 84–106.

Rupke, N. A. (2008). *Alexander von Humboldt: A Metabiography*. Chicago and London: University of Chicago Press.

Simon, M. (1890). Ueber das Parallelenaxiom. *Jahresbericht der Deutschen Mathematiker-Vereinigung, 1*, 39–41.

Sørensen, H. K. (2006). Niels Henrik Abel's political and professional legacy in Norway. In R. Siegmund-Schultze & H. K. Sørensen (Eds.), *Perspectives on Scandinavian science in the early twentieth century* (pp. 197–219). The Norwegian Academy of Science and Letters, I. Mat.-Nat. Klasse. Skrifter og avhandlinger 1. Oslo: Novus forlag.

Sørensen, H. K. (2012). Reading Mittag-Leffler's biography of Abel as an act of mathematical self-fashioning. In B. Wardhaugh (Ed.), *The history of the history of mathematics. Case studies for the seventeenth, eighteenth and nineteenth centuries* (pp. 115–144). Oxford: Peter Lang.

Swain, J. (1901). Daniel Kirkwood. *Publications of the Astronomical Society of the Pacific, 13* (80), 140–147.

Timmons, T. (2004). A prosopographical analysis of the early American mathematics publication community. *Historia Mathematica, 31*(4), 429–454.

Young, J. W. A. (1903). Concerning the bibliography of mathematics. *The American Mathematical Monthly, 10*(8/9), 186–191.

Greek Mathematics in English: The Work of Sir Thomas L. Heath (1861–1940)

Benjamin Wardhaugh

Abstract

This short note provides some basic information about the life and work of Sir Thomas Little Heath (1861–1940), the author of well-known English versions of ancient Greek mathematical works. It suggests some of the questions which a study of Heath might ask, and provides brief illustrations of some ways in which his work might be situated intellectually and biographically.

Keywords

Greek mathematics · Translation · Euclid · Archimedes

1 Introduction

Sir Thomas Little Heath (1861–1940): readers of this volume will certainly know his name and will very probably have used his work. Most prominent among that work is Heath's English version of Euclid's *Elements*, which remains widely available both in print and online more than a century after its first publication. Dover Books has kept the text of the 1926 edition in print in recent years (Euclid 1956, most recently reprinted in 2000), while the Green Lion Press has recently

B. Wardhaugh (✉)
All Souls College, University of Oxford, Oxford OX1 4AL, UK
e-mail: benjamin.wardhaugh@history.ox.ac.uk

V.R. Remmert et al. (eds.), *Historiography of Mathematics in the 19th and 20th Centuries*, Trends in the History of Science,
DOI 10.1007/978-3-319-39649-1_6

issued an English *Elements* amounting to Heath's translation without his commentary (Euclid 2002). On the web, David Joyce's popular English hypertext of the *Elements* (Euclid 1996) adopts a translation which "is similar to Heath's edition".[1]

In fact, over six decades Heath published English versions of several major Greek mathematical texts, as well as a monumental history of Greek mathematics:

1885: *Diophantos of Alexandria; a study in the history of Greek algebra* (second edition 1910)
1896: *Apollonius of Perga; treatise on conic sections*
1897: *The works of Archimedes, edited in modern notation*
1908: *The thirteen books of Euclid's Elements* (second edition 1926)
1912: *The Method of Archimedes*
1913: *Aristarchus of Samos, the ancient Copernicus*
1920: *Euclid in Greek, Book I*
1921: *A history of Greek mathematics* (2 volumes)
1931: *A manual of Greek mathematics*
1932: *Greek Astronomy*
1948 (published posthumously): *Mathematics in Aristotle*.

Many of these continue to be used in the twenty-first century.

Heath's contemporaries rated him very highly. D'Arcy Wentworth Thompson (1860–1948) (1941: 409), no mean judge of intellectual merit, ranked him "with Gino Loria, Paul Tannery and Zeuthen, next after Heiberg", among the greatest mathematical historian/philologists of his generation, and judged him "one of the most learned and industrious scholars of our time". Heath was a Fellow of the Royal Society, a Fellow of the British Academy, an honorary Fellow of Trinity College, Cambridge, and the holder of honorary degrees from Oxford and elsewhere. As David Eugene Smith (1860–1944) remarked, reviewing the second edition of Heath's *Elements*, its title page recorded "honors which this and other works upon the history of mathematics have brought to the author, and with the approval of the whole scientific world" (Smith 1936, 246).

Heath's achievements have endured despite drastic changes in the historiography —in the construction and use of the ancient mathematical past—since his lifetime. His influence remains pervasive, since most of the texts he translated have not been rendered subsequently into English (a new English *Elements* is under way at the time of writing). Thus even now in the English-speaking world only those who read Greek well can easily approach the canon of Greek mathematical writing without reading at least some of Heath's work.

Yet he and his agenda have received little explicit critical attention since the obituaries which followed his death in 1940. The bibliography on Heath, indeed, consists merely of obituaries in the London *Times*, the *Proceedings of the British Academy* and the Royal Society's *Obituary Notices*, and an article in the *Dictionary*

[1]http://aleph0.clarku.edu/djoyce/java/elements/aboutText.html.

of National Biography revised for the *Oxford Dictionary of National Biography* (Headlam 1940; Thompson 1941; Headlam and Thomas 2010).

This essay is a first attempt to sketch out some of the directions in which a study of Heath might proceed. It illustrates certain ways in which Heath and his achievement may be situated with respect to his circumstances and his intellectual agenda(s). It is based on Heath's published works, and the biographical notices and appreciations which appeared during his lifetime and shortly after his death. Archival research on Heath would doubtless turn up further valuable information. But it would be an enormously larger project to carry out a fully documented archival study of Heath's life and work, to fully understand the sources of his ideas and assess their importance, and to place him fully in a historiography of ancient mathematics.

2 A Self-made Man

A sense of hard, disciplined work and its rewards pervades much of what we know of Thomas Heath, both in the shape of his career, the testimony of contemporaries, and his own autobiographical hints in his work and letters. These values can be seen to have shaped both his historical work and its presentation.

Thomas Heath was the son of a butcher, Samuel Heath, in Lincolnshire. His education took place first in local schools; his first schoolteacher, the Rev. Anthony Bower, was a self-educated man, a tanner's son, and perhaps an important influence (Thompson 1941, 409). Heath does not appear to have discussed such matters in print, but it seems clear that a deliberate plan of self-education and self-improvement guided his trajectory both personally and professionally from quite early in his life. An example from his personal life: both Thomas and his brother Robert taught themselves to play the piano with what was reported to be very considerable proficiency, apparently without the benefit of tuition (Thompson 1941, 421; compare Heath 1948, preface).

Hard work at this stage bore fruit: Heath went up to Trinity College, Cambridge in 1879 (Venn and Venn 1922–1954, part II, vol. III, p. 313; Thompson 1941, 410). As an undergraduate he predictably acquired a reputation for efficiency (Thompson 1941, 411). Neither sport nor the debating union took up his time; his only reported hobby was music. At some stage, perhaps this, he became proficient in French, German and Italian, and acquired some knowledge of other modern languages (Smith 1936, vii). Crucially, he took honours in both Classics and Mathematics. In Classics he gained first-class marks in both Parts 1 and 2 of the tripos, in 1881 and 1883; in Mathematics he sat Part 1 of the tripos in 1882 and was placed as twelfth Wrangler.

At this stage a career in scholarship could surely have been thought of, but in fact Heath entered the civil service in 1884, and would spend his professional life there for more than forty years. He was first a clerk in the Treasury. From 1913 he was joint Permanent Secretary to the Treasury, from 1919 Comptroller General of the National Debt Office. He retired in 1926, aged sixty-five. For these services he was repeatedly rewarded with civil honours, becoming a Companion of the Bath in 1903, in 1909 a Knight Commander of the Bath, and in 1916 Knight Commander of the Royal Victorian Order (Headlam and Thomas 2010).

This pattern of important—if unprominent—work and visible reward obscures to some degree what seem to have been Heath's limitations as an administrator. In particular, the consensus among his obituarists was that he did not really thrive during the First World War, when admittedly unusual combinations of qualities and unusual flexibility were called for in civil servants. The move in 1919 to the position of Comptroller General, though it was not a demotion, amounted to a reduction of his responsibilities in a "less arduous office", as Thompson put it. Even the obituarist in the *Times* (Anonymous 1940) felt compelled to state that, although "his technique was perfect", "his mind was not, perhaps, sufficiently pliable or fertile in ideas to adapt itself" to wartime conditions. I do not know whether his wartime experiences may have affected his outlook in ways relevant to his historical publications.

Heath's work on Greek mathematics was done as, in his word, a "hobby" around the edges of this full and busy life. It, too, manifested a focus on unremitting hard work, perhaps to the exclusion or at least the diminution of other qualities. In his *Diophantos* Heath (1885, v–vi) claimed that he had "twice carefully worked out the solution of every problem from the proof-sheets". His books on Greek mathematics total something like five thousand printed pages, in addition to which he published articles (eleven are listed in Thompson's (1941) obituary), reviews and other short pieces such as contributions to the *Encyclopedia Britannica* (articles on Pappus and on porisms) and two brief popular books.

In both word and action Heath represented himself as an amateur of Greek mathematics, one who courted Euclid and others for love, not for gain: and thus perhaps as an outsider to the ranks of professional historians, with an outsider's privileges of needing to please no-one but himself and owing no reverence to the established pieties of the insider. In a letter to D.E. Smith in 1909 he called his historical work "My favourite hobby of Greek mathematics" (see Smith 1936, xix). Since Heath had no formal training specifically in historical scholarship, there was a sense in which his historical work was that of a hobbyist, although of course his study of classics will have acquainted him with ancient history through the lens of the classical authors. He would not be the only historian of mathematics to play the role of an outsider; arguably the discipline as a whole was (and remains) itself an outsider to mainstream historical work.

Others writing about Heath took up this theme of hard, besotted work. In her preface to his posthumous *Mathematics in Aristotle* (Heath 1948), Heath's widow Ada Mary Heath stated that "His eagerness to return to this work too soon after a serious illness in 1939 was probably instrumental in hastening his end." Thompson (1941, 423), typically, wrote of "sixty years of unstinted and unwearied work, all done for the love of it." Yet the danger that such an approach might lead to something less than brilliance surfaced here too: the same combination of technical perfection with a slight inflexibility to be found in Heath's civil service work.

If Thompson reckoned Heath "sound and careful to the last degree", he nevertheless noted that he had relatively little "critical taste and gift of brevity" and lacked "brilliant insight"; "nor does he show, or ever want to show, much imagination or speculative curiosity". "As a historian he was more sober than a man need always be". The consequence for his historical work was that "he was best as a biographer" (Thompson 1941, 413, 420, 423).

It is perhaps no surprise that a man who rose from humble origins through hard work should be found lacking in imagination or flair by some of those around him. Possibly a measure of resentment at his achievements might be detected in the responses of some contemporaries, although I do not believe that the published obituaries and appreciations can plausibly be read in such a way. It might be possible to interpret Heath's actions and words in terms of the brittleness of an acquired gentility; the less-than-perfect security of the butcher's son who became a knight. It would be equally possible to read them as Heath's deliberate self-fashioning, positioning himself as an outsider, an amateur, whose steady determination achieved without the luxury of showy brilliance.

A more oblique reference to the same characteristics—determination with stiffness—appeared in D.E. Smith's appreciation of Heath, published in *Osiris* in 1936. Struggling to account for Heath's achievements, Smith (1936, xvii) included the remarkable line of explanation "In the first place Sir Thomas Heath is an Englishman". He went on to point to his physical fitness (Heath had taken up mountaineering as another hobby) as an example of his characteristically British qualities. Elsewhere the connection between Britishness and the history of ancient mathematics was spelled out more clearly, insofar as that was possible. Reviewing Heath's Euclid in 1909 Smith (1909, 387f) had written "It may properly be said of Dr. Heath's latest contribution to mathematical literature… that it is characteristically British". That is, it was thorough: "when the English scholar does bring out a book, it is quite as when a British general takes possession of a conquered province—there is nothing more to say." It manifested tenacious hard work: "no continental writer has stayed with the problem with the tenacity of purpose that characterizes a Briton when he sets to work for his task". And it reflected the English love of Euclid (more: the common identification of English and Greek culture): "a work that no one but an Englishman could write in the true con amore spirit, one that appeals more to English education than to that of any other country". To this last point we will return.

3 "Mathematics… Is a Greek Science"

For whom was Heath writing? Several times he answered the question, and the answer was invariably, in the first place, mathematicians. In his Apollonius (1896, vii) he stated that he was writing for "mathematicians"; in his Archimedes (1897, v) likewise that he wished "to make the work of 'the great geometer' accessible to the mathematician of to-day". Thompson (1941, 417), too, believed Heath's books would “bring home to the modern mathematician the magnificence of the Greek achievement”.

Why should the modern mathematician be interested? For Heath a true understanding of mathematics necessarily began with its Greek foundations: “For the mathematician the important consideration is that the foundations of mathematics, and a great portion of its content, are Greek…. Mathematics in short is a Greek science” (Heath 1921, v). He elaborated elsewhere (Heath 1931, 1):

> Why should we study Greek mathematics? In the first place, it is true generally that, if we would study any subject properly, we must study it as something that is alive and growing and consider it with reference to its origins and its evolution in the past. In the case of mathematics, it is the Greek contribution which it is most essential to know, for it was the Greeks who first made mathematics a science.

Heath’s desire to make the Greek mathematical classics comprehensible, and his conviction that they were relevant, to the mathematician of his day, shaped some of the details of Heath’s reading and presentation of them, although this is not a subject I have investigated in detail. For example, Heath seems to have been quite ready to believe that the point of certain Euclidean constructions was as existence proofs: “The use of actual construction as a method of proving the existence of figures having certain properties is one of the characteristics of the *Elements*” (Euclid 1908, 234). (More recent scholars have often argued that the problems are in fact to show that certain figures can be constructed under the specified restrictions (Saito 2012; Harari 2003).) Heath’s position was one which had its roots in the late antique commentary of Proclus (c. 412–485), but it is surely no accident that it tended to make (some parts of) Greek mathematics comparable to the constructivist programmes of L.E.J. Brouwer (1881–1966) or David Hilbert (1862–1943).

From the view that the essence of mathematics was its Greek foundations there closely followed a sense of the pedagogical importance of the Greek mathematical classics, above all of Euclid. It was surely this which led Heath to devote his efforts to presenting Euclid and others in English rather than in their original language, the Greek he so loved.

The British love affair with Euclid’s *Elements* was drawing to a close by Heath’s time. The text had dominated school education since the eighteenth century, albeit never uncontroversially, but following the so-called Euclid debate of the later nineteenth century, the English universities had abandoned the compulsory teaching of Euclid in 1904 (Moktefi 2011). Heath’s work on Greek mathematics was thus mainly carried out during a period when the dependence of English mathematics pedagogy on Greek models was undergoing rapid and in his view profoundly ill-advised change. Introducing his English Euclid in 1908, Heath expressed the hope that “it would seem at least possible that… there will be a return to Euclid more or less complete [in geometrical teaching]” (Euclid 1908, vi).

In keeping with this hope, Heath also prepared—now much less well known—an edition in Greek of *Elements* Book I, a volume which he hoped would have a place in such a putative renewed pedagogical emphasis on Euclid. For Heath, direct contact with Greek geometry in its original Greek form could bring unparalleled benefits. "If the study of Greek and Euclid be combined by reading at least part of Euclid in the original, the two elements will help each other enormously." "How can any person who has only had such words as *theorem*, *problem*, *isosceles*, *parallelepiped* explained to him in English apart from their derivation get any such clear idea of their significance" as one who knows them as their Greek equivalents? (Euclid 1920, v, vi).

Heath himself acknowledged in the preface to his Greek Euclid (1920, v) that it might seem to some to represent "a wildly reactionary proceeding", and the same might be said of his larger project to present Euclid in English. Indeed, it is not clear that English readers had ever before possessed a version of Euclid's *Elements* which approached Heath's very high standards, and the school editions of the eighteenth and nineteenth centuries may be considered a tradition already debased long before. But he found some supporters, not least the faithful D.E. Smith, who hoped that the Greek volume would be widely taken up, as "one of the few books on geometry that no teacher can afford to be without, that is indispensable in the library of any well-equipped high school…" (Smith 1920, 266).

It is not clear whether Heath attempted to influence educational policy more directly than through these publications. But in any event, his hopes were not to be realised, and any plan he had of editing more material in the original Greek for these purposes did not come about. Heath's remarks about the desirability of a return to Euclid in the teaching of geometry were reprinted in the 1926s edition of his *Elements* (Euclid 1926), yet by this time Euclid's fall from his pedagogical pedestal had become still more irrevocable.

Heath's views on the interest of Greek mathematics to the modern mathematician, as on the vital importance of Euclid in the modern classroom, sprang from his ideas about the nature of mathematics. Mathematics was for Heath a single—albeit a growing—thing, not essentially different in classical Greece and in Victorian and Edwardian Britain. "Elementary geometry is Euclid, however much the editors of text-books may try to obscure the fact" (Euclid 1920, v). Smith, reviewing the second edition of Heath's *Elements*, characteristically went further (Smith 1927, 248): "no teacher of geometry can afford to be ignorant of the spirit of Euclid, since it is this spirit which constitutes the essence of all demonstrative geometry".

Naturally such a position on the nature, the philosophy, of mathematics, had consequences for Heath's handling of ancient mathematics and for his detailed textual practices. Like other scholars of his and the next few generations, Heath was interested—at least up to a point—in the notion of 'geometrical algebra': the idea that what Euclid and other Greeks expressed in geometrical language was in essence equivalent to certain algebraic statements and operations. The implication, taken in its most extreme form, was that Greek geometry could legitimately be translated into algebra without the loss of anything important: that the essence of mathematical ideas was independent of the technical language, the symbolism, or even the conceptual language in which it was expressed. The chapter in the present volume by Martina Schneider deals with some aspects of the controversial history of 'geometrical algebra'.

In fact Heath's own practices as an editor and translator varied. His rather different handling of different ancient authors reflected clearly the views of contemporaries—notably of Johan Heiberg (1854–1928)—about the nature of their writing. Netz (2012) has recently shown that with respect to Archimedes in particular, some work of what might be called unintended modernization was already performed by Heiberg in the Greek editions on which Heath relied, such as in the presentation of diagrams and the decision of which portions of the text to consider authoritative. "Very likely, this editorial policy reveals, therefore, a certain image of mathematical *genius*. Heiberg could well make his Euclid transparent and accessible; Archimedes had to be difficult" (Netz 2012, 202f).

Heath on the whole took up the view that Euclid was lucid and comprehensible and Archimedes precise but difficult. He also adopted the position that Apollonius was rebarbative in the extreme. Introducing his Apollonius (1896, viii–ix), Heath argued at some length for the necessity of modern notation to render the material comprehensible: his text was, as he put it (ix), "so entirely remodelled by the aid of accepted modern notation as to be thoroughly readable by any competent mathematician". Although it sprang from his beliefs about the nature and difficulty of the material, this approach illustrates Heath's belief that such matters as the detailed arrangement of material, its division into separate theorems, and of course its notation could be separated from essentials such as mathematical completeness, accuracy and coherence: that a text could be "Apollonius and nothing but Apollonius" (ix) despite being "entirely remodelled".

In his *Archimedes* the following year, Heath was a little more cautious, carrying out less compression (and presumably rearrangement) and less notational change (Archimedes 1897, vii–viii). Introducing the text, he was careful to leave open such questions as whether the Greek geometer "really" performed integrations (v–vi); although in his later edition of Archimedes' 'Method' he would claim without apology or explanation that "Archimedes' argument establishes" a result he expressed in modern notation involving two integrals (Archimedes 1912, 9). But once again Heath's "perfectly faithful reproduction of the treatises" (Archimedes 1897, viii) incorporated obtruded subheadings, modern-style fractions and even, often, algebraic notation within the main text. In fact only certain passages, selected for what Heath judged their historical importance, were translated literally, the remainder being given in modern notation and phrasing. Fidelity was to the original mathematical thought as Heath understood it, not to what he evidently regarded as the accidents of its presentation.

Even Heath's choice of titles is revealing. His edition of Diophantos (1885) bore the subtitle "a study in the history of Greek algebra" (Heath 1885); his account of the work of Aristarchus of Samos (Heath 1913) called him "the ancient Copernicus"; both reflect Heath's willingness to see ancient achievements as equivalent in their essence to those of more recent writers. Such claims were made explicit time after time, and in his astronomical works Heath became particularly fond of expanding on how (Heath 1932, iv) "The Copernican hypothesis was actually anticipated by Aristarchus of Samos".

When he turned to Euclid (1908), Heath adopted the view that less modification was needed. Euclid was for Heath a great expositor, a writer who required little editorial intervention in order to be comprehensible. The Euclidean text was therefore to be treated with a greater measure of self-conscious respect than was considered due to the works of other ancient mathematical writers. In striking contrast with his Apollonius, rewritten explicitly for the sake of accessibility, Heath promised to present (vii) "the real Euclid as distinct from any revised or rewritten version which will serve for schoolboys or engineers". That "real Euclid" was substantially the Euclid of Heiberg's critical text (although in fact Heath occasionally preferred other readings, notably those from Proclus, and his overall principles are not clear).

Unlike certain other Euclidean translators of his period (such as Clemens Thaer (1883–1974), editor of Euclid in German (Euclid 1933)), and unlike English predecessors as early as Isaac Barrow (1630–1677) (Euclid 1655), Heath refrained from introducing algebraic notation into the Euclidean text itself, even if his notes relied heavily on the translation of Euclidean ideas into algebraic language.

Possibly the change reflected increasing caution or deeper reflection on the issues involved. Certainly Heath's Euclid (1908, 372–4) contained a brief discussion of 'geometrical algebra', where he set out some of the issues as he saw them.

> The algebraical method has been preferred to Euclid's by some English editors; but it should not find favour with those who wish to preserve the essential features of Greek geometry as presented by its greatest exponents, or to appreciate their point of view (Euclid 1908, 373–4).

The contrast with Heath's handling of Apollonius could hardly be more striking.

4 To "Understand the Greek Genius Fully"

But Heath was not only writing for mathematicians, and his writings did not only embody mathematical agendas. Indeed, "Heath's interest in meeting the needs of cultured readers is manifest in all his works" (Smith 1936, xv). A much quoted line from his *History of Greek Mathematics*—"if one would understand the Greek genius fully, it would be a good plan to begin with their geometry" (Smith 1936, vi) —tells of his concern to write also for classical scholars, and indeed for cultured readers generally.

Heath's strategy for facilitating the attention of cultured people to the Greek mathematical classics included presenting them—or at least Euclid—in ways which made them look like works of classic literature. His Euclid followed in its layout the conventions then current for the presentation of canonical texts whether ancient or modern: it looked and felt like a work of great literature, with its combination of philological apparatus and explanatory commentary on pages which from a distance could have been those of Aeschylus or for that matter Jane Austen (1775–1817). Smith (1936, xi), too, believed that Heath's notes to Euclid should help to make "of elementary geometry something truly humanistic", noting that they "tell the story of geometry… over a period of more than two thousand years."

To make the works of Euclid and others into works of classical literature which any cultured person could (and should) read and learn from was only part of the agenda. More important was to embed them within a history that even the less scholarly could read and engage with.

In the *History of Greek Mathematics* Heath had in mind (Heath 1931, v)

> the requirements, on the one hand, of the classical scholar who might look for light on the interpretation of passages of mathematical content in Greek authors which came his way, and, on the other hand, of the expert mathematician who might wish so far to assimilate the whole argument of a particular treatise, say of Archimedes, as to be able, on occasion arising, to apply the same method to a different problem.

Thompson judged it a book "for every scholar". In the briefer *Manual of Greek Mathematics*, derived from the *History*, Heath set out the same historical sweep for a different class of reader: "the general reader who has not lost interest in the studies of his youth" and who wished to know something of the history behind Euclid and his *Elements* in particular (Heath 1931, v). In two still shorter works of just sixty pages each Heath (1920a, b) discussed the achievements of Aristarchus (once again "the Copernicus of Antiquity") and of Archimedes for a more general audience still, comprising readers of the SPCK series 'Pioneers of Progress: Men of Science'. These were accessible narrative accounts of their subjects' lives and works, with algebra and calculation kept to a reasonable minimum and introduced towards the end of each book, though by no means suppressed altogether. They were Heath's most 'popular' productions; nevertheless they were aimed at a reader who was not uncomfortable with a certain amount of mathematics.

The history all of these works provided was of a very particular kind, determined by Heath's own assumptions about the nature of mathematics and by the nature of the materials available to him. Heath's was a period in which a number of scholars turned their attention to ancient mathematics: Carl Anton Bretschneider (1808–1878); George Johnston Allmann (1824–1904); Hieronymus Georg Zeuthen (1839–1920); Paul Tannery (1843–1904); Johan Ludvig Heiberg; Gino Loria (1862–1954), to name only a selection. Most possessed, or acquired in the course of their work, a deep knowledge of classical philology. The philological apparatus provided by such scholars and made available to English readers by Heath may have done much to make the mathematical classics works of classical literature. It also supported—in part, determined—Heath's distinctive way of doing history. The characteristic methodological feature was the combination of historical and philological reconstruction. The evidence for Greek mathematics was and is patchy and problematic, and for the period before Euclid it is very scanty indeed. In a sense what Heath did was to apply a philological cast of mind to mathematical reconstruction. Mathematical argument, being logical, should in Heath's view be reconstructible from minimal evidence, and that minimum could often be provided by the philological evidence of the few surviving fragments of Greek mathematics before Euclid.

The resulting history was very much a product of its methodological assumptions, characterized by logical development and meticulously reconstructed proofs. The chain of reasoners was unbroken from Thales to Diophantus, and what Heath constructed might be characterized today as a history of the contents of Greek theoretical mathematics. That the forms, the practices, or indeed the circumstances might have changed was not part of Heath's project.

Heath did not altogether avoid occasional glances at the "practical utility" of Greek mathematics, (Heath, 1931, 1), nor at its value as a propaedeutic: "I am convinced that there is no subject which, if properly presented, is better calculated than the fundamentals of geometry to make the schoolboy (or the grown man) think" (Euclid 1920, viii). But his real aim in respect of non-mathematical readers was to improve and promote the study of the "Greek genius". Thus (Heath 1931, 2; cf. Heath 1920b, 2)

> the Greek genius for mathematics was simply one aspect of their genius for philosophy in general. Their philosophy and their mathematics both arose out of the instinct of the race, their insatiable curiosity, their passion for 'inquiry', and a love of knowledge for its own sake which the Greek possessed in a greater degree than any other people of antiquity.

As a result, Heath believed, (Euclid 1920, v) "Generation after generation of men and women will still have to go to school to the Greeks for the things in which they are our masters; and for this purpose they must continue to learn Greek." Hence Heath's Euclid in Greek, and hence both the importance and the validity of using mathematics as a route to the study and appreciation of the ancient civilizations.

5 Conclusion: A 'Body of Doctrine'

Heath's work on the classics of Greek mathematics can be read in a number of contexts. This brief note has suggested that we can usefully consider Heath's status as self-educated amateur or outsider; his views about the nature of mathematics and his determination to make ancient mathematics relevant to mathematicians of his own day; and his pedagogical agenda and his desire to promote the study of Greece and Greek in a time of what he felt was educational decline. A final thought may show one way these disparate contexts can be related to one another.

Heath's view of Greek geometry was an exalted one, and it occasionally broke forth in passages which hinted at something literally sacred about the texts and the ideas they set forth. Justifying his work on the *Elements* in 1908, Heath claimed that "the body of doctrine contained in the recent textbooks of elementary geometry does not… show any substantial differences from that set forth in the Elements" (Euclid 1908, v). The idea of geometrical truth as 'doctrine' to be handed on faithfully is one which seems to have influenced Heath throughout his career.

In 1920, as mentioned above, Heath contributed two volumes to a series produced by the Society for Promoting Christian Knowledge, on Archimedes and on Aristarchus. Just what the SPCK believed these volumes could do to promote specifically Christian knowledge I have not discovered, but there seems little doubt that for Heath setting forth faithfully the body of ancient geometry was in part a sacred duty for those on whom it fell. For the Greek origins were so sound that "in the centuries which have since elapsed, there has been no need to reconstruct, still less to reject as unsound, any essential part of their doctrine" (Heath 1931, 1). The task of the scholar was to re-present, not to replace, and indeed those who attempted to replace Euclid with new textbooks of their own devising came in for Heath's very sharp censure. Commenting on the pedagogical wasteland he found in modern geometry, Heath remarked that "Euclid can never at any time be more than apparently in abeyance; he is immortal" (Euclid 1920, v).

This is not easy to reconcile with the effort Heath himself expended to interpret and in some cases to rewrite the Greek texts. We can perhaps understand his project rather better if we see it in terms of transmission and exegesis. The essential ideas, the 'body of doctrine' were to be handed on faithfully and unchanged, any lapse or failure being a serious matter. Commentary, even voluminous commentary, was legitimate if it served the purpose of helping readers to understand that body of doctrine. Rephrasing and restructuring, too, could be justified as long as "the real Euclid"—the ideas, not the accidents of presentation—remained unchanged.

D.E. Smith seems to have concurred. Reviewing Heath's Greek Euclid for the *American Mathematical Monthly* (Smith 1920, 264) he wrote that Heath's was "like a voice from another sphere". He expanded (266): "to the teacher of elementary geometry… these notes will seem like the words of one having authority and not of those of the educational scribes and Pharisees." The review was reprinted in the *Classical Weekly*, and the passage was quoted in his appreciation of Heath for *Osiris* (Smith 1936).

Several factors indubitably contributed to the distinctive character of Sir Thomas L. Heath's work on the history of mathematics, and to its longevity. Possibly a sense that the transmission and faithful interpretation of the Greek mathematical heritage was literally a sacred duty underlay the different agendas he seems to have brought to the task. Whatever his agendas and idiosyncrasies, Heath was a remarkable man, exceptionally well-equipped in intellect and character to undertake the colossal labours of editing, translating and commenting to which he devoted much of his life.

Many questions can be raised on the basis of what has been said here.

- What exactly were Heath's philological principles? Can he be shown to have placed mathematical purity ahead of textual purity at specific points, or to have allowed his agenda—to construct a rational history of mathematics—to guide his textual choices?
- How far were Heath's practices (merely) a response to those of his predecessors, and how far did he (intend to) innovate in the study of ancient mathematics?

- The material and typographical dimensions are well worth exploring more: how far did Heath's editions participate in, and how far did they differ from, the common practices of presses such as Oxford and Cambridge in handling ancient or modern 'classic' texts in this period?
- How far did later editors and translators take up Heath's approach, and indeed how far, if at all, is the shape of ancient mathematics history today still a reflection in the English-speaking world of Heath's priorities and practices?

Heath and his books are one of the more prominent features of the landscape in mathematical historiography, and it is hoped that this brief note will stimulate serious work on them.

References

Anonymous. (1940). Sir Thomas Heath (obituary). *The Times*, p. 10. March 18, 1940.

Apollonius. (1896). *Treatise on conic sections* (T. L. Heath Trans.). Cambridge: Cambridge University Press.

Archimedes. (1897). *The works of Archimedes, edited in modern notation* (T. L. Heath Trans.). Cambridge: Cambridge University Press.

Archimedes. (1912). *The method of Archimedes* (T. L. Heath Trans.). Cambridge: Cambridge University Press.

Euclid. (1655). *Euclidis Elementorum libri xv. breviter demonstrati* (I. Barrow Trans.). Cambridge: Cambridge University Press.

Euclid. (1908). *The thirteen books of Euclid's Elements* (T. L. Heath Trans.). Cambridge: Cambridge University Press.

Euclid. (1920). *Euclid in Greek Book 1*. In T. L. Heath (Ed.). Cambridge: Cambridge University Press.

Euclid. (1926). *The thirteen books of Euclid's Elements* (2nd ed.) (T. L. Heath Trans.). Cambridge: Cambridge University Press.

Euclid. (1933). *Die Elemente von Euklid* (2nd ed., Vol. 2 1939) (C. Thaer Trans.). Leipzig: Oswald.

Euclid. (1956). *The thirteen books of Euclid's Elements* (T. L. Heath Trans.). New York: Dover.

Euclid. (1996). *Euclid's Elements* (D. E. Joyce Trans). http://aleph0.clarku.edu/djoyce/java/elements/elements.html. Online; Accessed December 2012.

Euclid. (2002). *The thirteen books of Euclid's Elements*. In D. Densmore (Ed.). Santa Fe: Green Lion Press.

Harari, O. (2003). The concept of existence and the role of constructions in Euclid's Elements. *Archive for History of Exact Sciences, 57*, 1–23.

Headlam, M., & Thomas, I. (2010). Heath, Sir Thomas Little (1861–1940). In L. Goldman (Ed.), *Oxford dictionary of national biography*. Oxford University Press: Oxford. Rev. Alan Booth, online edition http://ezproxy-prd.bodleian.ox.ac.uk:2167/view/article/33793. Accessed December 2012.

Headlam, M. F. (1940). Sir Thomas Little Heath, 1861–1940. *Proceedings of the British Academy, 26*, 424–438.

Heath, T. L. (1885). *Diophantos of Alexandria; A study in the history of Greek algebra*. Cambridge: Cambridge University Press.

Heath, T. L. (1913). *Aristarchus of Samos, the ancient Copernicus*. Cambridge: Cambridge University Press.

Heath, T. L. (1920a). *Archimedes*. London: Society for the Promotion of Christian Knowledge.

Heath, T. L. (1920b). *The Copernicus of Antiquity: Aristarchus of Samos*. London: Society for the Promotion of Christian Knowledge.

Heath, T. L. (1921). *A history of Greek mathematics*. Cambridge: Cambridge University Press.

Heath, T. L. (1931). *A manual of Greek mathematics*. Cambridge: Cambridge University Press.

Heath, T. L. (1932). *Greek astronomy*. London: J.M. Dent.

Heath, T. L. (1948). *Mathematics in Aristotle*. Cambridge: Cambridge University Press.

Moktefi, A. (2011). Geometry: The Euclid Debate. In R. Flood, A. Rice, & R. Wilson (Eds.), *Mathematics in Victorian Britain* (pp. 321–336). Oxford: Oxford University Press.

Netz, R. (2012). The texture of Archimedes' writings: Through Heiberg's veil. In K. Chemla (Ed.), *The history of mathematical proof in ancient traditions* (pp. 163–205). Cambridge: Cambridge University Press.

Saito, K. (2012). Reading ancient Greek Mathematics. In E. Robson & J. Stedall (Eds.), *The Oxford handbook of history of mathematics* (pp. 801–826). Oxford: Oxford University Press.

Smith, D. E. (1909). Heath's Euclid [review]. *Bulletin of the American Mathematical Society, 15*, 386–391.

Smith, D. E. (1920). Euclid in Greek [review]. *The American Mathematical Monthly, 27*, 263–266.

Smith, D. E. (1927). Heath's Euclid [review]. *Bulletin of the American Mathematical Society, 33*, 246–248.

Smith, D. E. (1936). Sir Thomas Little Heath. *Osiris*, *2*, iv–xxvii.

Thompson, D. W. (1941). Sir Thomas Little Heath. *Obituary Notices of Fellows of the Royal Society, 3*, 409–426.

Venn, J., & Venn, J. A. (1922–1954). *Alumni cantabrigienses: A biographical list of all known students, graduates and holders of office at the University of Cambridge, from the earliest times to 1900*. Cambridge: Cambridge University Press.

Otto Neugebauer's Vision for Rewriting the History of Ancient Mathematics

David E. Rowe

The common belief that we gain "historical perspective" with increasing distance seems to me utterly to misrepresent the actual situation. What we gain is merely confidence in generalizations which we would never dare make if we had access to the real wealth of contemporary evidence.
— Otto Neugebauer, *The Exact Sciences in Antiquity* (Neugebauer 1969, viii)

Abstract

Historians of mathematics have long exalted the achievements of the ancient Greeks as symbolized by a single name, Euclid of Alexandria. The thirteen books that comprise his *Elements* hold a place within Greek mathematics comparable to the Parthenon in its architectural tradition. Appreciation for Greek classicism was long reinforced by the formal ideal of Euclidean geometry, a style that persisted until well into the nineteenth century. Not until the early decades of the twentieth did a new picture of ancient mathematics emerge, advanced by the pioneering researches of Otto Neugebauer on Egyptian and especially Mesopotamian mathematics. Although grounded in detailed analysis of primary sources, Neugebauer's work was guided by a broad vision of the exact sciences in ancient cultures that predated the Greeks. He thereby broke with the traditional Greco-centric understanding of European science. Neugebauer's historical views and methodological approach, which elevated mathematical techniques while diminishing the importance of philosophical commentary, came under strong attack after he immigrated to the United States in 1939.

D.E. Rowe (✉)
Institut für Mathematik, AG History of Mathematics, Johannes Gutenberg Universität Mainz, Staudinger Weg 9, 55099 Mainz, Germany
e-mail: rowe@mathematik.uni-mainz.de

V.R. Remmert et al. (eds.), *Historiography of Mathematics in the 19th and 20th Centuries*, Trends in the History of Science,
DOI 10.1007/978-3-319-39649-1_7

Keywords
Greek mathematics · Historiography of ancient science · Göttingen mathematics · Geometric algebra · Pythagorean tradition

Otto Neugebauer (1899–1990) was, for many, an enigmatic personality. Trained as a mathematician in Graz, Munich, and Göttingen, he had not yet completed his doctoral research when in 1924 Harald Bohr (1887–1951), brother of the famous physicist, invited him to Copenhagen to work together on Bohr's new theory of almost periodic functions. Quite by chance, Bohr asked Neugebauer to write a review of T. Eric Peet's (1882–1934) recently published edition of the Rhind Papyrus (Neugebauer 1925). In the course of doing so, Neugebauer became utterly intrigued by Egyptian methods for calculating fractions as sums of unit fractions (e.g. 3/5 = 1/3 + 1/5 + 1/15). When he returned to Göttingen, he wrote his dissertation on this very topic. In 1927 he published the first of many researches on Babylonian mathematics and astronomy, a pioneering study on the evolution of the sexagesimal (base 60) number system (Neugebauer 1927). These works received high praise from leading Egyptologists and Assyriologists, helping to launch Neugebauer's career as a historian of ancient mathematics and exact sciences. Indeed, he would go on to revolutionize research in these areas, leaving a deep imprint on our understanding of these ancient scientific cultures to this very day.

Yet Neugebauer's general orientation as a historian seems strangely remote from today's perspective, so much so that even scholars who know his work well and respect it highly have great difficulty identifying with his methodological views. One who worked closely with him during his later career at Brown University, Noel Swerdlow (1941–), gave a most apt description of the "zwei Seelen" that dwelled within Otto Neugebauer and that colored all his work:

> At once a mathematician and cultural historian, Neugebauer was from the beginning aware of both interpretations and of the contradiction between them. Indeed, a notable tension between the analysis of culturally specific documents, whether the contents of a single clay tablet or scrap of papyrus or an entire Greek treatise, and the continuity and evolution of mathematical methods regardless of ages and cultures, is characteristic of all his work. And it was precisely out of this tension that was born the detailed and technical cross-cultural approach, in no way adequately described as the study of "transmission," that he applied more or less consistently to the history of the exact sciences from the ancient Near East to the European Renaissance.
>
> But if the truth be told, on a deeper level Neugebauer was always a mathematician first and foremost, who selected the subjects of his study and passed judgment on them, sometimes quite strongly, according to their mathematical interest. (Swerdlow 1993, 141–142)

Taking up this last point, one can easily appreciate why Neugebauer's approach to history persuaded few, while provoking some of his detractors to take a firm stand against his methodological views and what they felt was a deleterious influence on studies of the ancient sciences..[1] Neugebauer firmly believed in the immutable character of mathematical knowledge, which meant that his field of

[1]The sharpest attack against Neugebauer's methodological approach came from Sabetai Unguru in (Unguru 1975); for this text and reactions to it, see (Christianidis 2004). I discussed this within the larger context of the historiographical debates from that period in (Rowe 1996). See also the

historical inquiry, the exact sciences, differed from all other forms of human endeavor in one fundamental respect: in this realm there was no room for historical contingency. The methodological implications Neugebauer drew from this were simple and clear: once an investigator had cracked the linguistic or hieroglyphic codes that serve to express a culture's scientific knowledge he or she then suddenly held the keys to deciphering ancient sources. And since the content of these sources pertained to mathematical matters, one could, in principle, argue inductively in order to reconstruct what they originally contained, namely a fixed and determinable pattern of scientific results. Clearly, this type of puzzle solving held great fascination for Neugebauer, and he practiced it with considerable success in his research on Mesopotamian astronomy, beginning in the mid-1930s.

Neugebauer's work on Greek mathematics during these politically turbulent times was far scantier. Nevertheless, his views on Greek mathematics formed a central component of his overall view of the ancient mathematical sciences. When it came to purely human affairs, Neugebauer professed that he held no *Weltanschauung*, and he took pains to make this known to those who, like Oskar Becker (1889–1964), mingled ideology with science (see Siegmund-Schultze 2009, 163). Regarding historiography, on the other hand, Neugebauer adopted a rigorously empirical approach that worked well in some cases, but often led him to make sweeping claims based on little more than hunches. Not surprisingly, his views on historiography had much to do with the special context in which he first experienced higher mathematics (Rowe 2012).

1 Neugebauer's Cornell Lectures

In 1949, when Neugebauer delivered six lectures on ancient sciences at Cornell University, he was the first historian of mathematics to be given the honor of speaking in its distinguished Messenger lecture series. He did not waste this opportunity. Afterward, he went over his notes and gave the text its final, carefully sculpted form that we find today in the six chapters of Neugebauer's *The Exact Sciences in Antiquity*, published in 1951 with high-quality plates. His text begins by describing a famous work in the history of art: *Septembre* from the *Très Riches Heures du Duc de Berry* (see Plate 1, p. 127), one of the most famous works in the French Gothic tradition, which first gained public attention after 1856 when it was acquired by the Duc d'Aumale, founder of the Musée Condé in Chantilly. Neugebauer began by recalling the circumstances that brought this work to a close:

> When in 1416 Jean de France, Duc de Berry, died, the work on his "Book of the Hours" was suspended. The brothers Limbourg, who were entrusted with the illuminations of this

(Footnote 1 continued)
contribution by Schneider to the present volume "Contextualizing Unguru's 1975 Attack on the Historiography of Ancient Greek Mathematics".

book, left the court, never to complete what is now considered one of the most magnificent of late medieval manuscripts which have come down to us.

> A "Book of Hours" is a prayer book which is based on the religious calendar of saints and festivals throughout the year. Consequently we find in the book of the Duke of Berry twelve folios, representing each one of the months. As an example we may consider the illustration for the month of September. As the work of the season the vintage is shown in the foreground (see Plate 1). In the background we see the Château de Saumur, depicted with the greatest accuracy of architectural detail. For us, however, it is the semicircular field on top of the picture, where we find numbers and astronomical symbols, which will give us some impression of the scientific background of this calendar. Already a superficial discussion of these representations will demonstrate close relations between the astronomy of the late Middle Ages and antiquity. (Neugebauer 1969, 3)

Neugebauer went on to note four different types of writing for the numbers that appear in the *Book of Hours:* Hindu-Arabic as well as Roman numerals, number words (September through December for the seventh to the tenth months of the Roman calendar), and alphabetic numbers, here calculated modulo 19, the system used in connection with the Metonic lunar cycle. Regarding the latter, he noted that for a given year, the associated number between 1 and 18 was called the "golden number" in the late Middle Ages, after a 13th-century scholar wrote that this lunar cycle excels all others "as gold excels all other metals."

He then comments as follows about the state of scientific progress in the Latin West when seen against the backdrop of earlier developments:

> In the twelfth century this very primitive method [for calculating the date of a new moon] was considered by scholars in Western Europe as a miracle of accuracy, though incomparably better results had been reached by Babylonian and Greek methods since the fourth century B.C. and though these methods were ably handled by contemporary Islamic and Jewish astronomers. (Neugebauer 1969, 8)

Clearly, Neugebauer wanted his audience to realize that it was one thing to appreciate a magnificent work of art, quite another to think of it as a canvas for clues about the state of mathematical and astronomical knowledge in the culture within which it was produced.

For the second edition, he updated the material and added two technical appendices, but he still hoped to have "avoided… converting my lectures into a textbook" (Neugebauer 1969, ix). Evidently, he valued the less formal form of exposition associated with oral exposition, a hallmark of the Göttingen tradition (Rowe 2004). Still, Neugebauer grew up in Austria, not Prussia, which may help account for his playful sense of humor. A typical example comes in a passage where he comments on how astronomers took delight in harmonizing their science with anthropocentric religious views, whereas modern celestial mechanics teaches us to be humble creatures living in a solar system conditioned by accidental circumstances.

> The structure of our planetary system is indeed such that Rheticus [an early champion of the Copernican theory] could say "the planets show again and again all the phenomena which God desired to be seen from the earth." The investigations of Hill and Poincaré have demonstrated that only slightly different initial conditions would have caused the moon to travel around the earth in a curve [with small loops and]…. Nobody would have had the

Plate 1 "Septembre" from the Très Riches Heures du Duc de Berry, located today in the Musée Condé, Chantilly, France

> idea that the moon could rotate on a circle around the earth and all philosophers would have declared it as a logical necessity that a moon shows six half moons between two full moons. And what could have happened with our concepts of time if we were members of a double-star system (perhaps with some uneven distribution of mass in our little satellite) is something that may be left to the imagination. (Neugebauer 1969, 152–153)

2 Neugebauer and Courant in Göttingen

Significantly, Neugebauer dedicated this now classic book to "Richard Courant, in Friendship and Gratitude." Elaborating on that dedication in the preface, he wrote that it was Richard Courant (1888–1972) who enabled him to pursue graduate studies in ancient mathematics, and he went on to remark: "more than that I owe him the experience of being introduced to modern mathematics and physics as a part of intellectual endeavour, never isolated from each other nor from any other field of our civilization" (Neugebauer 1969, vii). Neugebauer was a man who chose his words carefully, and so we may be sure that this public acknowledgement of his debt to Courant was far more than just a friendly gesture. The last part of the quotation, that Courant saw mathematics and physics as fields of intellectual endeavor "never isolated from each other nor from any other field of our civilization" comes very close to capturing the essence of Neugebauer's own understanding of what it meant to study the history of mathematics as an integral part of human cultural life. Regarding Courant's personal outlook, he described this in connection with the Göttingen mathematical tradition they both shared and valued:

> …the real core of his work [consisted] in the conscious continuation and ever widening development of the ideas of Riemann, Klein, and Hilbert, and in his insistence on demonstrating the fundamental unity of all mathematical disciplines. One must always remain aware of these basic motives if one wants to do justice to Courant's work and to realize its inner consistency. (Neugebauer 1963, 1)

As a close ally of Courant, Neugebauer shared a positivist vision of mathematics as an integral part of scientific culture. In particular, both men were deeply influenced by the universalism advocated by Göttingen's two aging sages, Felix Klein (1849–1925) and David Hilbert (1862–1943) who broke with an older German tradition in which mathematical research was largely isolated from developments in neighboring disciplines, like astronomy and physics. Hilbert's strong epistemic claims for mathematics had also deeply alienated conservative humanists on the Göttingen faculty, many of whom feared a realignment of traditional disciplinary boundaries (Rowe 1986). Neugebauer's personal relationship with Courant reflects many of the broader mathematical and scientific interests the two men shared.

As director of the Göttingen Mathematics Institute during the Weimar years, Courant was faced with numerous challenges as he struggled to uphold its international scientific reputation. Part of his strategy was conservative in nature.

Through his connections with Ferdinand Springer (1881–1965), Courant launched the famed "yellow series," one of several initiatives that enabled Springer to attain a pre-eminent position as a publisher in the fields of mathematics and theoretical physics (Remmert and Schneider 2010). Courant was an innovator with a deep belief in the vitality of older traditions. His yellow series looked backward as well as forward; in fact, surprisingly few of its volumes betray a commitment to what came to be identified as modern, abstract mathematics. Far more evident was the way in which Courant and his co-editors built on the tradition of Klein and Hilbert, and with the yellow series he found a way to make local knowledge accessible well beyond the borders of Germany. For the history of the ancient exact sciences, Springer's short-lived *Quellen und Studien* series—launched in 1929 and edited by Neugebauer, Julius Stenzel (1883–1935), and Otto Toeplitz (1881–1940)—created a new standard for studies in this fast-breaking field.

Soon after Neugebauer arrived in Göttingen in 1922, Courant gave him various special duties to perform at the hub of operations, located on the third floor of the *Auditorienhaus*. There one found the famous *Lesezimmer* together with an impressive collection of mathematical models, long cared for by Klein's assistants. Now Neugebauer stood guard while Klein received nearly daily reports through those who were busy helping him prepare his collected works. Neugebauer's new interest in Egyptian mathematics also came to Klein's attention, along with a complaint that he had stuffed all the books on mathematics education tightly together on a high shelf, making them nearly inaccessible. By now Klein was an infirm old man who rarely left his home, which overlooked the botanical garden immediately behind the *Auditorienhaus*, but he still kept up a busy and tightly organized schedule. Neugebauer remembered how Klein called him over to be gently scolded. When he arrived, Klein greeted him by saying: "there came a new Moses into Egypt and he knew not Pharaoh!" (Reid 1976, 100) (a play on: "Now there arose up a new king over Egypt, which knew not Joseph", Exodus I.8). The young Neugebauer surely realized that watching over the *Lesezimmer* was no trifling matter.

3 Neugebauer's Revisionist Approach to Greek Mathematics

Neugebauer saw himself as a "scientific historian"; he had no patience for those who simply wanted to chronicle the great names and works of the past. George Sarton (1884–1956), who did little else, saw the history of science as a humanistic endeavor; nevertheless, he had the highest respect for Neugebauer's achievements. Sarton's views emerge clearly from correspondence during September 1933 with Abraham Flexner (1866–1959). At the time, Flexner was contemplating the possibility of founding a school for studies of science and culture at the Institute for Advanced Study. Sarton thought that Neugebauer was just the man for such an enterprise, a point he made by humbly contrasting the nature of their work:

> As compared with Neugebauer I am only a dilettante. He works in the *front trenches* while I amuse myself way back in the rear — praising the ones, blaming the others; saying this ought to be done, etc.–& doing very little myself. What Neugebauer does is fundamental, what I do, secondary. (Pyenson 1995, 268)

Neugebauer certainly did view Sarton as a dilettante through and through. When I interviewed him in 1982, he made a point of telling me this by lumping him together with Moritz Cantor (1829–1920), another encyclopedist of great breadth and little depth.

Although plans to bring Neugebauer to Princeton came to naught, Harald Bohr managed to arrange a three-year appointment for him in Copenhagen beginning in January 1934. Neugebauer managed to get most of his property out of Germany, but had to abandon a house with a partially paid mortgage. In Copenhagen, his research was supported in part by the Rockefeller Foundation. Almost immediately he began preparing a series of lectures on Egyptian and Babylonian mathematics that he would publish in Courant's yellow series as *Vorgriechische Mathematik* (Neugebauer 1934). According to Swerdlow this volume was "as much a cultural as a technical history of mathematics" and represents "Neugebauer's most thorough and successful union of the two interpretations" (Swerdlow 1993, 145) More striking still is the unfinished character of this work, which represents the first volume in a projected trilogy that remained incomplete. Neugebauer had planned to tackle Greek mathematics proper in the second volume, whereas the third would have dealt with mathematical astronomy, both in the Greek tradition culminating with Ptolemy as well as the largely unknown work of late Babylonian astronomers. Thus, his original aim, as spelled out in the foreword to the first volume, was to achieve a first overview of the ancient mathematical sciences in their entirety, something that had never before been attempted.

Swerdlow has offered compelling reasons to explain why Neugebauer dropped this project, one being that he simply found the rich textual sources for Mesopotamian mathematical astronomy far more important than anything he could ever have written about Greek mathematics. Nevertheless, we can trace a fairly clear picture of the line of argument Neugebauer originally had in mind by examining the summary remarks at the conclusion of his *Vorgriechische Mathematik* as well as some of his other publications from the 1930s. Neugebauer's writings from the 1920s contain few hints that his understanding of ancient mathematics was fundamentally opposed to older views. By the early 1930s, however, his analyses of Babylonian texts led him to a new conception, namely that the Greek penchant for geometrization represented a retrograde step in the natural development of the exact sciences. This did not mean, of course, that he held a low opinion of Euclid's *Elements*; he simply thought that historians and philosophers had distorted its true place in the history of mathematics. Thus, he once imagined how scholars in some future civilization might easily form a deceptive picture of mathematical knowledge *circa* 1900 if the only important text that happened to survived were Hilbert's *Grundlagen der Geometrie* (Neugebauer 1931, 132).

In the course of this transition, Neugebauer's assertions about the character of ancient mathematics often took on a strident tone. Particularly suggestive is an essay entitled "Zur geometrischen Algebra," published in 1936 in *Quellen und Studien* (Neugebauer 1936). Significantly, Neugebauer takes as his motto a famous fragment from the late Pythagorean Archytas of Tarentum, which reads: "It seems that logistic far excels the other arts in regard to wisdom, and in particular in treating more clearly what it wishes than geometry. And where geometry fails, logistic brings about proofs" (Neugebauer 1936, 245). Much has been written about this passage, in particular about what might be meant by the term "logistic", a matter Jacob Klein (1899–1978) discussed at great length in his study "Die griechische Logistik und die Entstehung der Algebra" (Klein 1936), which appeared alongside Neugebauer's article[2]. In fact, both scholars were chasing after the same elusive goal, though the similarity ends there.

Klein was a classical philologist who later became a master teacher of the "Great Books" curriculum at St. Johns College in Annapolis Maryland. Not surprisingly, he was intent on squeezing as much out of Plato as he possibly could. Thus he distinguished carefully between practical and theoretical logistic, offering a new interpretation of Diophantus' *Arithmetica* that placed it within the latter tradition. Neugebauer had no patience for the nuances of meaning classicists liked to pull out of their texts. Indeed, he had an entirely different agenda. His point was that rigorous axiomatic reasoning in the style of Euclid arose rather late, and that Archytas, a contemporary of Plato, was bearing witness to the primacy of algebraic content over the geometrical form in which the Greeks dressed their mathematics. With that, we can take another step toward attaining a closer understanding of Neugebauer's *Weltanschauung*.

Decades earlier, the Danish historian of mathematics H. G. Zeuthen (1839–1920) already advanced the idea that the Greeks had found it necessary to geometrize their purely algebraic results after the discovery of incommensurable magnitudes (Zeuthen 1896).[3] Neugebauer took up this by then standard interpretation, adopted by Heath (1861–1940) and nearly everyone else, but he then went much further, arguing that the algebraic content—found not only in Book II of Euclid but throughout the entire corpus of Apollonius' *Conica* (Neugebauer 1932) —could be traced back to results and methods of the Babylonians:

> The answer to the question what were the origins of the fundamental problem in all of geometrical algebra [meaning the application of areas, as given in Euclid's *Elements*, I.44 and VI.27–29] can today be given completely: they lie, on the one hand, in the demands of

[2]It was later translated into English by Eva Brann (1929) (Klein 1968).

[3]Ancient sources only hint at the circumstances surrounding this discovery, which probably took place during the latter half of the fifth century. Before this time, it was presumed that magnitudes of the same kind, for example two lengths, could always be measured by a third, hence commensurable. This is equivalent to saying that their ratio will be equal to the ratio of two natural numbers. This theory had to be discarded when it was realized that even simple magnitudes, like the diagonal and side of a square, have an irrational ratio because their lengths are incommensurable lengths. The discovery of such irrational objects in geometry had profound consequences for the practice of Greek geometry in the fourth century, see (Fowler 1999).

the Greeks to secure the general validity of their mathematics in the wake of the emergence of irrational magnitudes, on the other, in the resulting necessity *to translate the results of the pre-Greek "algebraic" algebra as well.* Once one has formulated the problem in this way, everything else is completely trivial [!] and provides *the smooth connection between Babylonian algebra and the formulations of Euclid.* (Neugebauer 1936, 250, my translation, his italics)[4]

The mathematical concepts underlying this argument are by no means difficult. It must be emphasized, however, that what may seem mathematically trivial (i.e. obvious) should hardly be thought of as historically self-evident. Since Zeuthen's time, it had been customary to interpret Greek problem-solving methods as manipulations closely related to techniques like "completing the square", used to solve quadratic equations. These Greek methods, called applications of areas, occupy a prominent place in Euclid's *Elements* as well as in his *Data,* a kind of handbook for problem solving. Neugebauer was struck by the parallelism between certain standard Babylonian problems and the Greek methods for solving very similar problems geometrically (Neugebauer 1969, 40–41, 149–150).

A typical algebra problem found in several cuneiform tablets from the Old Babylonian period requires that one find two numbers whose sum (or difference) and product are both given—Neugebauer called this the "normal form" leading to a single quadratic equation. This pair of problems, depending on whether the sum or difference is given, can also be found as Propositions 84 and 85 in Euclid's *Data.* Moreover, according to the neo-Platonic commentator Proclus—on the authority of Aristotle's student, Eudemus, author of a lost *History of Geometry* written just before Euclid's time—the three types of applications of areas (later used by Apollonius to distinguish the three types of conic sections: ellipse, parabola, and hyperbola) were discovered long before Euclid: "These things, say Eudemus, are ancient and are discoveries of the Muse of the Pythagoreans" (Heath 1956, 343).

Neugebauer would have been the last to argue that the Pythagoreans had anything to do with this ancient knowledge; for him, the key fact was merely that the original ideas were old, hence likely to have roots in still older cultures from which the Greeks borrowed freely. Having established that the mathematical content of the Babylonian texts was fundamentally algebraic, he now claimed that Mesopotamia was the original source of the algebra underlying the "geometric algebra" uncovered by Zeuthen at the end of the nineteenth century. Neugebauer was fully aware, of course, that his interpretation required a really bold leap of the historical imagination, since making a claim for the transmission of such knowledge over

[4]Die Antwort auf diese Frage, d. h. auf die Frage nach der geschichtlichen Ursache der Uraufgabe der gesamten geometrischen Algebra, kann man heute vollständig geben: sie liegt einerseits in der aus der Entdeckung irrationaler Größen folgenden Forderung der Griechen, der Mathematik ihre Allgemeingültigkeit zu sichern durch Übergang vom Bereich der rationalen Zahlen zum Bereich der allgemeinen Größenverhältnisse, andererseits in der daraus resultierenden Notwendigkeit, *auch die Ergebnisse der vorgriechischen "algebraische" Algebra in eine "geometrische" Algebra zu übersetzen.* Hat man das Problem einmal in dieser Weise formuliert, so ist alles Weitere vollständig trivial und liefert *den glatten Anschluß der babylonischen Algebra an die Formulierungen bei Euklid.*

such a vast span of time meant accepting that this mathematical linkage sufficed to fill a gap devoid of any substantive documentary evidence. Summarizing his position, he offered these remarks:

> Every attempt to connect Greek thought with the pre-Greek meets with intense resistance. The possibility of having to modify the usual picture of the Greeks is always undesirable, despite all shifts of view,… [and yet] the Greeks stand in the middle and no longer at the beginning. (Neugebauer 1936, 259, my translation)[5]

When we try to square this with Neugebauer's stated belief that we should be wary of generalizations about the distant past—the position quoted in the motto to this essay—the problems with such an argument only become more acute. Perhaps these evident difficulties help explain the intensely passionate language in the concluding parts of his text. The tone in *The Exact Sciences in Antiquity* is far milder, and yet his arguments remain substantively the same (Neugebauer 1969, 146–151). There is even brief mention of the same quotation from Archytas, and one senses what Swerdlow might have meant when he wrote that Neugebauer grew bored with Greek mathematics (Swerdlow 1993, 146).

Neugebauer's research represented part of a large-scale intrusion by mathematicians into a field that was formerly dominated by classicists. Before he entered the field the history of Greek mathematics was traditionally seen as strongly linked with the works and influence of Plato and Aristotle, a view that would later be contested by the prolific American historian Wilbur Knorr (1945–1997)[6]. Neugebauer's work thus struck a sympathetic chord among a younger generation of experts on Greek mathematics, even though he had left the field by the mid 1930s. Ever the anti-philosopher, he wanted to undermine the special German fascination with Greek philosophy, most particularly the Platonic tradition. In this respect, his work stood poles apart from that of Becker, or for that matter, Toeplitz, both of whom, like Neugebauer, published regularly in *Quellen und Studien.* These two older contemporaries combined fine-tuned mathematical analyses with careful philological readings of classical Greek texts. Neugebauer, on the other hand, showed very little interest in studies of this kind. Furthermore, he had an entirely different agenda: he aimed to overthrow the standard historiography that made mathematics look like the handmaiden of Greek philosophy.

Neugebauer's original vision thus entailed a radical rewriting of the history of ancient mathematics and exact sciences. One of his central theses was that rigorous axiomatic reasoning in the style of Euclid arose rather late. At the same time he liked to call on the testimony of Archytas, who—according to Neugebauer's reading—tells us that the Greeks of that era understood the primacy of algebraic content over geometrical form. If one probed the later Greek sources with a mathematically trained eye—as Neugebauer tried to show in his study of

[5]Jeder Versuch, Griechisches an Vorgriechisches anzuschließen begegnet einem intensiven Widerstand. Die Möglichkeit, das gewohnte Bild der Griechen modifizieren zu müssen, ist immer wieder unerwünscht, trotz aller Wandlungen … stehen also die Griechen in der Mitte und nicht mehr am Anfang.

[6]See, for example, the essays by Knorr in (Christianidis 2004).

Apollonius' *Conica*—what one found was a fundamentally algebraic style of thought. His revisionist stance also aimed to debunk the notion of a "Greek miracle" that sprang up during the sixth century from the shores of Ionia. Neugebauer was convinced that most of the sources that reported on the legendary feats of ancient heroes—Thales, Pythagoras, and their intellectual progeny—were just that: legends that had grown with the passing of time. So his constant watchword remained skepticism with regard to the accomplishments of the early Greeks, whereas Toeplitz, Becker, and others began to analyze extant sources with a critical eye toward their standards of exactness.[7]

4 Greek Mathematics Reconsidered

One can well imagine that for some experts on ancient Greek philosophy and early science, Neugebauer's views regarding the historical development of Greek mathematics were simply anathema. On the other hand, he published almost nothing that dealt with early Greek mathematics per se, partly no doubt in order to avoid controversy. Still, he had a number of notable allies in classics who shared his general skepticism. In fact, a debate was then underway in which these skeptics questioned the level of truly scientific activity among the followers of 6th-century physiologoi, particularly the early Pythagoreans. German classical philology had witnessed a very different type of debate when Friedrich Nietzsche (1844–1900) published his *Birth of Tragedy,* but in a sense the parallel holds true. Leading classicists saw themselves as *Kulturträger*, which meant that they were quite accustomed to playing for "high stakes" (or at least imagining they were). Owing to their spiritual affinity with the ancient Greeks, they did not think of themselves as mere scholars: their discipline and special expertise carried with it an implicit social responsibility, namely to explain the deeper meaning of Greek ideals to that special class of German society, its *Bildungsbürgertum,* who perhaps alone could appreciate the true mission of the German people, especially when faced with momentous "world-historical" events like the Great War.

After Imperial Germany collapsed following that calamitous struggle, it should come as no surprise that fresh fissures developed within the humanities and, in particular, the discipline of classical philology. This suggests that by the time Neugebauer brought forth his new vision for understanding the history of the exact sciences a quite general reorientation had long been underway among experts who specialized in classical Greek science and philosophy. At any rate, Neugebauer had plenty of good company. He could thus cite the work of classical scholars like Eva Sachs (1882–1936) and Erich Frank (1883–1949)—dubbed by their opponents as "hyper-critical" philologists—while defending his case for recasting the early history of Greek mathematics.

[7]See (Christianidis 2004) for a recent account of older as well as the newer historiography on Greek mathematics.

Thus, in a synopsis of (van der Waerden 1940) for *Mathematical Reviews*, he wrote:

> In the first paragraph the author shows that the famous paradoxa of Zeno (for example, of the tortoise and Achilles) are not at all directed against the infinite divisibility of geometrical magnitudes, but that their aim is simply to support the assumption of Parmenides that all movement is only a human fiction. The second part points out that *in Zeno's time no mathematical theory of importance existed* in which infinitesimal methods played a role. This fits in with the general concept of the development of Greek mathematics, which is familiar, at least since Frank's book "Plato und die sogenannten Pythagoreer" [Halle 1923]. The last paragraph emphasizes that the so-called "crisis" of the foundation of Greek mathematics did not originate in the problem of infinite divisibility but from the discovery of irrationals. (Neugebauer 1940)[8]

Bartel Leendert van der Waerden (1903–1996) was a distinguished Dutch mathematician who had taken a course on ancient mathematics with Neugebauer in Göttingen. They remained good friends and corresponded regularly about historical matters, but they also often disagreed. Only a year after he wrote the above, Neugebauer came back to the same issue while reporting on (van der Waerden 1941), a paper on Pythagorean astronomy:

> The author gives an outline of the development of Greek astronomy in its earlier phases. He seems to have overlooked the book of Frank, *Plato und die sogenannten Pythagoreer*, where essential points of his theory are already published. (Neugebauer 1941)

Neugebauer's persistent references to Frank's book appear to have made no impression on van der Waerden, who remained in Leipzig after the Nazis rose to power. This makes it highly unlikely, of course, that he knew of Neugebauer's printed remarks from 1940 to 1941, at least not until some time after the war had ended. Yet when he brought out the original Dutch edition of *Science Awakening* in 1950—a more popular account of the exact sciences in antiquity that drew heavily on Neugebauer's researches—van der Waerden presented the legendary Pythagoras as the founder of a scientific school, one in which the sage's teachings had a profoundly mathematical character as opposed to merely espousing the doctrines of a religious sect that practiced number mysticism.

Neugebauer, who was not Jewish, could have stayed on in Göttingen. After Courant's dismissal, however, he chose instead to leave for Copenhagen in January 1934. From this new outpost he continued editing Springer's *Zentralblatt* until 1938, at which point he resigned in protest of Nazi racial policies that had led to the removal of Jewish colleagues from its board. These events then paved the way for the founding of *Mathematical Reviews,* which Neugebauer co-managed beginning in 1940, after his arrival at Brown University. Courant, who was now teaching at New York University, had by this time severed his publishing connections with Springer. Ten days after the devastating blow to Jewish property and life during the *Reichskristallnacht,* he wrote to Ferdinand Springer informing him that he wished

[8]Neugebauer here alludes to the so-called "foundations crisis" that supposedly ensued with the discovery of incommensurable magnitudes. This interpretation became popular during the 1920s, but later fell out of favour (Christianidis 2004).

to resign as editor of the "yellow series" (Reid 1976, 208–209). Still, Courant continued to maintain his former contacts in Göttingen after the war. He often visited the Mathematics Institute, whose new director Franz Rellich (1906–1955) had earlier been part of the "Courant clique" that was forced to leave in 1933. Neugebauer, by contrast, refused ever to set foot in Germany again (he did visit Austria once or twice however).

Despite his loathing for the Nazis, Neugebauer steered clear of politics when commenting on the work of scholars whom he surely knew to be faithful followers of Hitler's brand of fanatical German nationalism. A striking example of this can be seen in his review of the German translation of the well-known *Commentary on Book I of Euclid's Elements*, written by the neo-Platonic philosopher Proclus in the fifth century. One should note that this rather large volume (Steck 1945) with extensive commentary by Max Steck (1907–1971), a hardcore Nazi from Munich, managed to get published in the year 1945. Neugebauer praised the work of the translator and then wrote this about Steck's contribution:

> The introduction [33 pp.] contains many words which fortunately have no English equivalent, e.g., "deutscher Geistraum," "Geistschau," "in- und ausstrahlen," etc. By means of this "denkanschauend" method Proclus is made a founder of the German Idealismus for which Cusanus, Copernicus, Kepler, Hegel, Gauss (!) and many others are quoted. On the other hand, Proclus is considered as the culmination of Greek mathematics. The author here follows [Andreas] Speiser with whom he shares the tendency to consider the last phase of Greek metaphysics as representative of Greek mathematics. The subsequent commentary on Proclus shows the same contempt for the chronological element of history. There is hardly a combination of any pair of famous names missing, however great their distance may be. (Neugebauer 1945)

Once he was located in the United States Neugebauer published regularly in English in the Danish journal *Centaurus* as well as in numerous American publications, including George Sarton's *Isis,* the official journal of the History of Science Society. By the early 1950s, however, a first wave of negative reaction began to swell up among émigré scholars now residing in the United States. In 1951 Neugebauer's revisionist interpretation came under strong attack in *Isis* in an article entitled "Philolaos in Limbo, or: What Happened to the Pythagoreans?", written by George de Santillana (1902-1974) and Walter Pitts (1923–1969). The first author, well-known for his book *The Crime of Galileo,* had fled fascist Italy to take up a post at the Massachusetts Institute of Technology. Thus, it was fitting that the authors began their essay by citing these famous words: "Several years ago there was published in Rome a salutary edict which, in order to obviate the dangerous tendencies of our present age, imposed a seasonable silence upon the Pythagorean opinion that the earth moves…." They then proceeded to explain their present purpose:

> These are the opening words of Galileo's preface to his Dialogue on the World Systems. One would be tempted to repeat them almost word for word today, apropos [sic] of certain contemporary philological research. The invisible edict or "trend" to which we refer has decreed that the whole development of Greek mathematics and astronomy must be condensed into a rather short interval of time around 400 B.C., so that almost all the

> mathematics, astronomy, and music theory of the "so-called Pythagoreans" becomes contemporary with Plato and his successors. (de Santillana and Pitts 1951, 112)

Three different groups were then identified as being responsible for this trend. The first of these is only vaguely named by alluding to "the massed power of Platonic and Aristotelian scholarship." Far more important for their critique was the role played by the aforementioned "hyper-critical philologists", especially Sachs and Frank, but also the American, W. A. Heidel, author of "The Pythagoreans and Greek Mathematics" (Heidel 1940). Frank, who had succeeded Heidegger in 1928 as professor of philosophy in Marburg, had been forced to flee Germany after losing this chair in 1935; he eventually came to Harvard as a Rockefeller Fellow. Unable to secure regular employment in the United States, he died in Amsterdam in 1949. His older study (Frank 1923) argued that when Aristotle spoke about "so-called Pythagoreans" he was referring to the circle around Archytas of Tarentum, who was a friend of Plato as well as a gifted mathematician. This argument supported Frank's larger thesis, according to which the early Pythagoreans were merely a religious sect and played no substantive role in early Greek science.

The third group of trend setters was "the recent school of scientific historians which has attempted to trace the connection between Babylonian and Greek mathematics." Several works are cited by three authors: Neugebauer, van der Waerden, and the mathematician Kurt Reidemeister (1893–1971). "Relying on Frank," it is charged,

> these authors have dismissed the entire tradition about early Greek mathematics, and supplanted it either with a most improbably late transference of Babylonian mathematics to Greece in the Vth century, or else have tried to fill the gap with speculations, conceived certainly in a true and subtle mathematician's spirit, derived from conjectural traces in Euclid and Plato. (ibid.)

Having identified Frank as the key culprit responsible for this hyper-critical treatment of sources on the Pre-Socratics—in the present case the authenticity of fragments attributed to Philolaos form the principal matter under dispute—de Santillana and Pitts proceed to demolish the arguments in his book. Since Frank was no longer among the living, there was small chance of a rebuttal, although they also chided the distinguished classicist Harold Cherniss (1904–1987) for having been duped by Frank's arguments regarding the authenticity of the Philolaos fragments (Cherniss 1935, 386).

The year 1951 also saw the publication of the original Copenhagen edition of *The Exact Sciences in Antiquity*. It was reviewed at length in *Isis* by Sarton, who noted that no one but Neugebauer could have written such a book. Sarton also paid tribute to Cornell University for its role in helping the author produce this idiosyncratic *synthesis* based on his six Messenger Lectures from 1949. This opportunity, Sarton felt sure, gave Neugebauer just the incentive he needed to address a broader set of historical issues, something he was otherwise loathe to do. In his review, Sarton put the matter this way: "as he does not like synthetic work and even affects to despise it, he would probably not have written this book without that flattering invitation, and we, his readers, would have been the losers" (Sarton 1952, 69).

One can easily read between the lines here, since Sarton, the doyen of American historians of science, certainly saw himself as a leading representative of that very genre of scholarship to which he here alluded (Sarton 1936a). Nor was this review altogether positive. The reviewer voiced skepticism, for example, when it came to Neugebauer's claims regarding the historical impact of Babylonian mathematics and astronomy. Noting that neither Hipparchus nor Ptolemy made mention of earlier Babylonian theoretical contributions, he wondered how historians could ever know that these Greek astronomers drew on such sources? As for Babylonian algebra, why should we assume that this knowledge survived long after the period of Hammurabi when there is no extant evidence *for a continuous tradition* of high mathematical culture in Mesopotamia? And if such mathematical knowledge persisted, how was it transmitted? After all, the complexity of the Babylonian algebraic and astronomical techniques required an expertise similar to Neugebauer's own. Sarton also took sharp issue with Neugebauer over the centrality of Hellenistic science, especially his claim that this melting pot of ancient science later spread to India before entering Western Europe, where it held sway until the time of Newton. In Sarton's view, the Hellenistic period marked the final phase of Babylonian science, though he admitted some slight influences on both the Indian and Islamic cultural spheres. For the most part, however, he contrasted the larger long-term impact of Greek science with the relatively meager legacy of the Babylonian tradition. For him, this was the gravest shortcoming of all; how could Neugebauer write a book called *The Exact Sciences in Antiquity* and virtually ignore the achievements of the Greeks? Doing that was comparable to writing a play entitled *Hamlet* while leaving out the figure of Hamlet himself. With that quip, Sarton could chide Neugebauer's Danish editors—identified as Zeuthen's countrymen—for allowing their distinguished friend to make such a blunder.

Sarton's criticisms reflect the views of a generalist who clearly found Neugebauer's overall framework far from convincing. He had the highest respect for the author's specialized contributions to research on the ancient exact sciences—work that required not only formidable mathematical abilities but also immense discipline—but this review makes plain that he saw Neugebauer's book as the product of a remarkable specialist. Sarton's overall verdict—seen from his personal vantage point of someone who hoped to open inroads for the history of science within the curriculum of American higher education—echoed Neugebauer's own forthright opinion that he "did not like synthetic work" (ibid.). *Exact Sciences*, Sarton opined, was of limited value for introductory courses; it should not and could not be taken as a model for teaching the history of ancient science. Though full of nicely chosen anecdotes and a good deal of general information, it simply did not pass muster as a global account of the history of the exact sciences in ancient cultures. Swerdlow later expressed a very different opinion when he wrote:

> Neugebauer here allowed himself the freedom to comment on subjects from antiquity to the Renaissance. The expert can learn something from it, and from its notes, every time it is read, and for the general reader it is, in my opinion, the finest book ever written on any aspect of ancient science. (Swerdlow 1993, 156)

Sarton saw himself as a champion of what he called a "synthetic approach" to the history of mathematics (Sarton 1936b, 11). What Neugebauer thought about this can well be surmised from the preface to the first edition of *Exact Sciences in Antiquity:* "I am exceedingly skeptical of any attempt to reach a "synthesis"—whatever this term may mean—and I am convinced that specialization is the only basis of sound knowledge" (Neugebauer 1969, vii–viii). Paging through Sarton's booklet, *The Study of the History of Mathematics,* one can easily understand Neugebauer's dismissive attitude. There one reads that:

> The main reason for studying the history of mathematics, or the history of any science, is purely humanistic. Being men, we are interested in other men, and especially in such men as have helped us to fulfill our highest destiny. As soon as we realize the great part played by individual men in mathematical discoveries — for, however these may be determined, they cannot be brought about except by means of human brains —, we are anxious to know all their circumstances. (Sarton 1936b, 12)

Sarton's humanistic approach to the history of mathematics thus derives from simple human curiosity, which he admits is the same instinct that feeds public fascination with murderers. Whereas newspapers skillfully exploit this "insatiable desire to know every detail of a murder case, those who are more thoughtful wish to investigate every detail of scientific discoveries or other creative achievements" (ibid.). This loftier interest apparently has much to do with Sarton's sympathy for hero worship: "One soon realizes that mathematicians are much like other men, except in the single respect of their special genius, and that genius itself has many shapes and aspects" (ibid.).

Not surprisingly, Neugebauer drew a sharp line between his work and that of dabblers like Sarton, though he never launched a frontal attack on the latter's own works. He did, however, occasionally publish critical responses to Sarton's opinions in *Isis*, one of which sheds much light on the intellectual fault lines that divided them. In a review of van der Waerden's *Science Awakening,* Sarton expressed dismay over the author's "shocking ingratitude" towards Cantor, whom he called "one of the greatest scholars of [the] last century, a man to whom every historian of mathematics owes deep gratitude." After citing this passage, Neugebauer went on to explain why he was writing this "Notice of Ingratitude" (Neugebauer 1956):

> Since I must conclude that this statement in its generality would also apply to myself, I should like to point out that I never felt a trace of indebtedness to Cantor's voluminous production. I do not deny, of course, the fact that it had a great influence, though in a direction quite opposite to what Professor Sarton's statement implies. I always felt that its total lack of mathematical competence as well as its moralizing and anecdotal attitude seriously discredited the history of mathematics in the eyes of mathematicians, for whom, after all, the history of mathematics has to be written. In methodological respects, Cantor's work might be of some value for historians of science since it contains so many drastic examples of how one should not approach a problem… If Cantor had not philosophized about a goose counting her young or about oriental mathematics, which was equally inaccessible to him, but instead had studied the texts themselves, he would have avoided countless misinterpretations and inaccuracies which have become commonplace. It was with good reasons that the *Bibliotheca Mathematica* for years ran a special column devoted

to corrections of errors in Cantor's *Geschichte der Mathematik*. But no amount of corrections can ever remedy consistent mediocrity. (Neugebauer 1956, 58)

Given that Neugebauer's academic career was decisively shaped by his training and background as a mathematician, one can easily understand his aversion to the writings of Cantor and Sarton. He was most definitely not a "synthetic" historian in the sense of Sarton, but we can say just as assuredly that his work was guided by a larger view of the history of mathematics. His was an approach to history deeply grounded in the mathematical culture he grew up in, and his sensibilities as a historian were from the very beginning guided by a grandiose vision. Neugebauer worked on details, but always with a larger landscape in mind. His attitude toward his own work seems to have also contained elements of playful irony. When he came to the end of his Messenger lectures on the exact sciences in antiquity, he offered a simile to describe the historian's craft:

In the Cloisters of the Metropolitan Museum in New York there hangs a magnificent tapestry which tells the tale of the Unicorn. At the end we see the miraculous animal captured, gracefully resigned to his fate, standing in an enclosure surrounded by a neat little fence.[9] This picture may serve as a simile for what we have attempted here. We have artfully erected from small bits of evidence the fence inside which we hope to have enclosed what may appear as a possible, living creature. Reality, however, may be vastly different from the product of our imagination; perhaps it is vain to hope for anything more than a picture which is pleasing to the constructive mind when we try to restore the past. (Neugebauer 1969, 177)

References

Cherniss, H. (1935). *Aristotle's criticism of Presocratic philosophy*. Baltimore: Johns Hopkins Press.

Christianidis, J. (Ed.). (2004). *Classics in the history of Greek mathematics*. Dordrecht: Kluwer.

De Santillana, G. & Pitts, Walter. (1951). Philolaos in Limbo, or: What Happened to the Pythagoreans? *Isis, 42*, 112–120.

Fowler, D. H. (1999). *The mathematics of Plato's academy. A new reconstruction* (2nd ed.). Oxford: Oxford University Press.

Frank, E. (1923). *Plato und die sogenannten Pythagoreer: Ein Kapitel aus der Geschichte des griechischen Geistes*. Halle: Niemeyer.

Heath, T. L. (Ed.). (1956). *The thirteen books of Euclid's elements* (Vol. 1). New York: Dover.

Heidel, W. A. (1940). The Pythagoreans and Greek mathematics. *American Journal of Philosophy*, 1–33.

Klein, J. (1936). Die griechische Logistik und die Entstehung der Algebra. *Quellen und Studien zur Geschichte der Mathematik, Astronomie und Physik, B: Studien, 3*, 18–105 and 122–235.

Klein, J. (1968). Greek mathematical thought and the origin of algebra. Trans. Eva Brann. Cambridge, MA: MIT Press.

[9]This refers to "The Unicorn in Captivity," one of seven tapestries dating from ca. 1500 located in The Cloisters in New York. In the pagan tradition, the unicorn was a one-horned creature that could only be tamed by a virgin; whereas Christians made this into an allegory for Christ's relationship with the Virgin Mary.

Neugebauer, O. (1925). Litteraturanmeldelse (T. E. Peet, The Rhind Mathematical Papyrus). *Matematisk Tidsskrift A*, 66–70.
Neugebauer, O. (1927). Zur Entstehung des Sexagesimalsystems. *Abhandlungen der Akadamie der Wissenshaften zu Göttingen, 13*, 1–55.
Neugebauer, O. (1931). Zur vorgriechischen Mathematik. *Erkenntnis, 2*, 122–134.
Neugebauer, O. (1932). Apollonius-Studien. *Quellen und Studien zur Geschichte der Mathematik, Astronomie und Physik, B: Studien, 2*, 215–254.
Neugebauer, O. (1934). Vorlesungen über Geschichte der antiken mathematischen Wissenschaften, Vol. 1. Vorgriechische Mathematik. Berlin: Verlag Julius Springer.
Neugebauer, O. (1936). Zur geometrischen Algebra (Studien zur Geschichte der Algebra III). *Quellen und Studien zur Geschichte der Mathematik, Astronomie und Physik, B: Studien, 3*, 245–259.
Neugebauer, O. (1940). Review of (van der Waerden 1940). *Mathematical Reviews*.
Neugebauer, O. (1941). Review of (van der Waerden 1941). *Mathematical Reviews*.
Neugebauer, O. (1945). Review of (Steck 1945). *Mathematical Reviews*.
Neugebauer, O. (1956). A Notice of Ingratitude. *Isis, 47*, 58.
Neugebauer, O. (1963). Reminiscences on the Göttingen Mathematical Institute on the Occasion of R. Courant's 75th Birthday. Otto Neugebauer Papers, Box 14, publications vol. 11.
Neugebauer, O. (1969). *The exact sciences in Antiquity* (2nd rev ed.). New York: Dover.
Pyenson, L. (1995). Inventory as a Route to Understanding: Sarton, Neugebauer, and Sources. *History of Science, 33*, 253–282.
Reid, C. (1976). *Courant in Göttingen and New York: The Story of an Improbable Mathematician*. New York: Springer Verlag.
Remmert, V. & Schneider, Ute. (2010). Eine Disziplin und ihre Verleger. *Disziplinenkultur und Publikationswesen der Mathematik in Deutschland* (pp. 1871–1949). Bielefeld: Transkript.
Rowe, D. E. (1986). 'Jewish Mathematics' at Gottingen in the Era of Felix Klein. *Isis, 77*, 422–449.
Rowe, D. E. (1996). New trends and old images in the history of mathematics. In R. Calinger (Ed.), *Vita mathematica: historical research and integration with teaching* (pp. 3–16). Cambridge, UK: Cambridge University Press.
Rowe, D. E. (2004). Making mathematics in an oral culture: Göttingen in the Era of Klein and Hilbert. *Science in Context, 17*, 85–129.
Rowe, D. E. (2012). Otto Neugebauer and Richard Courant: On exporting the Göttingen approach to the history of mathematics. *The Mathematical Intelligencer, 34*, 29–37.
Sarton, G. (1936a). *The Study of the History of Science, with an Introductory Bibliography*. Cambridge: Harvard University Press.
Sarton, G. (1936b). *The study of the history of mathematics*. Cambridge: Harvard University Press.
Sarton, G. (1952). Review of Otto Neugebauer, The exact sciences in antiquity. *Isis, 43*, 69–72.
Siegmund-Schultze, R. (2009). *Mathematicians Fleeing from Nazi Germany: Individual fates and global impact*. Princeton: Princeton University Press.
Steck, M. (Ed.). (1945). *Proklus Diadochus, Kommentar zum ersten Buch von Euklids „Elementen"* (trans. by Leander Schönberger). Halle: s.n.
Swerdlow, N. M. (1993). Otto E. Neugebauer (26 May 1899-19 February 1990). *Proceedings of the American Philosophical Society, 137*, 138–165.
Unguru, S. (1975). On the need to rewrite the history of Greek mathematics. *Archive for History of Exact Sciences, 15*, 67–114.
van der Waerden, B. L. (1940). Zenon und die Grundlagenkrise der griechischen Mathematik. *Mathematische Annalen, 117*, 141–161.
van der Waerden, B. L. (1941). Die Astronomie der Pythagoreer und die Entstehung des geozentrischen Weltbildes. *Himmelswelt, 51*, 97–103 and 113–119.
Zeuthen, H. G. (1896). *Geschichte der Mathematik im Altertum und Mittelalter*. Kopenhagen: Verlag A. F. Hoest.

The 'Mathematization of Nature': The Making of a Concept, and How It Has Fared in Later Years

H. Floris Cohen

Abstract

The concept of 'the mathematization of nature' arose in close connection with the equally novel concept of 'The Scientific Revolution of the seventeenth century'. The pioneers were Dijksterhuis (*Val en worp. Een bijdrage tot de geschiedenis der mechanica van Aristoteles tot Newton*. Noordhoff, Groningen, 1924), Burtt (*The Metaphysical Foundations of Modern Physical Science. A Historical and Critical Essay*, Routledge & Kegan Paul, London, 1924), and Koyré (*Etudes Galiléennes*. Hermann, Paris, 1939/1940). Thanks in good part to Koyré's agenda-driven perseverance, these became highly productive concepts. Nonetheless their analytically sharp delineation quickly gave way to an unceasing process of meaning dilution. In the 1970s Kuhn and Westfall sought (but to little avail) to halt the process by suggesting two ingenious 'two-current' accounts of the Scientific Revolution, with the mathematization of nature figuring prominently in each. The idea of the mathematization of nature being key to the Scientific Revolution has kept informing a 'master narrative' that professionals have ever since the 1980s tended to regard as hopelessly obsolete. It is certainly true that, in their original guise, these concepts are incapable of capturing fundamental aspects of the pursuit of knowledge of nature in seventeenth century Europe. But is that good enough a reason to give them up altogether? Two efforts at fresh reconceptualization have recently been undertaken, one by Gaukroger, one by the author of the present chapter. Having compared the two elsewhere, he outlines here in what ways he has partly reinstated, partly revised, partly expanded, but in all cases built forth upon, the still vital notion of the mathematization of nature. In this partly altered guise, it marks three of the six revolutionary transformations that in his view comprise the Scientific Revolution of the seventeenth century.

H. Floris Cohen (✉)
Descartes Centre, Utrecht University, Trans 10, 3512 JK Utrecht, The Netherlands
e-mail: h.f.cohen@uu.nl

V.R. Remmert et al. (eds.), *Historiography of Mathematics in the 19th and 20th Centuries*, Trends in the History of Science,
DOI 10.1007/978-3-319-39649-1_8

Keywords
Scientific Revolution · Mathematization of nature

The mathematization of nature, in the sense of a historical event, has its starting point around the year 1600. As an expression, it was coined in 1935. As a historical concept (my principal concern here), it first emerged eleven years earlier, in 1924. It did so in Tilburg (The Netherlands) and in New York simultaneously, at the hands of a mathematics high school teacher by the name of E.J. Dijksterhuis and of a promising philosopher by the name of E.A. Burtt. The account that follows derives in good part from my book *The Scientific Revolution. A Historiographical Inquiry* (1994). It is a book about ideas—among them the leading ideas of those historians who first enunciated and then elaborated the concept of 'the mathematization of nature' together with the closely related concept of 'The Scientific Revolution of the 17th century'. Between the 1920s and the 1970s these and numerous later scholars created what is still regarded (albeit no longer adopted with much if any enthusiasm) as the 'master narrative' of the early modern period in the history of science. Among those who jointly produced and then worked out those original ideas about the Scientific Revolution were both 'internalist' and 'externalist' historians. The former focused on the broad conceptions and the specific theories of thinkers like Galileo or Descartes or Newton, whereas the latter sought to derive these from some contemporaneous socio-economic substructure—an effort that later fused into the now prevalent attention given to socio-cultural context.

What I did not do in that book, was to place the ideas of these historians themselves in such a wider context. This was a point that, soon upon its publication, a colleague from abroad confronted me with. 'What a pity', he said in effect, 'that you, Floris, have not contextualized those ideas in their turn! Take, for instance, Dijksterhuis. I have noted that he was a mathematics teacher in Tilburg, and I know the Netherlands well enough to realize that that town lies in the heartland of Catholicism in your otherwise Calvinism-stamped country. Would it not be wonderful, then, for you to show how Dijksterhuis' ideas, as for instance his sympathetic treatment of medieval thought about nature, emerged from his Catholic background?' I responded that it might well have been a good thing to contextualize all those captivating ideas that I had assembled, summarized, analyzed, and compared with each other, but that to do so properly would at least have trebled the length of an already fairly massive book. 'Moreover', I told him, 'it so happens that Dutch national school law furthers the presence in every town of schools of three kinds, Catholic, Protestant, and confessionally neutral, and that Dijksterhuis, who himself was mildly Protestant, happened to teach at a high school of the latter variety. Should we not, then, take the utmost caution in attributing certain ideas to their supposed local context?'

The moral of this little anecdote will be clear. I am happy to comply with the setup of the present book and address the local situatedness and the respective agendas of the three men who in the 1920s—1930s created the concept of the mathematization of nature. But I do so only while keeping firmly in mind that that situatedness and those agendas do not necessarily explain much about the many terrific, albeit no doubt meanwhile partly superseded ideas of these protagonists, whom I shall now introduce, starting, indeed, in the mathematics classroom of a 32-year old teacher at the State High School of Tilburg in the year 1924.

1 Dijksterhuis' *Val En Worp*

Indeed, 1924 is the year when Dijksterhuis' first book came out, *Val en worp* ('Free Fall and Projectile Motion'). Why did he select for in-depth historical treatment this seemingly narrow subject? His objective was to present an overview of theories of free fall and projectile motion, from their original status as fitting, although slightly irregular, portions of Aristotelian natural philosophy, through their becoming the highly problematic focus of increasingly unorthodox thought about nature, up to and including their final codification in Newton's Laws. Indeed, he concentrated on free fall and projectile motion because in his view these were the two key phenomena around which the seminal transition from Aristotelian natural philosophy to modern mechanics had taken place.

About the precise nature of this transition Dijksterhuis had clear-cut ideas. Here, in his own words, is the book's basic thesis:

> The wide-spread idea that the roots of modern science are always there where we find more distinct attention being paid to the phenomena of nature and a higher estimation of experience as a source of knowledge than occurred in Scholasticism, is generally erroneous. Precisely in the field of mechanics, which after all is the foundation of modern physics, the above opinion loses its force. We saw already how in the very depths of the most abstract scholastic dialectics the first kinematic results were reached and the outlines were sketched of the most important dynamical concepts. As our inquiry continues, we shall learn how, time and again, the same influences as were operative there continue to cause every important progress in mechanics—abstraction made from the disturbances which, in the reality of nature, obscure the ideal phenomenon; the deepening and sharpening of those vague and intuitive conceptions which already from the earliest times had been called forth by experience; in one word: the introduction of natural science in the sphere of mathematics.[1]

Already as a student, Dijksterhuis had adopted the Plato-like conception of mathematics that he was to keep upholding until his death in 1965. The two things that he enjoyed most in the astronomy seminar taught in Groningen by J. C. Kapteyn (the great cartographer of the Southern sky) were the professor's vast excursions in the history of astronomy and his knack for deriving the various heavenly phenomena by means of little else but abstract algorithms. Once Dijksterhuis became a mathematics teacher, he got involved in a running debate over

[1]See for all passages by Dijksterhuis, Burtt, and Koyré here quoted Chap. 2, 'The Great Tradition', of my historiographical book of 1994.

whether mathematics should be taught chiefly in empirical garb, with much attention being given to real-world phenomena and their cautious quantification, or in the pure, unadulterated, really Platonist-Euclidean vein so dear to Dijksterhuis and the perennial minority that felt and feels like he did. The main thesis of *Val en worp*, then, reflects those struggles of the early 1920s over how best to teach Dutch high school kids mathematics and mechanics.[2]

Mechanics, indeed! In Dijksterhuis' view mechanics really is a branch of mathematics. The principal thesis of *Val en worp* governs, basically unaltered, Dijksterhuis' later, much more encompassing and also better known *The Mechanization of the World Picture*. This book appeared in the original Dutch in 1950, to be translated fairly well in German in 1956 and fairly poorly in English in 1961. Here is its final conclusion:

> A complete characterization is not reached, nor is the true contrast between classical [that is, modern] and medieval science made perfectly clear, until the definition of mechanics as the science of motion is made to include the feature of mathematical treatment as well. And this statement must be made even more precise. Classical mechanics is mathematical, not only in the sense that it makes use of the tools of mathematics for the sake of convenience in abbreviating arguments which, if necessary, could also be expressed in everyday speech; but it is so in the far stricter sense that its basic concepts are mathematical concepts, that mechanics itself is a mathematics. In fact, it is only thus that the cardinal difference from medieval physics is revealed; as has been shown, at one stage of its development the latter, too, liked to make use of mathematical methods, but only a few of its representatives—Oresme in particular—tried, as a precursor to a principle later to be formulated by Galileo, to use mathematics as the language of physics.

In order to do full justice to the book's fundamental thesis, then, it should rather have been entitled 'The *Mathematization* of the World Picture'. After extensive treatment of Greek, medieval, and Renaissance science, Part IV covers 'The Birth of Classical Science',[3] which opens with Copernicus' *De Revolutionibus* and ends with Newton's *Principia*. However, in his discussion of Copernicus Dijksterhuis shows at length to how very large an extent Copernicus still operated in hoary, Ptolemaean fashion, and how ambiguous his message of a moving Earth appears once, in Books II–VI of *De Revolutionibus*, the business of building models for the various planetary trajectories truly begins. The real protagonists of the 'birth of classical science' in Dijksterhuis' treatment are neither Copernicus nor Descartes (whom quite in keeping with *Val en worp* he portrays as, above all, a know-all philosopher in the speculative mold), but Kepler and Galileo. These two are the men with whom the historical process of the mathematization of nature truly begins in Dijksterhuis' apparent view.

[2]I have taken the biographical details mentioned here and further down from the biography by Klaas van Berkel (1996).

[3]The translator saw fit to render Dutch 'geboorte' = English 'birth' (German 'Geburt') as 'evolution'.

2 Burtt's *Metaphysical Foundations*

It so happens that the mathematization of nature—surely the *concept*, albeit not yet quite literally the *expression*—came into the world by way of a twin birth. Quite unbeknownst to the two authors, in the same year 1924 when *Val en worp* came out, the firm Routledge and Kegan Paul in London published the doctoral dissertation of a likewise 32-year old philosopher at Cornell University, E.A. Burtt. Curiously, he did not actually earn the doctor's hat until a year later, when he had finally scraped together the funds needed to procure the one hundred copies that Cornell's stern rules required for attaining the title of Doctor of Philosophy.[4] The title of the dissertation was *The Metaphysics of Sir Isaac Newton*; of the book's trade version, *The Metaphysical Foundations of Modern Physical Science*. Just like *Val en worp* it is a seminal work, not only in mature history of science writing as such, but also by way of an early case made for the rise of modern science coming down, in essence, to a conception of the natural world as governed, at bottom, by mathematical rule and order.

Unlike Dijksterhuis, who regarded mathematization as the only viable pathway toward proper scientific method, Burtt was profoundly unhappy with the surely intellectually brilliant, yet from a philosophical point of view hopelessly uncritical and needlessly overpowering episode of which he set out to analyze the history:

> Wherever was taught as truth the universal formula of gravitation, there was also insinuated as a nimbus of surrounding belief that man is but the puny and local spectator, nay irrelevant product of an infinite self-moving engine, which existed eternally before him and will be eternally after him, enshrining the rigor of mathematical relationships while banishing into impotence all ideal imaginations; an engine which consists of raw masses wandering to no purpose in an undiscoverable time and space, and is in general wholly devoid of any qualities that might spell satisfaction for the major interests of human nature, save solely the central aim of the mathematical physicist.

The leading idea of Burtt's book may be summed up by the paradox that concurrent with, and as a direct result of, one of the highest achievements of the human mind, the autonomy of that very same mind was downgraded and banished from the supposedly real universe of atoms moving according to mathematical laws across geometrical space. Consequently, he saw it as his ultimate aim to build a metaphysics surely compatible with modern mathematical science, yet capable of reinstating "man with his high spiritual claims" to a more fitting position than that of an entity reducible to the mathematical, atomic categories of modern science. So not the achievement of science, but its metaphysical foundation is to be repudiated and to be replaced by something better, of which Burtt characteristically admitted that he had as yet very little idea. Even so, Burtt argued, it was an indispensable first step to turn to history. Not, to be sure, to the history of philosophy. After all, philosophers have ever since Newton struggled largely in vain to accomplish such a desired reinstatement. The very failure of their attempts testifies to the

[4] I have taken the biographical details mentioned here and further down from a study by D.D. Villemaire (2002).

extraordinarily powerful grip the metaphysical foundation of modern science has exerted upon intellectual thought ever since—it has disabled philosophers "to rethink a correct philosophy of man in the medium of this altered terminology." No, the history to turn to is the history of how this brilliantly innovative yet, from a human perspective, depressing scientific world-view managed against many odds to come into being. Burtt saw the process first emerging unambiguously in the work of Kepler, who identified

> the underlying mathematical harmony discoverable in the observed facts as the cause of the latter, the reason, as he usually puts it, why they are as they are. This notion of causality is substantially the Aristotelian formal cause reinterpreted in terms of exact mathematics… Thus we have in Kepler the position clearly stated that the real world is the mathematical harmony discoverable in things. The changeable, surface qualities which do not fit into this underlying harmony are on a lower level of reality; they do not so truly exist.

Even more radical in this tendency to regard, and analyze, the real world as at bottom mathematical is, in Burtt's treatment, Galileo, who for the first time makes "the clear distinction between that in the world which is absolute, objective, immutable, and mathematical; and that which is relative, subjective, fluctuating, and sensible."

With Burtt, then, Galileo and Kepler are the great pioneers of the mathematization of nature, whereas Newton caps the historical process thus set on its relentless course. With Dijksterhuis in *The Mechanization of the World Picture* it is quite the same. I shall now address the views of the third historian to come forward with, at bottom, the same conception of the mathematization of nature as the core event to separate for good pre-modern from modern science, the Russian-French philosopher A. Koyré, born just as Dijksterhuis and Burtt in 1892.

3 Koyré's *Etudes Galiléennes*

Koyré turned from philosophy of religion to history of science as, at the very same time but at the other side of the Atlantic, Burtt made the exact reverse move—*The Metaphysical Foundations* was to remain the only work he contributed to our field. Koyré developed an interest in the history of ideas in the context of their own time rather than as links in some evolutionary chain, due to his work in the 1920s on the German mystic Jakob Böhme. Inspired by authors like P. Duhem, P. Tannery, and É. Meyerson, he devoted to the history of science his *Etudes Galiléennes* (poorly translated into English in 1978). Made up of three closely intertwined essays that began to appear in 1935, it came out as a book in Paris a few months before the city was overrun by Nazi armies and its author joined the self-proclaimed leader of the Resistance, General de Gaulle, in London, moving on to the USA later in the war.

Koyré's book not only enunciates the two closely related concepts of 'the mathematization of nature' and of 'the Scientific Revolution of the 17th century', but also comes up with literally these two expressions. The book's core thesis, elaborated yet not substantially altered in all his later works until his death, one year ahead of Dijksterhuis, in 1964, has been summarized very ably by Koyré himself. Here it is, in his own English that still betrays an essentially French sentence structure:

> I shall therefore characterize this revolution by two closely connected and even complementary features: (a) the destruction of the cosmos, and therefore the disappearance from science—at least in principle, if not always in fact—of all considerations based on this concept, and (b) the geometrization of space, that is, the substitution of the homogeneous and abstract—however now considered as real—dimension space of Euclidean geometry for the concrete and differentiated place-continuum of pre-Galilean physics and astronomy.
>
> As a matter of fact, this characterization is very nearly equivalent to the mathematization (geometrization) of nature and therefore the mathematization (geometrization) of science.
>
> The disappearance—or destruction—of the cosmos means that the world of science, the real world, is no more seen, or conceived, as a finite and hierarchically ordered, therefore qualitatively and ontologically differentiated, whole, but as an open, indefinite, and even infinite universe, united not by its immanent structure but only by the identity of its fundamental contents and laws; a universe in which, in contradistinction to the traditional conception with its separation and opposition of the two worlds of becoming and being, that is, of the heavens and the earth, all its components appear as placed on the same ontological level; a universe in which the *physica coelestis* and *physica terrestris* are identified and unified, in which astronomy and physics become interdependent and united because of their common subjection to geometry.
>
> This, in turn, implies the disappearance—or the violent expulsion—from scientific thought of all considerations based on value, perfection, harmony, meaning, and aim, because these concepts, from now on *merely subjective*, cannot have a place in the new ontology.

It is hard to overlook, in the final passage, a dash of Burtt's core message—Koyré, in footnoting Burtt in somewhat deprecating, even patronizing terms, was not overgenerous in acknowledging all that *The Metaphysical Foundations* had quite apparently meant to him.

Whereas Galileo is the principal protagonist for all three, and likewise Newton caps the story with all three, Koyré diverges from both Dijksterhuis and Burtt in that he took not Kepler but Descartes as, so to say, Galileo's co-protagonist. This is so chiefly because, to Koyré, the principle of inertia that Galileo first enunciated and that, in his view, Descartes first systematized, became as it were the embodiment of the mathematization of nature. To Koyré the principle of inertia was the central event in the revolution he was analyzing, in that it stands for the very 'geometrization of space' which was at the heart of the revolution. This is so for two different, yet closely connected reasons. The principle served Galileo, in the *Dialogo*, as the principal weapon to defend Copernicanism—the doctrine that did more than any other to usher in the destruction of the Greek Cosmos. But, so Koyré argued, there is a still deeper ground. In stating the persistence of rectilinear motion in empty space, the principle of inertia asserts something that, in everyday reality, is just not the case—in the world that surrounds us day by day, motion invariably comes to a stand-still. The principle therefore explains what is the case by means of what is not the case, and cannot even be. Only in a fully idealized situation (in Euclidean space) is the principle valid, and yet it governs every motion in our real world.

So much for the three men who, in ways much more striking for their both vast and deep similarities than for their (on the whole subordinate) differences, jointly created the immensely influential concept of The Scientific Revolution as marked by the mathematization of nature.

Nonetheless, the name that comes to mind at once when retracing these two productive historical concepts is not Burtt or Dijksterhuis but Koyré alone. How, then, is it is that the chronologically third of the three has come to be personally identified with it, rather than the two men who preceded him by at least a decade?

4 Contextual Factors in the Balance

For possible answers, we turn to local situatedness and to agendas, on the look-out for significant respects in which Koyré differed from the two others.

In terms of *academic background*, the major difference is between Dijksterhuis, the mathematician, and Burtt and Koyré, both philosophers. Koyré's education was with Edmund Husserl, the German phenomenologist, Burtt's with American pragmatism—clearly, neither philosophical stance provides particularly ready explanations for what got both men into history, let alone into history of science. Prior to the Second World War, history of science was hardly a field at all, let alone a 'profession'—no one got there but along some convoluted pathway strictly of his or her own making.

In terms of the subject of the present book, the *historiography of mathematics in a somewhat narrower sense*, none of the three was then or later to put his mark on the field, with the partial exception of Dijksterhuis' concern in the 1930s with Euclid's *Elements* and with Archimedes' works overall.

In terms of *language*, it would be a mild anachronism to say that Burtt, whose native language was of course English, started with an advantage over the other two. The languages in which major, albeit incidental, contributions were on roughly equal footing being made to the history of science were English, German, and French. Paris in the 1930s even saw something of a proto-professionalization, run by the Italian émigré A. Mieli as head of a self-styled bureau of scientific priority, which was nipped in the bud by the advent of the Second World War, and in terms of which Koyré, albeit connected with the 'Ecole Pratique des Hautes Etudes', found himself at the sidelines throughout the period.[5] Dijksterhuis, with Dutch for sole language, was obviously at a disadvantage in this regard. It is conceivable that, if the publisher of *Val en worp* had not refused to go along with Dijksterhuis' wish to have the book published in German, such a feat might have furthered the book's renown and distribution. More significant is that the wife of D. Struik (the later historian of mathematics, who had just moved from the Netherlands to the US) considered but ultimately failed to translate *Val en worp* into English. Given that

[5] I take these facts about A. Koyré from notes made by P. Redondi (1986).

the triumphal march of the mathematization of nature began right after the Second World War in the Anglo-Saxon world, thus giving a strongly American/British flavor to the new wave of professionalization, the presence by then of an English version of *Val en worp* might well have made a difference. However, consider Burtt once again. *The Metaphysical Foundations* was lying ready for that wave, both instigated and successfully ridden by mostly young academics like Gillespie, Kuhn, Butterfield, and the Halls. But Burtt's book (albeit read and respected by some) hardly served as the obvious vehicle to spread the exciting message of how nature was first mathematized in the early 17th century, thus bringing about science essentially as we still know it. Instead, Koyré's work provided the vehicle, and our question still is why this was so.

For a possibly more satisfactory answer we now turn to *agendas*. I already emphasized that no ready-made agendas for work in the history of science lay ready prior to the Second World War—no one who engaged with the field did so for other than very much personally stamped reasons, not for any pre-established research agenda. Even Dijksterhuis, who comes closest to having one, did not really have an agenda when he wrote *Val en worp*—the book is in line with the Platonic conception of mathematics he was busily defending against numerous fellow high school teachers, yet hardly dictated by it. Koyré did not have any identifiable agenda at all, at least until completion of *Etudes Galiléennes*. But this changed with the publication of that book in 1940. Or rather, since hardly a more unlucky date for publication in Paris of a book of that nature is at all conceivable, Koyré's arrival in the USA by mid-war and his subsequent association with the famous New York School of Research did it. This is what led to publication, in English translation, of two of the three essays of which the book is composed in slightly rewritten guises, in the newly founded *Journal of the History of Ideas* and another journal of high academic standing. From then on, Koyré dedicated himself unrelentingly to spreading his message, through books like *From the Closed World to the Infinite Universe* and by many more means. That is, Koyré turned the mathematization of nature into the core item on his new-found agenda.

What about Dijksterhuis and Burtt in this regard? As for the former, my guess is that even early translation of *Val en worp* into English by Ruth Struik would not have made much of a difference. Koyré was a true cosmopolitan. Born in Russia, he fought for his country in the First World War, then moved to Husserl's Freiburg, then to Paris where he picked up flawless French, intermittently spent years in Egypt, after which he returned to Paris, crossed the Atlantic during the Second World War and upon its conclusion kept moving back and forth between the US and France until he died in 1964. Dijksterhuis, in stark contrast, was happy to deliver one lecture after another for the benefit of a variety of non-specialist audiences in the Netherlands, and even after his late appointment to a professorial chair in the 1950s locked himself up in his own study, not even bothering about the quality of the English into which a none too qualified translator sought to put his masterpiece, *The Mechanization of the World Picture*. Surely 'the mathematization of nature' had become part of Dijksterhuis' agenda, only, it formed just one item on a wider list and, even more importantly, he failed entirely to pursue this agenda with anything like the single-minded dedication of Koyré, blessed as the latter was with a far more outgoing personality.

As for Burtt, the most curious thing about him is that, for all the novelty and brilliance of his first book, he never mustered sufficient confidence in himself to come up with any agenda at all, let alone stick to it. Within ten years of writing *The Metaphysical Foundations* he abandoned that wonderful book for good, leaving it in effect to its own devices. He also effectively dropped the ambition that follows from the book's own core message to think up a philosophy both in line with modern mathematical science and capable of expressing a responsible view of us human beings and our 'high spiritual claims'. Henceforth he wrote, as it were, in the margin of other thinkers, producing clear-cut and instructive, yet essentially service-based books on comparative religion or on contemporary philosophical currents, seeking everywhere for what a large diversity of thinkers had in common rather than what held them apart.[6] It is a rare and, in my view, admirable posture for a philosopher to take, but not one conducive to the pursuit of an agenda of one's own. So Burtt remained over his very long career (he died in 1989 at the age of 97) an even lonelier bird, academically speaking, than Dijksterhuis.

Here, then, we seem to have the decisive difference sought for—the personal creation of a powerful agenda *plus* the driven, outgoing personality required for its relentless pursuit—that serves to explain why it was Koyré's 1940 *Etudes Galiléennes* rather than Dijksterhuis' (1924) *Val en worp* or Burtt's (1924) *Metaphysical Foundations* that inaugurated the decades-long triumph of this fertile conception of the mathematization of nature as key to the Scientific Revolution of the 17th century.

5 Historical Concepts as They Come and Go

The expression 'decades-long' implies, of course, that the triumph was not to endure forever. Indeed, the unique prominence accorded to the idea that the 17th century witnessed a unique rupture with long-standing conceptions of the constitution of the natural world—a rupture that ushered in modern science essentially as we know it and that was marked above all by making science forever mathematical—has faded and has kept fading since the 1960s, or possibly even earlier. This has much to do with the unmistakeable one-sidedness of the concept.

Up to a point, its one-sidedness was of course a boon—a very large number of historical events, hitherto unconnected, now appeared bound together in one powerful, unifying concept. The achievement of great innovators like Galileo, Descartes, Kepler, and Newton, never neglected of course but so far examined each in its own right as a rule, suddenly appeared in a new light, capable (or so it seemed) of revealing a fundamental break at a retrospectively crucial point in the

[6]Some later books by Burtt that impressed me in particular are *In Search of Philosophic Understanding* (1965) and *The Human Journey* (1981).

story of nothing less than the human adventure itself. Here was the seemingly tailor-made intellectual core around which, in the first decade or so after the Second World War, a budding profession could organize itself, and did, with the USA in a leading position and Britain and the European Continent soon following in its wake. After all, the new concept came with an appealing, radically novel way of doing history of science. Dijksterhuis, Burtt, and most polemically so Koyré presented their novel method as the antithesis of the customary search for the 'primitive roots' of present-day insights—of a positivist-triumphalist celebration of how, starting from some simple-minded notions, we got ever smarter so as in the end to attain our present-day scientific truths. Instead, these men were determined to examine the past guided by viewpoints taken in that past, doing so with ever increasing understanding of a problematic such as it appeared at the time in the living experience of contemporaries, themselves; also with an awareness of the numerous twists and windings along which their protagonists managed to attain a possibly sharper insight. A further, likewise vast implication of the new concept was that the history of science could boast a significance far beyond the boundaries of its proper domain. As nature began to be mathematized full-scale, with Newton's work as the prime exemplar, what happened in and with the new science began to have an ever increasing impact on what happened in the world at large, whether regarded from a social, a political, an economic or a cultural perspective.

And yet, historical concepts tend to come and go—fresh and with an illuminating one-sidedness when they come; stale and diluted on their seemingly inexorable pathway to the exit. Take the concept of the Renaissance. If you go back to Jacob Burckhardt and make an effort to read his *Kultur der Renaissance in Italien* with the fresh eyes of his contemporaries in 1860, it is not hard to be thrilled by the sheer daring with which the author took a well-known expression and turned it into a veritable historical concept. What so far was just a piece of convenient periodization with rather vague, mostly personal associations hanging over it now became a clear-cut idea of what, in a radical break with the medieval period, the event at bottom was about. By stark contrast, in present-day usage the 'Renaissance' has become a catch-all expression for the most divergent notions, few of which still recall those 14–16th century trends which Burckhardt singled out as somehow standing for the whole—as, so to speak, expertly selected *partes pro toto*. Right from the start, the idea of a radical break with the preceding Middle Ages found itself challenged with particular force as medievalists began to point out how much continuity there really was between their own period and that of the Renaissance, while (as they warmed to the subject) claiming for themselves nothing less than a 'Renaissance of the 12th century'. With the mathematization of nature, bound up as it was with the equally novel concept of the Scientific Revolution, we may discern a comparable process of ongoing meaning dilution, albeit held in check at times by efforts at salvaging what could (and can) be salvaged still. It is these twin processes of dilution and salvaging of the concept of the mathematizaton of nature that I shall now address, doing so in even less detail than spent above on how the concept came into being, and also without providing more than some passing additional context.

Not only these two processes of meaning dilution, but the underlying dynamics, too, are quite comparable. As earlier with Burckhardt's Renaissance, in the case of the mathematization of nature, too, an original sense of exciting freshness gave way to a mostly tacit yet quickly prevailing sentiment that history may not, after all, be the right place for clear-cut concepts. Also, the claim, central to both concepts, of a radical break with the immediate past was countered by medievalists in both cases. Finally, and above all, one-sidedness turned from an asset into a liability. I shall from here on discuss the latter aspect in particular—the problem of increasingly apparent complexity and how to handle it properly.

6 Apparent Complexity Variously Dealt with

Dilution on the grand scale, then, set in by the early 1950s with three British historians, Butterfield and the Halls.[7] The time span of the Scientific Revolution itself is now being widened by large stretches. Both Burtt and (in the *Mechanization*) Dijksterhuis needed a century and a half (from Copernicus' *De Revolutionibus* to Newton's *Principia*); so did Koyré once he expanded the half century of his pristine concept (Galileo plus Descartes, c. 1600–c. 1645) to comprise likewise the century and a half covered by the other two. Butterfield even extended the scope of his Scientific Revolution to half a millennium, as following Duhem he included impetus theory (c. 1300) while also insisting on a 'postponed revolution in chemistry' that, alas, did not come about until Lavoisier (c. 1800). Over such a lengthy period the process of mathematization of nature inevitably lost the singular significance it had gained in the works of the three pioneers, in each of which it had provided the central story line. It is, however, possible to discern in Butterfield's very accessible and vastly popular *The Origins of Modern Science* (1949) a tacit distinction being made between what he presents as The Scientific Revolution and, as it were, an *inner* Scientific Revolution of far shorter duration (Copernicus to Newton). In the latter, prominence is still relegated to the mathematization of nature as key to all those significant, profound changes that marked this shorter period, in particular.

With A. Rupert and Marie Boas Hall, who between the mid-1950s and the mid-1980s dedicated four volumes to the Scientific Revolution, the picture looks rather similar. Periodization changes somewhat from one book to another, but in each case it starts prior to 1543 and ends beyond 1700. More significantly, little if any clear-cut conceptualization remains, with one sole exception—key to each of their stories is the ever increasing *rationality* of how between c. 1500 and c. 1750 natural phenomena and processes were conceived by successive pioneers. In the view of the Halls, rational conceptions and methods gradually take over as the unholy triad of magic, mysticism, and superstion recedes to the background. As a

[7]Strictly speaking, Rupert Hall's wife, Marie Boas, was an American, but their joint work was chiefly located in Great Britain.

consequence of all this, the mathematical treatment of natural phenomena no longer provides the main story line, now standing rather for a particularly noteworthy and important exemplar of rationality victorious.

It is not hard to guess what primarily drove this first wave of meaning dilution. How, in a responsible survey, to handle all those numerous investigations undertaken between Copernicus and Newton that were not marked by mathematics, or hardly so? What place in the story line to give to Bacon, to Harvey, to Boyle, to Hooke? How to deal with medicine or chemistry or even (God forbid) alchemy? What about the practice of experiment?

Surely a prime place given to the mathematization of nature did leave some room for answers to such questions. These, however, invariably entailed either a flat denial of any relevance for the Scientific Revolution (e.g., Koyré about Bacon) or none but a subsidiary role assigned to these non-mathematical investigators and these non-mathematical fields. Thus came into being what it became customary to call the 'master narrative of the Scientific Revolution', with the mathematization of nature hanging over it in fairly diluted fashion, yet preventing everything non-mathematical from being treated as more than, at best, a secondary by-product of that Revolution.

Suppose for a moment that, if the twin concepts of the mathematization of nature and of the Scientific Revolution had not entailed the vast methodical reform just mentioned, which after the 1940s took the budding profession by storm and thoroughly and forever transformed the responsible writing of history of science, then the viability of these twin concepts might not have survived this first wave of meaning dilution. Instead, two remarkable efforts were undertaken to salvage the mathematization of nature, by maintaining it in by and large its original guise inside (and this was the new thing) a partially reconceptualized Scientific Revolution. Building forth upon passing remarks by Burtt and by Koyré themselves, both T. Kuhn and R. Westfall came up in the 1970s with a two-current account of the Scientific Revolution. While in substantial agreement over the presence throughout the Scientific Revolution of penetrating and enduring efforts at the mathematical treatment of natural phenomena, they differed in how they defined the parallel current. With Kuhn, it was Baconian, natural-historical, craft-linked, empiricist, and perfused with increasing amounts of heuristic experimentation. With Westfall, the parallel current was the mechanical philosophy in the sense of an all-encompassing conception of natural phenomena as produced by various law-like motions of tiny particles. Unlike with Kuhn, Westfall's two currents were not so much static entities as, rather, subject to a dynamics of tension between the Galilean and the Cartesian approaches to motion and of ultimate merger in Newton's work.

For a long time these promising, albeit partly contradictory efforts at reconceptualization (very sketchy in Kuhn's one article on the subject, less so in Westfall's books)[8] failed to be taken up in a conceptual sense. The principal reason for this probably rested in another radical transformation of the method for history

[8]Whereas Kuhn did no more than sketch his two-current account in one surely brilliant article, 'Mathematical versus Experimental Traditions in the Development of Physical Science' (1977),

of science, away from the history of ideas toward investigation of their local situatedness, that is, toward a socio-culturally contextual approach to the history of science. One aspect of this transformation, which in the 1980s took the profession by storm, was a great deal of attention henceforth being given to practice, be it in connection with empiricist approaches to nature or with how in scientists' everyday activities experiments were actually being conducted. With local situatedness now at the heart of history of science writing, the universal claims for science as it was customarily held to have emerged in (and due to) the Scientific Revolution were routinely denied, or ignored, or at the very least regarded as irrelevant to the everyday, locally determined process of science. In tandem with a rising influx in the profession of more historians and fewer scientists, most often with decreasing mastery of the *lingua franca* of mathematical science prior to c. 1800, Latin, the concept of the mathematization of nature was bound to lose more and more ground. Just as the pioneers, Dijksterhuis, Burtt, and Koyré, but also still Butterfield and the Halls had in effect handled the mathematical treatment of natural phenomena as a *pars pro toto* for the process of the Scientific Revolution, just so its 'natural history' aspect became the new *pars pro toto* from the 1980s onward.[9] Only, this new *pars pro toto* treatment was accompanied by a strong tendency to get rid of the notion of the Scientific Revolution as such. One more driving force in the same direction was an unceasing stream of continuist views, most upholders of which kept making a large variety of cases for strong links between conceptions and approaches of 17th century protagonists and certain medieval (or in a rare case Renaissance period) predecessors. Also, as historians kept delving into more and more intricacies of 17th century thought about, and practice in, the examination of natural phenomena, more and more complications became manifest. Galileo undertook many more heuristic experiments than Koyré had been prepared to give him credit for; the bulk of innovative 17th century inquiry was in a natural-historical rather than a mathematical vein; experimentation in the frame of the Royal Society turned out to be linked up in important ways with Restoration politics; alchemy appeared to be practiced by just about every individual concerned at the time with problems of the constitution of matter, such as, notably, Robert Boyle and Isaac Newton; and so on, and so forth. In short, the big picture became ever more blurred, and the apparent simplicity of the historical formula 'Scientific Revolution = mathematization of nature' came to look (as indeed I am sure it is) quite untenable.

Convoluted as this brief sketch of historiographical developments since the 1950s may look, it still oversimplifies these in many ways. Yet the sketch suffices for my present purpose, which is to make clear the trouble in which almost a dozen historians of science found themselves who between the early 1990s and the present day accepted an invitation by some publisher (or decided on their own initiative) to write a textbook about the Scientific Revolution. After all, such textbooks are meant

(Footnote 8 continued)
Westfall's is spread over two books, his pioneering *Force in Newton's Physics* (1971a) and a shorter textbook based on it, *The Construction of Modern Science* (1971b).

[9]Just one example of this tendency is H. Cook's *Matters of Exchange* (2007).

as as rule for class-room usage in the first place. Feeling uneasy, albeit on partly different grounds, with the still prevailing 'master narrative', these men (and one woman) were not as textbook writers in a good position to make an effort at thorough reconceptualization of the Scientific Revolution—certainly not in an intellectual climate where the very notion was routinely being pooh-poohed. Consequently, in each of these textbooks we encounter some toned-down version of the mathematization of nature placed in a large variety of settings, as for instance treatment of mathematical science in the 17th century as wholly or partly subservient to 'the mechanical philosophy' of moving particles.[10]

7 The Mathematization of Nature Revived in a Partly Altered Setting

In two recent, much lengthier books, however, numerous problems assembled over past decades in connection with the very viability of the concept of the Scientific Revolution have been taken up with far larger ambition than a textbook allows for, one by Stephen Gaukroger and one by myself. Guided (or so I hope) by an effort at impartiality I have compared elsewhere our respective ends and approaches. In that article I emphasized that the apparent overlap between his work on the subject and mine is partial only, in that he aims not so much (the way mine does) at understanding how modern science came into the world as rather (even more ambitiously) how between the High Middle Ages and the present day science has managed to turn from a marginal cultural phenomenon into one central to just about all our modern concerns. So let me confine myself here to a very brief exposition of how the mathematization of nature has fared in my own conception of the Scientific Revolution, which appeared in 2010 under the title *How Modern Science Came Into the World. Four Civilizations, One 17th Century Breakthrough.*[11] In many ways this book may be regarded as picking up again and extending further those illuminating yet underexploited efforts at reconceptualization undertaken in the 1970s by Westfall and by Kuhn.

Key to my consistently comparative account in this regard is the presence of certain 'modes of nature-knowledge' in ancient China and in Greece. The latter civilization even provided two such modes—one speculative-dogmatic, all-encompassing, and causal; the other partial, philosophically non-committal, and, above all, abstract-mathematical. The former I call 'Athens' for short; it comprises the schools of Plato, Aristotle, Epicurus, and the Stoa. The latter I call 'Alexandria', exemplified by Euclid, Archimedes, Ptolemy and dozens more mathematical

[10]I have written brief summaries of these textbooks in a 'Postscript' to my historiographical book of 1984, available in a pdf file on my website www.hfcohen.com under 'Books'.

[11]A shorter version, directed at a wider audience, came out in Dutch in 2007, then in German translation in 2010 and in Chinese translation in 2012. It came out in English as *The rise of modern science explained* (Cambridge UP, 2015).

scientists. Now these two distinct, both geographically and intellectually separate modes of nature-knowledge found themselves in later times transplanted to three culturally distinct environments in succession. By 'transplanted' I mean that in every single case texts were being translated and appropriated, and their contents substantially enriched. One such transplantation is to the Islamic world in the aftermath of its first civil war, instigated and won by the Abassid caliphs. The next, a side-effect of the Reconquista, brings the Greek corpus in its Arabic version to medieval Europe. The third and final transplantation, which is occasioned by the fall of Byzantium, brings the original Greek texts preserved there to Renaissance Italy and then to the rest of Europe. As in ancient Greece originally, in each of these three cases the two distinct modes of nature-knowledge (the natural-philosophical and the mathematical one) experience together a flourishing period that culminates in a Golden Age, that fairly soon gives way to steep decay. The decay, however, is punctuated by bouts of incidental high-level achievement, as with Ptolemy in Alexandria, or with Nasir ed-Din al-Tusi in the Islamic world.

The history of nature-knowledge in Renaissance Europe (c. 1450–c. 1600) fits into this standard pattern in several major ways. Even so it is marked by two big exceptions. Besides the recovery and renewed flourishing of the 'Athenian' and the 'Alexandrian' modes of nature-knowledge, a third one comes up at the same time. This third, home-made mode of nature-knowledge is not intellectualist like the two originally Greek ones, but rather empiricist and marked by vigorous efforts to make accurate description yield certain practical ends, as with Vesalius or Leonardo da Vinci or Paracelsus. Further, quite unlike its predecessors the Golden Age of Renaissance nature-knowledge does not give way to decay but (on the contrary, and quite unexpectedly so) to each mode of nature-knowledge apart undergoing revolutionary transformation. This three-pronged episode c. 1600–c. 1645 is what I regard as the onset of the Scientific Revolution. At the hands of Kepler and Galileo hyper-abstract mathematical science in the vein of Ptolemy and Copernicus, of Archimedes and Benedetti, is turned into *realist*-mathematical science. That is, 'Alexandria' is, wholly unforeseen, and for bystanders at first quite enigmatically so, turned into what I call 'Alexandria-plus'. Mathematical science of course remains an in many ways abstract affair, as it cannot be otherwise, but the point is that around 1600 it begins to be linked up in unprecedented fashion with the reality of natural phenomena, as notably in Kepler's 'celestial physics', in Galileo's rules for falling and projected bodies, and most fundamentally so in Galileo's conception of motion overall. Within decades, Beeckman and Descartes begin to transform the ancient philosophy of atomism such as to make the motions and mutual configurations of invisible particles, not their shapes, central to their philosophical account of how the natural world is at bottom constituted. Also in these first four decades of the 17th century Bacon, Gilbert, Harvey, and van Helmont proclaim and/or effect a slightly less revolutionary transformation, in that experimentation becomes key to ongoing efforts at accurate description undertaken with a view to practical benefits.

Rather than continuing unabated, the advance of these three efforts at revolutionary transformation is halted by a profound crisis of legitimacy due to apparent strangeness and apparent sacrilege. In the 1640s the pace of discovery slows down due to censorship, self-censorship, and other side effects of the war of all against all that threatens more and more to rip Europe completely apart. But the crisis is surmounted in the nick of time, and the innovative achievement of the pioneers can now be pursued, elaborated, and tested in a second wave of revolutionary transformation (c. 1660–c. 1685). This episode is marked above all by four pioneers, Huygens, Boyle, Hooke, and young Newton, all of whom turn meanwhile customary thinking in terms of mechanisms of particles in motion from speculative, all-encompassing *dogma* into conceiving of such mechanisms as heuristically fruitful *hypotheses*. This crucial move makes it possible for Huygens (and independently for young Newton) to discover and partially to resolve the tension they alone perceive to exist between Galileo's and Descartes' conceptions of motion. Even though the two distinct conceptions overlap in part, they appear to both men to be in their fundamental constituents mutually incompatible (the one being mathematical, the other philosophic). At the third and final stage of the historical process of the Scientific Revolution, mature Newton revolutionizes what he and Huygens have meanwhile achieved in this regard, in once again radically novel ways which find their culmination point in the *Principia*.

In this utterly sketchy outline, then, the mathematization of nature remains one vital constituent of the entire process of the Scientific Revolution. At its first stage, intensely abstract 'Alexandria' is turned into realist-abstract mathematical science, validated where at all possible by means of carefully thought-out experiments. In the revolutionary transformations that follow upon a potentially lethal interval of ideology-driven crisis, newly realist mathematical science co-constitutes the fusion processes that mark the second and the third stages. Whereas from Euclid, Archimedes, and the other pioneers of the 'Alexandrian' approach onward up to and definitely including Kepler and Galileo mathematical science has stood apart from all other, natural-philosophical and empiricist undertakings, starting with Huygens in the late 1650s it is taken up in an increasingly intricate fabric of once again radical innovation.

Obviously, there is no way I could have reached these conclusions about the mathematization of nature without prior acquaintance with, and without reaping immense benefits from, what the pioneers and their successors came up with starting in 1924, now ninety years ago. These men, on whose shoulders I have felt myself standing all the time, were not stupid (to put the point at its mildest). It is a priori most unlikely that their by and large shared conception of nature mathematized in the 17th century, for all its one-sidedness, might have failed entirely to capture some deep truths about how modern science came into the world. The illumination it once provided has not necessarily faded to the full. True, the writing of history has not stood still—ongoing innovation in nature-knowledge over a period which (however demarcated exactly) certainly includes the 17th century has turned out to be a far more complex affair than the pioneers could possibly suspect. Even so it is our job as historians not just to take note of the boundless complexity of any historical process whatsoever, but to conceptualize and reconceptualize it until at least a rough fit with those complex events has been attained. In most current thinking about the history of science, the mathematization of nature has quietly receded to the background. In my view the concept deserves a better fate.

References

Berkel, K. van. (1996). *Dijksterhuis. Een biografie*. Amsterdam: Bert Bakker.

Burckhardt, J. (1860). *Die Cultur der Renaissance in Italien. Ein Versuch.*. Schweighauser: Basel.

Burtt, E. A. (1924). *The metaphysical foundations of modern physical science. A historical and critical essay*. London: Routledge & Kegan Paul.

Burtt, E. A. (1965). *In search of philosophic understanding*. Indianapolis: Hacket.

Burtt, E. A. (1981). *The human journey*. Calcutta: University of Calcutta.

Butterfield, H. (1949). *The origins of modern science 1300–1800*. London: Bell.

Cohen, H. F. (1994). *The scientific revolution. A historiographical inquiry*. Chicago & London: University of Chicago Press.

Cohen, H. F. (2010). *How modern science came into the world. Four civilizations, one 17th century breakthrough*. Amsterdam: Amsterdam University Press.

Cohen, H. F. (2013). 'Two new conceptions of the scientific revolution compared'. Contribution to: 'Scientific culture in the modern Era: A forum'. *Historically speaking. The Bulletin of the Historical Society, 14*, 24–26.

Cook, H. J. (2007). *Matters of exchange. Commerce, medicine, and science in the Dutch golden age*. New Haven & London: Yale University Press.

Dijksterhuis, E. J. (1924). *Val en worp. Een bijdrage tot de geschiedenis der mechanica van Aristoteles tot Newton*. Groningen: Noordhoff.

Dijksterhuis, E. J. (1950). *De mechanisering van het wereldbeeld*. Amsterdam: Meulenhoff.

Gaukroger, S. (2006). *The emergence of a scientific culture. Science and the shaping of modernity 1210–1685*. Oxford: Oxford University Press.

Koyré, A. (1939–1940). *Etudes Galiléennes*. Paris: Hermann.

Kuhn, T. S. (1977). Mathematical versus experimental traditions in the development of physical science. In Idem, *The Essential Tension* (pp. 31–65). Chicago: University of Chicago Press.

Redondi, P. (ed.). (1986). *Alexandre Koyré, De la mystique à la science. Cours, conférences et documents 1922-1962*. Paris: Editions de l'école des hautes études en sciences sociales.

Villemaire, D. D. (2002). *E.A. Burtt, Historian and philosopher*. Dordrecht: Kluwer.

Westfall, R. S. (1971a). *Force in Newton's physics. The science of dynamics in the seventeenth century*. New York/London: Elsevier/MacDonald.

Westfall, R. S. (1971b). *The construction of modern science. Mechanisms and mechanics*. Cambridge: Cambridge University Press.

Histories of Modern Mathematics in English in the 1940s, 50s, and 60s

Jeremy Gray

Abstract

The theme of the meeting on the historiography of mathematics in the 19th and 20th centuries invited participants to reflect on "the cultural contexts in which the history of mathematics was written". It noted the change in the field of history of mathematics, away from an initial concentration on "major internal, universal developments of certain sub-disciplines of mathematics" towards "a focus on contexts of knowledge production involving individuals, local practices, problems, communities, and networks". The expressed hope was that by "analysing the often hidden agendas of former historians of mathematics we [will be] led to reflect upon our own professional objectives as well as on the methods and tools we employ today". In that spirit this paper offers some reflections on the principal texts written in English on the history of modern mathematics in the 1950s and 1960s, concentrating on the ones that have had a lasting influence on the field, and draws out some thoughts about how place and audience have exercised, and continue to exercise, a marked effect on the growth and shape of the subject. The six authors I have considered, all of them American, wrote, in the main, for students of mathematics, and their approaches were adapted to the prevailing mathematical syllabus. Only one, Carl Boyer, made a serious effort to keep open links with contemporary history of science, and this has doubtless contributed to the present uneasy relations between history of mathematics and history of science. New initiatives are needed in the history of mathematics, such as the current attention to mathematical practice might supply, along with a broader perspective, and a renewed attention to methodology (attending, for example, to communications networks and corpuses of texts).

J. Gray (✉)
Department of Mathematics and Statistics, The Open University,
Milton Keynes MK7 6AA, UK
e-mail: j.j.gray@open.ac.uk

V.R. Remmert et al. (eds.), *Historiography of Mathematics in the 19th and 20th Centuries*, Trends in the History of Science,
DOI 10.1007/978-3-319-39649-1_9

Keywords
Carl B. Boyer · Julian Lowell Coolidge · Morris Kline · Historiography of geometry

1 The Authors and the Field

There is only one study of the topic of this paper that is of any scope, and that is *Writing the history of mathematics: its historical development*, edited by J.W. Dauben and C.J. Scriba and published in 2002 (Dauben and Scriba 2002). It contains a short article by Grattan-Guinness (2002) on the British historians of the 19th and early 20th centuries, of whom only James Glaisher (1848–1928) and Edmund Whittaker (1873–1956) could be relevant to the present article, and a longer essay by Joseph Dauben on the Americans (D'Ambrosio et al. 2002). This disparity reflects the relative strength of the subject in the younger country. Among the historians that Dauben considered are Raymond Clare Archibald (1875–1955), E.T. Bell (1883–1960), Carl B. Boyer (1906–1976), Florian Cajori (1859–1930), Julian Lowell Coolidge (1873–1954), Howard Eves (1911–2004), Louis C. Karpinski (1878–1956), Morris Kline (1908–1992), George Sarton (1884–1956), Dirk Struik (1894–2000), David Eugene Smith (1860–1944), and Raymond Wilder (1896–1982). He also looked at the work of some emigrés: Max Dehn (1878–1952), Wilhelm Magnus (1907–1990), and Ernst Hellinger (1883–1950). In keeping with the structure of the book, Craig Fraser wrote about Kenneth O. May (1915–1977) in the chapter on the Canadian story (D'Ambrosio et al. 2002). Interestingly, the Newton historian I. Bernard Cohen (1914–2003) is only mentioned elsewhere.[1]

For the purpose of this short article there are only six historians of mathematics in the above list who made significant contributions to the development of the discipline by writing on the modern period: Boyer, Coolidge, Eves, Struik, Kline, and Wilder.

A fuller account, which cannot be attempted here, would take on board the cultural context as it was manifested in the simultaneous story in the history of science and examine the place of history of mathematics within (or outside) it. This would be a significant project, but one that would have to confront the vexed relationship of the history of mathematics and the history of science, and therefore the complicated history of the history of science itself, and that too cannot be dealt with properly here. It is worth noting, however, the history of science itself was once a new specialism in history, as for example economic history also was. These

[1] It would be a separate, and important, story to trace the ways historians of science and historians of mathematics have dealt with Newton, but at least there is an extensive literature. The situation for a comparable figure such as Euler, and 18th century science, is much worse, and the disciplinary divide is surely not helping.

disciplines conform to the general rule that, as historians enlarge their discipline, when science comes in it it does so with no more mathematics than is strictly necessary. This is true of the study of Egypt and Mesopotamia, the Islamic world, and China, with only one notable exception, the work done at Brown University, enriched as it was with the presence of Otto Neugebauer (1899–1990). So what follows is a narrower analysis of what was done only within the history of mathematics, which, it must be said, was largely independent of the history of science in the period considered.

Any analysis of the work done by the six authors selected above would have to propose a methodology that would help us consider what sorts of things were done under some systematic headings. The books and articles they wrote can profitably be considered under the headings of research, record, and instruction. A book or article may fall under the heading of research if it brings forward new information or analysis. Articles of record are such things as obituaries and commemorative articles that put in one place information generally available but scattered. Instruction, which will be the largest category, includes articles and books written for teaching purposes or for the general education of the mathematical readership.

Such work may also be analysed under the headings of depth, 'width', and breadth. Depth applies most readily to work offering a new analysis or insight, perhaps based on fresh textual evidence, that aims to change and improve our understanding. I shall also use it for works that bring unusually large amounts of new information to light, such as editions of works and correspondence. At the other extreme, works of significant breadth offer good coverage of a topic, saying much or all of what needs to be said at a given level but without any real claim to novelty. Typically, such works are for classroom use. The intermediate category of width is for works that combine both depth and breadth, but make a compromise between novelty and generality.

It is also helpful to think of a book or paper in terms of its actual, intended, or ideal audiences. Sometimes these can be readily identified by the author, and subsequently by his readers, for example as college students, or a peer group of colleagues many of whom the author can name. Beyond these groups stand the larger audiences of similar but unknown people who are presumed to have generic backgrounds, skills, knowledge, and interests. The designated audience often has a distinctive effect on the book or paper. This is clearly so with structured textbooks, but is often apparent in the way the content is presented and the compromises made with contemporary student education. In the history of mathematics these compromises show up in the frequently limited use of sources, the modernised mathematics, the absence of quotations, and in numerous other ways.

A major book or paper is different. A book that offers depth may place unusual demands on its readers in the way it handles sources, linguistic issues, technical mathematics, and other fields of knowledge, and it may well carry a deeper and more sophisticated argument, all of which the author regards as the price of scholarship. In the 1950s and 1960s there was very little literature in the history of mathematics of this kind in English. Indeed, it can be argued that the influence of Neugebauer's *The exact sciences in antiquity* (Neugebauer 1951) well outside the field of ancient mathematics and science rests in part on its being an original, scholarly, and yet readable book with a clear argument at a time when not much like it was being written. Of course it rests on a prodigious amount of difficult work, but the book was visible, and showed what could be done if one immersed oneself in primary sources, at a time when very little of that was being done.[2]

The six authors wrote the books and articles that shaped the subject in the years from 1950 to 1970 and beyond, and so we can ask, and I believe we should ask, the subjective questions: What of this work is good? What has lasted?

2 Professionalisation

The training, academic positions, and likely audiences of the six historians of mathematics identified above also exhibit characteristic features that say a lot about their work. Indeed, the most obvious fact about them is that while all six were trained in mathematics, only Boyer had any professional training in history of mathematics. In this respect they were unlike the Newton scholars Bernard Cohen or Rupert Hall (1920–2009) in England. This is indicative of an important way in which history of mathematics differed markedly from the history of science: it was always oriented more towards mathematics than the history of science ever was towards science, and this is apparent when looking at the Faculties to which the subjects belonged. The history of science has almost always belonged within a Faculty of history or humanities; history of mathematics within a mathematics Faculty.

All six of the men we are considering taught in mathematics Departments. Most, one could say, wandered in from mathematics, some continued to do mathematics as they took up the study of its history, some turned to history only after they had given mathematics up. Many of them had a strong commitment to education. Morris Kline, who had a good career as an applied mathematician, became notorious for views on the teaching of modern mathematics that derived from his sense of what mathematics should be about.

Four among them had had careers as research mathematicians: Coolidge, Wilder, Struik, and Kline. Coolidge had a long career in algebraic geometry at Harvard, working in a style that was becoming exhausted by the end when a generation of younger mathematicians were bringing in the new methods of modern algebra and paying more attention to rigour. Wilder was a student of R.L Moore (1882–1974) in Texas who became a distinguished topologist, Struik a differential geometer, and Kline, as already noted, an applied mathematician. None were full-time, professional historians of mathematics however deep their commitment to it at some, usually late, stage in their lives. None has a successor, someone trained by them

[2]As this book discusses elsewhere, it was, of course, a continuation of a strong German tradition in the history of mathematics that was broken in the Nazi time and which Neugebauer was helping to transplant to the United States.

who continued to work in the field. None of these men taught in Departments of the history of science, and this brings up an important point: there never was a discipline of the history of mathematics in the United States in the sense that there was and is a history of science discipline. Indeed, there was not much of a sense of community, or communal practice, among these historians. They very seldom if ever cited each other's work, let alone discussed it, and I have not found an occasion where anyone explicitly criticised another's work or attempted to correct it explicitly.

Who, then, formed their audience? The content of their work make it clear that the answer is in the main mathematics students of various kinds, mathematicians of various kinds, and teachers of mathematics. Only Boyer tried to keep links open with historians of science, and only Wilder made links with philosophy of mathematics in the period.

Furthermore, only Coolidge, Boyer, and Struik published repeatedly on the history of mathematics; Eves, Wilder, and Kline wrote their book (or, in Kline's case, two: Eves 1964; Wilder 1969; Kline 1953, 1972). The relative absence of research papers and monographs by these authors is indicative of their commitment to the field, but in Kline's case his commitment to mathematics education led him to supervise a number of theses that became long and valuable papers in the early issues of *Archive for history of exact sciences*. Eves' book is famous for its rather mathematical problems—most of his other publications are problem-related. For many years he was mainstay of the Mathematical Association of America, where he emphasised the importance of problems and problem solving in the training of a mathematician, and his book is better regarded as belonging to the tradition of teaching mathematics, through its history than as book in the history of mathematics. As a work of history it is wholly derivative and short of attention to primary sources.

3 Methodologies

As for their approaches to the writing of history, it is clear that Coolidge, Eves, Boyer, and Kline were descriptive. In the main, what they describe is how one discovery followed another, one researcher did well or not so well with the topics he or she took up. They wrote in the tradition of history of ideas, in which actors were described almost entirely through their ideas, and the roles of places, institutions, patrons, and states were discussed peripherally if at all. Accordingly, these four authors wrote about little more than mathematics and physics.

Struik, and Wilder in their different ways were argumentative. Struik wrote in a manner influenced by old-fashioned Marxism, in which the influences on a mathematician could include the priorities of a nation, political and military events, and reference to the means and forces of production. But Struik's *A concise history of mathematics* (Struik 1948a) was not a social history of mathematics; the social-historical material functions as a scene-setting exercise, and the bulk of the pages Struik wrote were descriptive in the manner just indicated.[3] Wilder made the most marked change, and embraced philosophy, the foundations of mathematics, anthropology and culture, becoming a Fellow of the American Anthropological Association. It would be interesting to examine his influence on what became the field of ethno-mathematics—unfortunately I cannot do so here.

Much of what these people wrote stayed in print for a long time, and editions and re-editions have been produced until recently. Although I shall argue later that historians of mathematics today need to do things differently, and in some respects better, it is clear that much of what these people wrote met and meets a need in the mathematical community. Eves successfully combined the teaching of mathematics and the history of mathematics to a broad audience. Struik's *A concise history of mathematics* is a good complement: better on narrative, short on (mathematical) problems, and more readable. Before finding fault with either of these works we should ask ourselves, given the fact that almost all of modern mathematics is unknown to first-year students and difficult to understand, what would we do in less than 300 pages?

Kline's *Mathematical thought from ancient to modern times* (Kline 1972) is a different matter. It is best regarded, and generally used, as a reference work. It is largely accurate but not many people will have read it from cover to cover. It remains, over 40 years after its first publication, the best place to orient oneself when looking into topics one does not know, and it is still a good benchmark when refereeing papers. Although it contains many references to original sources, it reads as if it is establishing, or perhaps confirming, the official verdict. The generally major mathematicians are brought forward in conjunction with their generally agreed major works. Historical matters are discussed in terms of priorities and influences, and these generally rely, when not routine, on previous scholarship [such as Whittaker's work on complex function theory and analytical dynamics, (Whittaker 1917; Whittaker and Watson 1915)]. The chapters on applied topics are fresher than the ones on 19th and 20th century geometry or number theory, topics that Kline found less congenial, and in keeping with the view of the mathematics of his day, statistics is largely excluded for the book.

Much the same can be said of Boyer's well-regarded *A History of Mathematics* (Boyer 1968), which is a textbook, replete with Exercises (of a heavily mathematical kind). It aspires to be reasonably complete, it fills some of the gaps to be mentioned below in his more specialised works, it runs from before the Greeks to the start of the 20th century, with a chapter on China and India and one on the Arabic Hegemony, and it is a readable, descriptive account that does not differ in

[3]Whereas *Yankee science in the Making* (Struik 1948b) was an original study of science and technology in a local setting, but it deals with science and technology from the colonial era to 1861, and so is not relevant here.

approach from his other works. Indeed, precisely because it is a textbook it is in some ways more limited, and it can be set aside for the purposes of this paper.[4]

To sum up in the three categories of breadth, width, and depth, Eves (1964) and Struik (1948a) offered breadth of coverage, and of the two Eves did less and Struik did more. Two problems stand out with this literature, as with any account to provide the readers' first acquaintance with the field, and they are exacerbated when the books are used for teaching. The first problem is the size of the field. Any significant subject has this problem; there is so much to cover that many details are flattened, and a tendency to fit the account to what the readers may be supposed to know comes to dominate the selection. The second problem is the mathematics itself, which may not fit the exigencies of the readers' knowledge. If it does not, the author must decide if some mathematics teaching must be done, if the details be given but left for the reader to learn about elsewhere, or if the details are simply dropped. If the book forms part or all of a course then the material must be examinable, and this further biases the presentation in favour of the mathematics. Students may be asked to use a method of Newton's to find a tangent to a curve, rather than to write about the method in the context of its time.

Of the two, Eves embraced the problems with a view to making them into virtues, and Struik stepped back from them with a view to minimising their impact. Eves wanted his students and his other readers to learn about the mathematics of the past by doing something like it, with an eye on how that would help them with the mathematics they were studying generally. Struik had in mind an evolutionary development of mathematics that led, reasonably enough, to some of the topics of importance in the education of a mathematician. This allowed him a greater sensitivity to the past, and a degree of dynamism to his account. But the genre, and the brevity of the books, imposed its limits on both men, and it is likely that in their ways they are, and will remain, useful books that, for a topic such as this book addresses, are of marginal significance. It might be, for example, that a good short book focussing "on contexts of knowledge production involving individuals, local practices, problems, communities, and networks" could be written that would usefully supplement either of Eves's or Struik's books, but it is less clear that such a book could replace them for their intended audiences.

Kline, and Boyer in his *History of mathematics* (Boyer 1968), gave their readers width: a broad coverage with some detail about what was done and how it was done. Boyer's book drew on the research that had produced his two earlier books, and they will be considered more thoroughly below because of their originality. But the deadening hand of works of reference falls upon both *Mathematical thought from ancient to modern times* (Kline 1972) and Boyer's *History of mathematics*—a certain safeness, of fairness, of addressing a consensus flows beneath the occasional

[4]The first edition was published in 1968, a second edition, with a Foreword by Isaac Asimov, came out in 1989, and a valuable third, extensively revised edition edited by Uta Merzbach was published in 2011. A glance at the last two chapters, which have been added to the book, show very clearly how the deliberate attempt to conform to Boyer's approach results in a picture of 20th century mathematics that is overwhelmingly pure.

sharp opinion or authorial preference. The fact that these books are still in print attests to their usefulness, and it is for a new generation of historians to discover new ways of being as valuable and pertinent.

But the call that this book attempts to answer concerns original research, and of the six authors considered, only Boyer and Coolidge stand out as people who wrote serious, readable books based on a detailed immersion in the primary published sources. Only Boyer and Coolidge provided depth of coverage and found original things to say that could change the received view of the past of mathematics that was current in their day—Boyer in his *History of analytic geometry* (Boyer 1968) and Coolidge in most of his books. I will argue below that Boyer offers a better, and still valuable, model for the historian of mathematics today, but I would like to state a personal opinion: Boyer and Coolidge wrote books that I still consult and I'm glad I've read. For these reasons, the second half of this paper will be confined to an analysis of the work of Boyer and Coolidge alone.

4 Coolidge and Boyer

Julian Lowell Coolidge was born in Brookline, near Boston, on September 28, 1873 and studied mathematics at Harvard and then at Oxford [see Struik (1955) and the introduction to the second edition of Coolidge (1949)]. He became an instructor in mathematics at Harvard, and in 1902 a member of the Faculty. He then took the opportunity to study abroad two years, and took a Ph.D. from Bonn under the direction of Eduard Study (1862–1930). He then travelled to Turin to work under Corrado Segre (1863–1924), and his mathematical work (and as a result, his later historical work) shows the decisive influence of these two men. Study was an expert on line geometry in non-Euclidean spaces, a topic that derived from work on rigid body mechanics and projective geometry in equal measure and in particular from the work of Julius Plücker (1801–1868), Felix Klein (1849–1925), and Ferdinand Lindemann (1852–1939). Segre was an expert in the birational geometry of curves and a formidably well-read scholar. In 1918 Coolidge became a full professor at Harvard, where he stayed until he retired in 1940. He died on 5 March 1954 (Fig. 1).

He was regarded as a good teacher, but judged by the standards of Study and Segre he was not a leading mathematician, and he often seems to have thought that his own field of study had entered into an irreversible decline. Starting in his sixties he wrote three historical books, of which the first was his *A history of geometrical methods* (Coolidge 1940), the second his *A history of the conic sections and quadric surfaces* (Coolidge 1945), and the third his *The mathematics of great amateurs* (Coolidge 1949).

As the opening words of the preface to his *A history of geometrical methods* makes clear, Coolidge saw himself as following Michel Chasles, whose *Aperçu historique* (Chasles 1837) he much admired but could see was by then out of date, and even in its day had lacked a discussion of the contributions of German

Fig. 1 Julian Lowell Coolidge (1873–1954)

geometers. D'Ambrosio et al. (2002, 267) observes that Coolidge did not write with teachers in mind nor for pedagogical reasons, but to see whether any general conclusions can be drawn from the 5000 years of mathematics, and whether any general tendencies stood out (his unremarkable answers were a growing tendency towards generalisation and abstraction). What interested him were the methods mathematicians had used to come to their results, and he arranged his findings, drawn from the excellent resources of the Harvard libraries, chronologically and by topic.

In many ways the book resembles a tourist guide book from the same period. Like such books it is scholarly, accurate, full of information and oddly dry. It is a reliable introduction to the many mathematical topics that it treats, but perfunctory about people and places. There is more than a hint of the school master about some of the judgements, and it was surely not meant to be read from cover to cover. The same is true of his other books, and even on their own terms they fail to give a sense of historical drama because they lack an interesting historical perspective, but, like the books for tourists, they introduce the reader to many fascinating things.

In terms of their intended audiences, the first of Coolidge's books was aimed at an audience of professional mathematicians, and the later two at a broader audience that could include mathematics students. His *A history of geometrical methods* (Coolidge 1940) could only have been written by a well-trained mathematician familiar with a range of demanding sources in several languages, and it offers itself as a summary or work of record.

When we consider Boyer's work from this perspective we find an interesting comparison (Fig. 2).

Among the historians of mathematics considered in this paper, Boyer has the best (and indeed the only) claim to being a full-time, professional historian of mathematics. He studied mathematics at Columbia University, and took his undergraduate and masters degrees there before going on to teach at Brooklyn

Fig. 2 Carl Benjamin Boyer (1906–1976)

College from 1928 until his death in 1976 at the age of 69. In the late 1930s he returned to Columbia and took his PhD there in the history of mathematics in 1939. Kline, in his obituary of Boyer (Kline 1976), dismissed the idea that Boyer might have been influenced by David Eugene Smith, pointing out that Smith had already retired, and credits Boyer with making this move on his own. Moreover, Kline pointed out, there was no course in the history of mathematics at Columbia and no historian on the mathematics faculty. It is clear from other sources that after the second world war Boyer was unique in being heavily involved in promoting the history of mathematics within the Metropolitan New York section of the History of Science Society.

Two of Boyer's books are particularly worth discussing. His *The concepts of the calculus: a critical and historical discussion of the derivative and the integral* is essentially his PhD thesis (Boyer 1939). It was first published in 1939, republished in 1949, and became a Dover publication in 1959, when it changed its title to *The history of the calculus and its conceptual development*. His *History of analytic geometry* began life in 1956 as volumes 6 and 7 in the *Scripta Mathematica Studies* series of Yeshiva University, and strictly speaking only became a book when it was published by Dover in 2004 (Boyer 1956). I shall argue that they convey distinctly different lessons to the historian of mathematics today.

We can learn a lot about them by looking at the key words and phrases they employ. In the introduction to his *The concepts of the calculus* Boyer explained that he will deal with the elementary calculus only. The book will be neither authoritative nor comprehensive, it will offer a development of the basic concepts, and aim at clarity of exposition, not confusingly elaborate all-inclusiveness or meticulously precise erudition. It will reach only as far as Weierstrass and the rigorous epsilon-delta calculus. Some of this is merely prudent PhD speak, whereby original work is overlaid with a sheen of modesty as the apprentice takes his first steps into the world of experts. The truth is that there was no book like this in English, whatever their might have been in German to guide him.

The key words and phrases of Boyer's *Analytic geometry* are subtly different. Here he said he was offering an integrated treatment as opposed to existing accounts of special aspects, albeit one that was limited (for reasons of size) to what appears in elementary general college courses in analytic geometry, and to work done before the 1850s.

It might seem that the first book is the more vigorous and the second one rather more of a teaching text, but in fact the reverse is the case. We can see this in one way by considering how the books look from the perspective of 2014.

The concepts of the calculus (Boyer 1939) is a book of some 330 pages, in which Newton and Leibniz get 43 pages between them. It was written well before Tom Whiteside (1967–1981) got to work, and is now inadequate on Newton, and well before whatever scholarship will be done that will make it inadequate on Leibniz and the calculus—in which respect, see for example Bos (1974) and De Risi (2007). On the other hand, it has generous accounts of pre-calculus ideas in antiquity and the medieval period (80 pages) and 90 more pages (in the chapter entitled 'Anticipation') on work done before Newton and Leibniz. Thereafter only 70 pages take us down to Weierstrass, where the book ends.

Boyer's *Analytic geometry* (Boyer 1956) gives more or less equal weighting to the material in its chapters from antiquity to 1850, thereby offering a considerable amount of material. Page after page covers the work of this or that mathematician in detail, for by the 1950s Boyer had immersed himself much more thoroughly in original sources, in particular the extensive collection of old journals in the New York Public Library.

In both books, the original mathematics has been modestly rewritten or modestly suppressed, so that the book is 'user-friendly' to students and mathematicians. Both books are clear, and in each historical specifics (such as concepts, and ways of writing mathematics) are played down and the similarities with the subject (as it was done in the 1930s–1950s) are played up. Whether this is overdone or inevitable, the result is to make the book readable for a general audience. But in the end —and again this is a personal opinion—*The concepts of the calculus* does not surprise me, but *Analytic geometry* does. The first book sounds familiar and safe throughout, whereas the second book discusses mathematicians whom I am either reassured to see or who were new to me.[5]

It is not just the by-now superseded accounts of Newton and Leibniz that make me feel as I do. I think that Boyer's choice of audience, and the compromises with the sources he made to reach that audience, worked well for his *Analytic geometry* but less well for his *The concepts of the calculus*. This has to do with what can be called the received narrative of the calculus.

[5]Is this perhaps because Boyer's *The concepts of the calculus* defined the field, and my personal opinion is just that a statement about how I learned the subject in the period after the book was written and what I make of topics I do not personally research?

Anyone writing a history of the calculus, and any such author thinking of the audience for such a book, would conclude before lifting their pen that Archimedes should get a mention, that there must have been people in the build-up to the discovery of the calculus by Newton and Leibniz, that what Newton and Leibniz did was not rigorous (and that Bishop Berkeley's criticisms should be mentioned), that therefore it would be necessary to look at how the calculus was rigorised (and here is was 'well-known' that Lagrange failed, Cauchy succeeded, and Weierstrass finished the job).

In this respect, D.E. Smith's *A source book in mathematics* (Smith 1929) is interesting. The entries on the calculus are taken from Cavalieri, Fermat on maxima and minima, Newton on fluxions, Leibniz's 1684 publication of the differential calculus, Berkeley, Cauchy on differentials and derivatives, Euler on differential equations, Bernoulli on the brachistochrone problem, Abel on integral equations, Bessel 'on his functions', and extracts of a different kind from the work of Möbius, Hamilton, and Grassmann.[6] And much though historians of mathematics now wince at his name, E.T. Bell's *Men of mathematics,* first published in 1937, discusses Archimedes, Descartes, Fermat, Pascal, Newton, the Bernoullis, Euler, Lagrange, Laplace, Monge, Gauss, Cauchy, and Weierstrass among others (Bell 1937).

Bell's highly opinionated, no-nonsense take on his chosen men became a publishing success, and his account of them was by no means Boyer's. But insofar as his protagonists enter into the history of the calculus, Bell's selection of their achievements is often close to Boyer's. Where Boyer scores is in the time after the Greeks and before 1600: Bell leapt from Archimedes to Descartes; Boyer discussed medieval writers in the Aristotelian tradition.

When we get to the 17th century, however, the preferences of Smith, Bell, and Boyer in the history of the calculus begin to converge. Indeed, Boyer's begin to seem rather more narrow, as we shall see. It is true that anyone writing a history of the calculus would find it hard to leave any of it out. To this extent, the familiarity of Boyer's material was forced on him by the subject. That said, what Boyer's *The concepts of the calculus* leaves out is telling. The topic of differential equations is almost entirely omitted, and with it many of the reasons why the calculus was done. One consequence is that Euler is much less visible than he should be.

It is true that *The concepts of the calculus* (Boyer 1939) is the *first* history of the subject in English, and quite reasonably Boyer wanted it to be read. He took his audience to be young mathematicians, and his narrative to be what, educationally, was taken to be of most importance to them. He presented the calculus in the form that mathematicians of his time said it was, and not engineers, and this gave him his rationale for selecting material. The calculus was presented as a conceptual enterprise, one with a narrative in which brilliant intuitions and routine techniques struggle to acquire rigorous foundations until success is attained in the 19th century. Because the calculus has been defined as a conceptual exercise, it is not necessary

[6]Extracts that say quite a bit about what was taken to be important in mathematics in 1929, such as integral equations and special functions.

to discuss very many technicalities. This helps with the further aim of the book, the attempt to tell the curious non-mathematician why they should care about the calculus—which is not an unworthy aim.[7]

That said, the received narrative narrows the book (and in so doing makes it possible for Boyer to write the book), but at the cost of inhibiting the dialogue with the sources. Boyer's sources for Newton and Leibniz were poor (much of what Whiteside was able to draw on was inaccessible to someone in New York in the 1930s and Leibniz's unpublished manuscripts were in Germany and little known) and therefore Boyer stuck with published material, as he had to. But Lagrange and Euler were easier to reach, and yet Euler in particular was scantily treated, in keeping with the policy that mathematicians stumbling around with limits fit the narrative, but mathematicians with different attempts to vindicate the calculus do not.

The absence of the uses of the calculus not only led to the marginalisation of the calculus of several variables (partial differentiation is not mentioned), it pandered to the view that the calculus is all about *rigour*, to the exclusion of everything else, and that mathematics is really (very) pure mathematics. But to argue that mathematics is done for its own sake, albeit to very high standards, however much that might have been the opinion of the mathematical community, is false to the history of the calculus, and omits much of the reason the calculus was done at all.[8]

If we turn to Boyer's other major book, his *Analytic geometry* (Boyer 1956), the comparison is clear. First of all, there is no received narrative for analytic geometry. The most that can be said is that Descartes should get a mention, and something should be said about the conic sections (which have a long history). Otherwise, there is no pre-existing consensus among mathematicians as to what the history of the subject should contain. For example, should differential geometry be included? Boyer's reasonable answer was 'no', but he could just as well have answered 'yes'. Or, to pose another question, how much of the projective geometry of algebraic curves should be included?

An interesting comparison is with Coolidge's treatment in his *A history of the conic sections and quadric surfaces* published about a decade before (Coolidge 1945). Coolidge's account is very dry. Coolidge described the work of each author in his (Coolidge's) own terms and not without some judgemental remarks, and arranged these descriptions in themes—often a precise mathematical topic such as foci, singular point, or abridged notation—that he arranged chronologically. Although the linking commentary is often intelligent, the result is an accumulation of results with indications of the accompanying methods and proofs.

[7]How different it would have been had Boyer set out to write a history of solution methods to ordinary differential equations. We still lack a history of differential equations.

[8]One can ask the extent to which this was the opinion in Columbia University or in the big American Universities of the time just before the influx of leading European mathematicians fleeing the Nazis began to change hearts and minds.

Boyer's treatment overlaps Coolidge's extensively, but with much more of a sense of people doing geometry, and of them being led first this way and then that. He was keener to present the authors, if not in their own words then at least in the spirit of their words, and by comparison with his account of the history of the calculus, in his *Analytic geometry* he was free of these defects an agreed narrative could impose. Now he had less of a point to make, and wrote about how *more* mathematics was created rather than *better*. His account up to 1850 was reasonably complete, and his terminus more natural, because after the 1850s the subject became much harder (with the introduction of complex coordinates and the acceptance of higher dimensions). So his stopping point a sensible one, but not a climax to which the shape of the book must conform.

Even before we attempt a comparison of the situation for historians of mathematics in the 1950s and today we can see that Boyer worked in an entirely unstructured environment. Boyer was no Joseph Needham (1900–1995) or Otto Neugebauer, but he came to his chosen field as the only historian of mathematics with a formal training and had to work out what to do.

The model he adopted was the history of ideas model. This is the one that mathematicians most easily recognise, the one people in mathematics education can most easily use, and the one still commonly used today. It's not a bad model. It's not shallow, and it's not pernicious. It's probably unfashionable in history of science, but it is entirely compatible with other models such as ones dealing with social and institutional factors, or journals and networks of correspondence. In particular, it allows for considerations of the uses of mathematics, although Boyer chose not to go there.

His *A history of mathematics* (Boyer 1968) and his *The concepts of the calculus* (Boyer 1939) offered breadth rather than depth, but achieved width: a wide chronological range is covered, with topics and individuals looked at in some depth, and his *Analytic geometry* (Boyer 1956) is rich enough to be considered a book of some depth. That said, it is not narrow. Boyer did not try to write the definitive book on Fermat, or Monge, and outside the field of history of mathematics it may look very narrow, but that would be unfair (it is not a book someone could usefully write today that could be entitled *French analytic geometry 1680–1730*).

The reason for this, and the justification for it, is that Boyer was trying to create the subject of history of mathematics, to give it a place in academic life, and an audience. That task today is confronted by people writing on important subjects that have at best a scattered presence in the literature, not by those who have come after him on the topics he wrote about.

There are risks in such an enterprise, notably the perils of giving the audience what the author believes (or thinks, or knows) that they want. And although *The concepts of the calculus* (Boyer 1939) was his first book and *Analytic geometry* (Boyer 1956) came 17 years later, and the sources for the first book were lacking in ways they were not for the second book, *Analytic geometry* is a better book because it does not presume to know what the audience has to be told.

The concepts of the calculus fails and is an out-of-date book not just because its scholarship on Newton, Leibniz, and even Euler and Lagrange, has been surpassed, but also because its approach to its chosen subject was already tired. It was written by someone who knew in advance what to look for, and who knew directly how much calculus his audience could take. His *Analytic geometry* survives, in contrast, because it was written by someone with only a rough and ready idea of what the subject 'should be'. Its originality and lasting value is that Boyer was surely surprised by the diversity of what he found in the old journals and books, and if someone wrote a history of the same subject today they should have Boyer's book by their side.

The *Calculus* (Boyer 1939) knows what the calculus is, but *Analytic geometry* (Boyer 1956) is not so sure what geometry is, and it is the better book for it. Boyer improved as a historian of mathematics the more independent-minded he became, and that is why of all the authors discussed he remains the best model for historians of mathematics today.

5 The History of Mathematics Today

And what of today? It is one thing to unmask the not-so-hidden agendas of the historians of mathematics of the 1950s and 1960s, and it is always pleasant to think that we have moved on and improved, even that we have left the old ways behind. But there are reasons to think that it is not so simple.

A large branch of the history of mathematics today is archival. The eight volumes of Whiteside's monumental *Mathematical papers of Isaac Newton* were largely based on new evidence that changed the story we had of Newton's creation of the calculus (Whiteside 1981). Sometimes, and Neugebauer's work is a case in point, archival work may even start the story. These are examples of historians bringing forward wholly new documents, editing them well, and giving them a context. Another branch of historical work is more a matter of interpretation. Little new information may be presented, but a large amount of already available information is assembled in a new way. Mostly, historical work is a bit of both. The journals in the history of mathematics that began in the 1960s and 70s, *Historia Mathematica* and *Archive for history of exact sciences* typically publish a mixture of new texts and new interpretation.

The situation in history of science seems to be different. History of science journals carry far fewer articles on Western science in the modern period that fall under the heading of "Here is a forgotten paper by X" than do journals in the history of mathematics, and more that could be entitled "Here is an argument to change our views about Y". There is more attention to how what was once called 'internal' history of science fits into existing historical debates, and these are often of an outward-looking kind that form parts of much larger, more general-historical, topics such as popularisation, the legitimation of science, imperialism, and so on. This is surely because there is a recognised discipline in the history of science, with training, a professional structure from undergraduate work to professorships, and the result of the creation of a near monopoly of production by those trained in the field. Standards are taught in courses and maintained both by the publication side of work and the hiring and promotion side, and the impression one has is that if books and articles are read by people who are not historians of science then they are historians more generally. There is even a sense that history of science may have gone too far towards history: Hasok Chang's BSHM Presidential address to the International Congress in the History of Science Technology and Medicine was entitled 'Putting Science Back in History of Science'.[9]

Matters are still different in the history of mathematics. There is a number of beginners, people without years of training and who are working out for themselves what history of mathematics means to them. This includes a stream of professional mathematicians who are drawn into the history of their subject, a number of whom do well and whose work is welcomed and used by others. Standards are maintained by the journals, and to some extent by books, but hirings and promotions are conducted within mathematics departments, and this is reasonable when the work of historians of mathematics is also read by mathematicians. For better or worse, there are no strong 'schools' in the history of mathematics, no groups pushing historians towards or away from comparable groups or larger ones.[10]

We should consider the reasons for this disparity. One lies in the subject of mathematics itself. George Sarton, who is nowadays a rather neglected figure, called the history of mathematics a secret within a secret because its important ideas were intelligible only to those who had already studied them.[11] Whatever that says about the appetite for the relatively elementary mathematics of the Greeks, it cannot be disputed as a remark about the history of the mathematics of the 19th and 20th centuries. The decades since 1970 have seen an unparalleled profusion of attempts to popularise mathematics, and several national mathematical societies have explicitly recognised the need, as well as their social duty, to attempt to reach the general public. Some of the work of historians of mathematics, indeed, fits into that endeavour. But it is an endlessly difficult task, and the reasons are worth analysing.

It may be supposed that it is as difficult to do top-level physics as it is to do top-level mathematics, but it cannot be doubted that it is easier to explain to a general audience the outline of the work of a Nobel prize-winner in physics than that of a Fields Medal winner in mathematics. One reason is that there is a much stronger prior interest in what physicists do than in what mathematicians do: interesting (if baffling) new physics is more interesting than just plain baffling new

[9]Hasok Chang, BSHS Presidential Address, delivered at the International Congress of the History of Science, Technology and Medicine, 22 July 2013, Manchester, U.K.

[10]I speak only of the Anglo-Saxon world, and of the relatively weak relations with, for example, mathematics education, or philosophy, or even history of science itself.

[11]His explanation was: "For the growth of mathematics is unknown not only to the general public, but even to scientific workers" (Sarton 1936, 8f). Sarton is one of the founding fathers of the modern discipline of then history of science, and he wrote, somewhat gingerly, about the history of mathematics as part of his work. But, although he had a secure position at Harvard between the wars it seems agreed that he did not shape the discipline that largely grew up despite him.

mathematics. Another is that there are social implications in physics that mathematics lacks: nuclear power, space research, even the sheer cost and huge scale of modern particle physics. A third is that it seems easier to explain in vague terms what a physicist is trying to do than a mathematician—the technicalities can be held at bay for just significantly longer.[12]

Therefore, to write the history of modern physics offers certain advantages. In particular, the rise of big science—expensive science tied to industrial and military needs, with a large and growing influence on education—can rightly and profitably be addressed in ways that the rise of mathematics cannot be. Attention to the consequences of science produces interesting and important work even if the science is barely described, be it the advent of the computer, a new drug, or the aeroplane.

There are exceptions to the rule, and they are interesting. Turing's work on the foundations of mathematics and then at Bletchley Park, and the Turing test, together form one; Gödel's theorems, remarkably enough, are another. There was a lot of excitement in the popular press with the proof of Fermat's last theorem, as there was with the resolution of the Poincaré conjecture. Personalities played a part, but then they do in the popular interest in Einstein. But the truth is that these examples do not alter the basic picture. On the one hand, if historians of mathematics go beyond a certain level of mathematical difficulty their work will only be read by people who have studied mathematics to that level and mostly likely are in some way still actively involved in mathematics; there are very few major topics in mathematics that have any appeal to any but specialists. Histories of mathematics that go in these directions will only be read, and most likely can only be written, by mathematicians and historians of mathematics. On the other hand, there are relatively few topics in pure mathematics, and not many more in applied mathematics, that will appeal to any but specialists either.

There is a sense in which this is a prison made by the historians of mathematics. Consider, for example, a history of the aeroplane. From the first investigations of why planes fell out of the sky to the design of the Airbus this is a story with a strong mathematical aspect, and the early years of it are now being written about. But one has the strong feeling that if the partial differential equations aspect of the story declined and the aspect of flying, or the first world war, took over then historians of mathematics might feel that the resulting account was no longer part of their subject. This is part of the vexed relationship between history of mathematics and history of science that has already been commented upon.

It seems likely, therefore, that the issue of place—the professional place of the historian of mathematics—will remain as it has been for many years, with historians of mathematics mostly in mathematics departments, because that is where the core of the audience is.

[12]The comparison with modern biology is, of course, a hands-down victory for biology, as it always will be for a science with a handle on human health and even on what it may be to be human.

6 Reflections

Can the history of mathematics move away from this core, and, indeed, should it? In what directions should it move, and towards which professional groups and which audiences? This short essay is confined to the history of modern mathematics, and histories written in English in the 1950s and 1960s, but we can at least notice that these questions have long since been confronted by historians of other kinds of mathematics: the mathematics of the ancient Near East, of the Arabic and Islamic world, of India, and of China, for example. However, it is not so comforting to examine the situation facing historians of the mathematics of the 16th, 17th, and 18th centuries, who may or may not feel comfortable with their colleagues in history of science departments.

It may be valuable to recall some more remarks by George Sarton from his (Sarton 1936). Very early on he remarked: "the history of mathematics should really be the kernel of the history of culture. Take the mathematical developments out of the history of science, and you suppress the skeleton which supported and kept together all the rest" (Sarton 1936, 4). Later, after a discussion of the inevitable compromises any historian must make with the complexities of the past, and of the more immediate benefits that a history of mathematics might offer mathematicians that would not, however, satisfy the philosopher and the humanist (Sarton 1936, 21), he claimed that the main reason for studying the history of mathematics is purely humanistic: we are interested in other people and, even if they are geniuses (his term), in how they saw themselves and how they were affected by their circumstances (Sarton 1936, 23). On his view, the main factor in the development of mathematics, much more than external circumstances, was "the availability of creative genius, which cannot be controlled". And the genius, it would seem, can see into the future rather better than the rest of us, but nonetheless also imperfectly (Sarton 1936, 17).

Words like 'genius' and 'humanist' need careful handling these days, but there is much in Sarton's remarks from another era for us to measure ourselves against: the intimate relationship between mathematics and science, the balance of broadly external and internal factors, the call to "full and honest biographies" (Sarton 1936, 23). If it is true that recent history of modern mathematics has focussed on "major internal, universal developments of certain sub-disciplines of mathematics" then we have already slipped away from answering Sarton's call.

The leaderless move towards "a focus on contexts of knowledge production involving individuals, local practices, problems, communities, and networks" is one attempt to reframe the debate and allow the history of mathematics to diversify productively. The language is closer to that of history of science, and would certainly accommodate a shift away from the history of mathematics, where 'mathematics' is interpreted as the sole kind of knowledge being produced, to the production of various kinds of inter-related knowledge, some of which was in mathematics. To go in this direction, we have been invited "to reflect upon our own professional objectives as well as on the methods and tools we employ today".

It is always valuable to reflect on priorities from time to time. What was bold and innovative can become solid work, then predictable and finally stale. Faced with a call for new directions, nothing is easier than to counsel caution and the virtues of the familiar. It would be churlish to respond to a call for the study of local practices by observing that Chang indicated a widespread agreement that history of science filled up with "an aimless profusion of microstudies, in the aftermath of the demise of grand narratives".

The thrust of the call, as I see it, is to find some way of engaging again with the demon at the heart of mathematics: its claim to offer timeless truths. This claim is at the heart of the belief that mathematics is valuable, and it the deeply held faith of many mathematicians. A history of mathematics that was only account of mathematicians stumbling towards the light (one might almost say the 'Light' because there is no dispute that it is the true light) would be a tiresome thing, perhaps a good source of cautionary tales for research mathematicians but little more.

One way to confront the demon would, of course, be to deny that mathematics does consist of timeless truths—or even, when it comes to modern mathematics, timeless valid deductions from axioms. But this doesn't seem likely to succeed; we have excellent reasons for regarding many of the deductions in mathematical arguments as valid, whatever one makes of the premisses.

A better way is to take more seriously the struggling mathematicians. There is a lot of steady progress in mathematics, good work that is folded into the story although its author is forgotten. But there are numerous occasions when much more was required: much if not all of what is called modern mathematics is different in significant ways from what came before it in the first half of the 19th century. And when we look at these events, we see a wide range of mathematical practices in the modern period. There are mathematicians intimately concerned with the mathematics of contemporary physics, be it special and general relativity or quantum mechanics.[13] At the same time there are the mathematicians around Emmy Noether (1882–1935) and, only a little later, Bourbaki. There is classical real and complex analysis, largely unaffected by either of these developments, and one could go on. What one often sees is that these mathematical topics have geographical centres, and conversely that many geographical centres make particular choices of mathematical subjects. The extreme case of Poland, where between the wars mathematicians concentrated on real analysis, point-set topology, and logic is often mentioned, but the Soviet Union was different from Germany even before the vicious politics of the 20th century took its toll.

Moreover, there is a sense in which many mathematical communities were small and therefore had to concentrate their efforts of a limited number of fields. The historian therefore seeks to know why those priorities were chosen, and not others, and why they were pursued in these ways and not those.[14]

---

[13]See Schneider (2011) for an account of van der Waerden's involvement in quantum mechanics. The literature on relativity theory is happily too large to be indicated here.

[14]Even today each mathematician and every mathematics department must make such choices; there is every reason to suppose that focussing the history of mathematics on such issues will interest professional mathematicians.

Equally, there is a sense in which many mathematical communities were large. Soviet priorities and Nazi purges were mass events, whole mathematical communities were mobilised, moved, even wiped out. Lives of surviving mathematicians were transformed, mathematical fields invigorated or left to wither. The histories of modern mathematics in America, Japan, and China are all very different, and cannot properly be told by concentrating on the elite mathematicians, however important they were.

Unexpected developments complicate and enrich any history. The landscape of mathematics in 2014, by comparison with fifty years before, sees a great deal of combinatorial mathematics, statistical mechanics, and probability. The wider landscape sees the explosion of computing and communications in which mathematics has played a major, but still largely hidden, role. Contexts, priorities, even the criteria for success vary from the professorial seminar to industrial mathematics, from applied mathematics to algebraic geometry. Fields such as mathematical logic rise and, it would seem, decline in importance. There are extensive highly technical areas involving a lot of mathematics where rigour is perforce absent (chemistry beyond the few solvable cases of Schrödinger's equation, computer simulations to study cosmology or, today, global warming) and other criteria of adequacy and theoretical grasp have been developed.

All these issues direct us back to mathematical practice. What mathematicians do, how and why they do it, what counts as valid as well as what is valuable, what makes for understanding, the constraints and stimuluses they perceive as well as those they do not are a better focus for the history of mathematics than the 'timeless truths' that mathematicians discover.

But I would like to conclude these reflections with a remark about "the often hidden agendas of former historians of mathematics". Although more could be said about that agenda (how elitist it was, how male-oriented, how Western-focussed, and so on) the risk with such analyses is that we too easily persuade ourselves that we are somehow better.

An agenda is a plan or list of things that must be done. A hidden agenda may simply be undisclosed (the author could have stated it but did not) or one of which the author claims to be unaware, or indeed is unaware of, and conscious decisions may rest on hidden agendas. Hidden agendas are usually supposed to have had a distorting, if not outright pernicious influence on the work. If we are to look for our own hidden agendas, we must collectively try to distinguish between these various cases, bearing in mind that an agenda is revealed by what is there and what is not, but a hidden agenda is not necessarily revealed by a gap. Any agenda, explicit or hidden, puts audiences in the position of having to respond: do they want to add to the agenda, to reject it, or to do something in between?

Boyer stopped short of the 20th century, surely from a sense of his abilities and of what he thought he could usefully do, not from some hidden view about mathematics. Kline went much further, and plainly struggled with the pure mathematics of the late 19th and early 20th centuries. But was this from an agenda? Perhaps, indirectly, one he was quite explicit about elsewhere: he liked applied mathematics and thought that too much of contemporary pure mathematics was valueless and having a harmful effect on the subject as a whole.

Boyer, Coolidge, and Kline excluded statistics from their work. I am not aware of an explicit statement from any of them as to why, but it is easy to see that they were responding to a consensus that they shared with their audience, one that they could easily have articulated. Put simply and for brevity, this consensus said that statistics was not part of mathematics, although it overlapped somewhat with it, because it lacked the high conceptual aspect of core mathematics, and because it was (accordingly) taught mostly outside mathematics departments. If you do not share that view (which was already difficult to hold in the 1960s) then a possible response would be to write a history of statistics in the same spirit as Boyer, Coolidge, and Kline and set it along side their work, thus filling a gap as it were. They had a hidden agenda, if you like, but not one to be confronted so much as outflanked.

All three historians concentrated on mathematicians who lived and worked in ancient Greece or Western Europe, with little more than a nod to mathematics in Mesopotamia, or in the Islamic world. This may largely have been a matter of sources, but it did also reflect an agenda that the ancient Greeks and modern Western Europeans had, as it were, ‘got it right’ while those elsewhere could only make modest contributions. The result, however much it might rightly annoy someone in the Islamic tradition today, is a distortion that says more about the authors’ idea of what mathematics is, I believe, than about some belief in Western supremacy.[15]

We can, and should, read earlier historians with an eye to what they are not saying as well as what they do say, and with attention to their reasons, disclosed or hidden, for making the choices they did. But there is surely a hope that we can conduct a more closely examined practice today.

If we “reflect upon our own professional objectives as well as on the methods and tools we employ today” we might well disclose our undisclosed agendas. We might admit to sharing some hidden agendas (in my own case, to the high value I attach to elite pure mathematics). But we would presumably remain unaware of the hidden agendas that shape us unconsciously.

The hope in such cases is that our methodologies will compel us to examine our hidden beliefs. A commitment to establishing, say, an exhaustive communications network, or an explicitly defensible corpus of texts, might compel us to examine sources we might otherwise have ignored. I would hope that a wider view of what mathematics is, attention to actual mathematical practice, and less certainty before

[15]This is an opinion, but one can instructively compare their work with remarks by earlier historians and mathematicians.

we set off on a large piece of research that we know what to find, will produce histories of mathematics that are exciting and interesting to mathematicians of all kinds, to historians of mathematics and, who knows, even to historians and beyond them to the public at large.

References

Bell, E. T. (1937). *Men of mathematics*. New York, Simon and Schuster. Reprint 1965.

Bos, H. J. M. (1974). Differentials, higher-order differentials and the derivative in the Leibnizian calculus. *Archive for History of Exact Sciences, 14*, 1–90.

Boyer, C. B. (1939). *The concepts of the calculus: A critical and historical discussion of the derivative and the integral*. New York: Columbia University Press. Repr. 1949, and, as *The history of the calculus and its conceptual development*, Dover 1959.

Boyer, C. B. (1956). *History of analytic geometry, volume 6, 7 of Scripta Mathematica Studies*. New York: Scripta Mathematica. Repr. 2004, Dover. Yeshiva University.

Boyer, C. B. (1968). *A history of mathematics* (3rd ed.). New York: Wiley.

Chasles, M. (1837). *Aperçu historique sur l'origine et le développement des méthodes en géométrie... suivi d'un Mémoire de géométrie, etc.* Hayes, Bruxelles.

Coolidge, J. L. (1940). *A history of geometrical methods*. Oxford: Oxford University Press, Dover reprint, New York 1963.

Coolidge, J. L. (1945). *A history of the conic sections and quadric surfaces*. Oxford: Oxford University Press, Dover reprint, New York 1968.

Coolidge, J. L. (1949). *The mathematics of great amateurs* (2nd ed.). Oxford: Oxford University Press. Oxford: Clarendon Press, 1990; Dover reprint of 1st ed., Oxford 1963.

D'Ambrosio, U., Garciadiego, A. R., Dauben, J. W., & Fraser, C. G. (2002). The Americas. In *Writing the history of mathematics: Its historical development, volume 27 of historical works, historical studies* (pp. 249–287), Birkhäuser: Basel, Boston, Berlin.

Dauben, J. W., & Scriba, C. J. (Eds.). (2002). *Writing the history of mathematics: Its historical development, volume 27 of historical works, historical studies*. Birkhäuser: Basel, Boston, Berlin.

De Risi, V. (2007). *Geometry and monadology: Leibniz's analysis situs and philosophy of space*. Basel: Birkhäuser.

Eves, H. W. (1964). *An introduction to the history of mathematics*. New York: Holt, Rinehart and Winston. 1966. (6th ed. 1990).

Grattan-Guinness, I. (2002). The British Isles. In *Writing the history of mathematics: Its historical development, volume 27 of historical works, historical studies* (pp. 161–178). Birkhäuser: Basel, Boston, Berlin.

Kline, M. (1953). *Mathematics in western culture*. Oxford: Oxford University Press

Kline, M. (1972). *Mathematical thought from ancient to modern times*. Oxford: Oxford University Press.

Kline, M. (1976). Carl B. Boyer—In Memoriam. *Historia mathematica, 3*, 387–394

Neugebauer, O. (1951). *The exact sciences in antiquity*. Munksgaard, Copenhagen (2nd ed.). Providence, R.I.: Brown University Press, 1957, Dover, New York 1969.

Sarton, G. (1936). *The study of the history of mathematics, and the study of the history of science*. Cambridge, Mass: Harvard U.P. New York: Dover, 1957.

Schneider, M. R. (2011). *Zwischen zwei Disziplinen: B.L. van der Waerden und die Entwicklung der Quantenmechanik*. Berlin: Springer.

Smith, D. E. (1929). *A source book in mathematics*. New York: McGraw-Hill. New York: Dover, 1959.

Struik, D. (1948a). *A concise history of mathematics*. New York: Dover. (2nd ed., 1967).

Struik, D. (1948b). *Yankee science in the making*. Brown, Boston: Little.

Struik, D. (1955). Julian Lowell Coolidge, in memoriam. *American Mathematical Monthly, 62*, 669–682.

Whiteside, D. T. (Ed.). (1967–1981). *The mathematical papers of Isaac Newton*. Cambridge: Cambridge U.P.

Whittaker, E. (1917). *A treatise on the analytical dynamics of particles and rigid bodies* (2nd ed.). Cambridge: Cambridge U.P.

Whittaker, E. T., Watson, G. N. (1915). *A course on modern analysis*. Cambridge: Cambridge U.P. (2nd, completely revised edition).

Wilder, R. (1969). *Evolution of mathematical concepts. An elementary study*. New York: Wiley. Open University Press, Milton Keynes 1978.

Polycephalic Euclid? Collective Practices in Bourbaki's History of Mathematics

Anne-Sandrine Paumier and David Aubin

Abstract

In this paper, we argue that Bourbaki's historiography, which has been extremely influential among mathematicians and historians of mathematics alike, reflected the special conditions of its elaboration. More specifically, we investigate the way in which the collective writing practices of the members of the Bourbaki group in both mathematics and the history of mathematics help to explain the particular form taken by the *Elements of the History of Mathematics* (1960). At first sight, this book, which has been seen as an "internalist history of concepts," may seem an unlikely candidate for exhibiting collective aspects of mathematical practice. As we show, historical considerations indeed stood low on the group's agenda, but they nevertheless were crucial in the conception of some parts of the mathematical treatise. We moreover claim that tensions between individuals and notions related to a collective understanding of mathematics, such as "*Zeitgeist*" and "mathematical schools," in fact structured Bourbaki's historiography.

Keywords

Bourbaki · Mathematical schools · Collective practices of mathematicians · Historiography of mathematics

D. Aubin (✉)
Institut de mathématiques de Jussieu-Paris rive Gauche, Sorbonne Universités, Université Pierre et Marie Curie (CNRS, Sorbonne-Paris-Cité, Sorbonne universités), 4 place Jussieu 75252, Paris Cedex 05, France
e-mail: david.aubin@upmc.fr

A.-S. Paumier
Fondation Mathématique Jacques Hadamard, Institut Des Hautes Études Scientifiques, 35 Route de Chartres, 91440 Bures-sur-Yvette, France
e-mail: anne-sandrine.paumier@ens-lyon.org

V.R. Remmert et al. (eds.), *Historiography of Mathematics in the 19th and 20th Centuries*, Trends in the History of Science,
DOI 10.1007/978-3-319-39649-1_10

1 Introduction

Incessant chronological difficulties, which arise when we suppose the physical existence of a single Euclid, lessen, without vanishing, if we accept to take his name as the collective title of a mathematical school.[1]

That the French historian of mathematics Jean Itard (1902–1979) retained the idea of Euclid as a collective mathematician, and more precisely as "the collective title of *mathematical school*," was anything but innocent. In the early 1960s, the "polycephalic" author Nicolas Bourbaki and the members of this group of mathematicians were dominating the mathematical scene in France and beyond. Many, at the time, saw in Bourbaki the latest incarnation of what constituted a true "mathematical school" (or in French *une école mathématique*), wholly original in nature, with clear methods, objectives, and ideology.[2]

Only in the context of the Bourbakist experience might the idea of a collective Euclid have had some sort of appeal.[3] As explained by the historian Fabio Acerbi, the lack of historical and biographical data about Euclid had produced yet another fantasy about the author of the *Elements*. A polycephalic Euclid indeed seemed highly implausible to scholars who were well aware of the staunch claims for authorship that characterized Greek literature, and mathematics especially (Acerbi 2010).[4] That Bourbaki saw itself as a new Euclid is rather obvious.[5] We would like to conjecture that such possibility was entertained only as a consequence of changing mathematical practices at the time of Bourbaki and the rise of collective undertakings. More generally, as was stressed recently by one of us (Paumier 2014), collective practices came to characterize more and more the mathematical life of the period.

[1]"Les difficultés qui surgissent à chaque instant dans la chronologie lorsque l'on admet l'existence physique d'un seul Euclide s'atténuent sans disparaître lorsque l'on accepte de prendre son nom comme le titre collectif d'une école mathématique." (Itard 1962, 11). Unless when taken from a published translation, all translations are our own.

[2]For historical accounts of Bourbaki's domination in the 1960s, see Corry (1996) and Aubin (1997). It was in a paragraph referring to the "école Bourbaki" that André Delachet first revealed to the world that Bourbaki was a "polycephalic" mathematician (his term) (Delachet 1949, 113–116).

[3]Note that following conventions established by Liliane Beaulieu, we will refer to members of the group as *Bourbakis* and their followers as *Bourbakists* (Beaulieu 1989).

[4]For a recent examination of the polycephalic Euclid hypothesis, see the corresponding article in the MacTutor History of Mathematics archive (http://www-history.mcs.st-andrews.ac.uk/Biographies/Euclid.html, last viewed on March 16, 2013). Three possibilities are listed for Euclid: individual author, team leader, or mere pen name ("The 'complete works of Euclid' were written by a team of mathematicians at Alexandria who took the name Euclid from the historical character Euclid of Megara who had lived about 100 years earlier").

[5]This is made especially clear in Beaulieu's analysis of a poem written by some member of the Bourbaki collective (Beaulieu 1998, 112).

Now, Bourbaki was not only a mathematician, but also a highly influential historian of mathematics. In 1960, the "historical notes" appended to each volume of Bourbaki's monumental treatise were gathered in a single volume titled the *Elements of the History of Mathematics*. Interestingly, therefore, this volume on the history of mathematics was the (collective) product of mathematicians who themselves felt that they were experimenting with original ways of (collectively) writing a mathematical treatise, while at the same time developing a new (collective) understanding of mathematics as a whole. Were these collective writing practices at all reflected in Bourbaki's historiography? Although written collectively, Bourbaki's *Elements of the History of Mathematics* in fact appear as an unlikely candidate for exhibiting collective aspects in the practice of mathematics. According to Jeanne Peiffer (Dauben and Scriba 2002, 40), it is an "internalist history of concepts" which has only little to say about the way in which mathematics emerged from the interaction of groups of people in specific circumstances.

In the following, we will try to unpack this intricate relation among various ways of seeing and not seeing the importance of collective practices in the history of mathematics. We first introduce some aspects of the collective life of mathematics during the twentieth century, and then focus on the practice and content of the history of mathematics written by the Bourbakis. Although the Bourbakis have individually written many texts concerning the history of mathematics, we will focus on the collective volume of the *Elements of the history of mathematics*.[6] We then examine the concrete practices involved in the writing of Bourbaki's historical notes, as far as it is possible to determine them. We investigate the place of collective practices in Bourbaki's historiography and examine in particular the role played by the notion of "mathematical school."

2 Collective Practices in 20th-Century Mathematics

> The form of collaboration we have adopted is new; we did not limit ourselves to chop the topic into various pieces and to distribute among ourselves the writing of the diverse parts; on the contrary, after having been discussed and prepared at length, each chapter is assigned to one of us; the text thus obtained is seen by all, it is once gain discussed in details, and it is

[6]Jean Dieudonné and André Weil are the most prolific authors in that field. Let us just mention here that Dieudonné has written histories of functional analysis (Dieudonné 1981), algebraic and differential topology (Dieudonné 1989) and directed a collective volume (Dieudonné 1978), and that Weil has written a history of number theory (Weil 1984) and expressed his ideas about the way history of mathematics should be practiced and written (Weil 1980), see the contribution by Schneider in this volume. About the way Weil used historical mathematical texts, see Goldstein (2010).

always reviewed at another time, and sometimes more. We are thus embarked on a truly collective enterprise which will have a deeply unified character.[7]

In this letter, which has become famous, Szolem Mandelbrojt (1899–1983) described Bourbakis' working rituals to several officials, in order to obtain funding for their congresses. There is little need now to insist on the originality of the enterprise.[8] The group of mathematicians who would adopt N. Bourbaki as a pen name met for the first time in Paris, in December 1934. At first, they sought to write a treatise of analysis that would serve as a university textbook for the next ten years. In the first months of its existence, the Bourbaki group is therefore referred to as the "Committee for a treatise of analysis" [*comité du traité d'analyse*]. As is also well known, the scope of the project widened considerably and the treatise, which started to appear under the title *Elements of mathematics* in 1939–1940, was announced to cover vast areas of mathematics. For mathematicians, Bourbaki soon came to represent a highly recognizable mathematical style, systematically relying on the axiomatic method to introduce mathematics in the most rigorous, abstract manner. The members of the Committee slightly varied from one meeting to the next, but more or less stabilized at Besse-en-Chandesse, in July 1935, at the first of what would become a series of annual, or semiannual, Congresses. Over the years, newsletters, meeting reports, and drafts of various parts of the chapters circulated among members and are now accessible online, thanks to the work of Liliane Beaulieu and her collaborators.

Although Bourbaki's experience may well have been unique, the period we will study here, roughly from 1934 to 1960, is characterized by great changes in mathematicians' working practices. This corresponds to a change of scale, which in other fields has been captured by the expression "Big Science." To quote from a book about the phenomenon, "Seen from the inside—from the scientists' perspective—big science entails a change in the very nature of a life in science" (Galison 1992, 1). In mathematics, the change of scale was translated into an anxiety about keeping abreast with the rapid development of a field where the number of practitioners skyrocketed. Bourbaki expressed this in a famous article on "The Architecture of Mathematics," first published in 1948:

---

[7]"La formule de collaboration que nous avons adoptée est nouvelle; nous ne nous sommes pas bornés à partager le sujet en tranches et à nous distribuer la rédaction de ces diverses parties; au contraire, chaque chapitre après avoir été longuement discuté et préparé, est confié à l'un d'entre nous; la rédaction ainsi obtenue est vue par tous, elle est à nouveau discutée en détails, elle est toujours reprise au moins une fois, et quelques-fois plusieurs. Nous poursuivons ainsi une oeuvre véritablement collective, qui présentera un profond caractère d'unité." Mandelbrojt to Mme Mineur, Borel, and Perrin (1936); repr. in "Journal de Bourbaki No. 6; 27/11/1936," Archives Bourbaki, http://purl.oclc.org/net/archives-bourbaki/70. All were last accessed on September 27, 2016.

[8]Historical literature on Bourbaki is rather extensive. We refer especially to (Beaulieu 1989). In English, one is referred to, among others, (Beaulieu 1993, 1994, 1996; Aubin 1997, Mashaal 2006).

> The memoirs in pure mathematics published in the world during a normal year cover several thousands of pages. […] No mathematician, even were he to devote all his time to the task, would be able to follow all the details of this development. [Bourbaki, 1950, 221]

This was not a mere marketing ploy to sell Bourbaki's treatise as an efficient solution to the fear of disunity in mathematics. As we shall see later, André Weil (1906–1998) expressed the same idea in his private correspondence. Some early drafts of Bourbaki's treatise also put this in a blunter way:

> There is a single *mathématique*, [that is] one and indivisible: this is the *raison d'être* of the present treatise, which aims at introducing its elements in the light of a 25-century-old tradition.[9]

But, of course, Bourbaki was just one response to this anxiety. Another was the significant change in mathematicians' collective working practices. In her recent Ph.D. dissertation, Paumier has focused on the emergence, diffusion, and growing importance of various forms of collective organization such as the seminar, the specialized international conference, and the research center.

In their formative years at the École normale supérieure in Paris, the first generation of Bourbakis attended Jacques Hadamard's (1865–1963) seminar at the Collège de France. Weil underscored how the seminar played a crucial role in his own training: "the *bibli* [library] and Hadamard's seminar […] are what made a mathematician out of me" (Weil 1992, 40). At the time, the seminar represented a novelty for mathematicians in Paris.[10] In 1933, a small group among the future Bourbakis clearly understood the benefits to be reaped from this form of organization and set up a seminar under the moral authority of Gaston Julia (1893–1978) who was a full professor at the Sorbonne. To quote Beaulieu, Julia's seminar was "the laboratory of a restricted team" of mathematicians working together on definite topics (Beaulieu 1989, 133–137). This kind of collective organization for mathematical research would quickly explode: in the 1960s, more than 30 seminars met regularly in Paris and its surroundings. As a group or individually, the Bourbakis played a significant part in this development. Launched in 1945–1946, the Bourbaki Seminar quickly became a major social event for mathematicians in France and abroad. Soon, Paul Dubreil (1904–1994), Henri Cartan (1904–2008), and Laurent Schwartz (1915–2002), among others, also organized their own seminars.

Another form of collective organization that spread after WWII was the international conference that focused on a research subfield. Looking for an efficient way to help the reconstruction of French mathematics, the Rockefeller Foundation gave money to the CNRS to sponsor such meetings. The series of the "*colloques internationaux du CNRS*" were supposed to be "small and informal" so as to foster effective collaboration. In Warren Weaver's own term, "the attendance of mature contributors [was] restricted to say 15; with provision, however for additional

[9]"Il y a une mathématique, une et indivisible: voilà la raison d'être du présent traité, qui prétend en exposer les éléments à la lumière de ving-cinq siècles." "Introduction au Livre I (état 3) [ou état 2]" (n.d.), p. 1. Archives Bourbaki, http://purl.oclc.org/net/archives-bourbaki/563.

[10]On Hadamard's seminar, see [Beaulieu, 1989, 60–65] and [Chabert and Gilain, ming].

listening and observing audience of young men" (Zallen 1989, 6). They were to include two to five non-French speakers. In the field of mathematics, the Bourbakis were again greatly involved in those conferences. The conference on Harmonic Analysis held in Nancy in 1947 for instance involved Schwartz, Mandelbrojt, and Roger Godement (1921–2016). Cartan, Charles Ehresmann (1905–1979), and Jean Leray (1906–1998) took part in a conference on Algebraic Topology in Paris, also in 1947. Dubreil was one of the organizers of a conference on Algebra and Number Theory in 1949, at which Weil and Jean Dieudonné (1906–1992) spoke. About the Nancy conference, a report stated:

> Beyond scientific results [that were] improved, clarified, or established, beyond the long-lasting personal contact that will result from it, this scientific event has shown that it was possible for a small number of qualified people to work very fruitfully on a well circumscribed topic. The material format tried out on this occasion proved as useful as it was pleasant.[11]

In their mathematical work, members of the Bourbaki group very consciously explored new forms of collective organization. Before we try to assess whether this had an effect on the content of their historiography, let us examine the collective practices that went into the elaboration of Bourbaki's historical writings.

3 Bourbaki's Historical Notes as Collective Work

As far as editorial fortune goes in the history of mathematics, the *Eléments d'histoire des mathématiques* by N. Bourbaki was immensely successful: first published by Hermann in 1960 and reprinted in 1964, a second and a third editions were published, respectively, in 1969 and 1974 with corrections and additions; it was then reissued by Masson in 1984 and later reprinted by Springer as recently as 2007.[12] This book was also translated into Italian and Russian (1963), in German (1971), Spanish (1972), and Polish (1980). The English translation, the rather poorly rendered *Elements of the History of Mathematics*, only came out later (Bourbaki 1994b). This editorial success is all the more surprising when one realizes that most of these "elements" had been written much earlier, sometimes in the early 1940s, if not before. Indeed, nothing was original in the volume when it first appeared in 1960. It merely gathered most, but not all, of the "historical notes,"

[11]"Outre, les résultats scientifiques améliorés, éclaircis ou établis, outre, les contacts personnels durables qui en résulteront, cette manifestation scientifique a montré qu'il était possible de travailler très utilement sur un sujet bien délicité entre un petit nombre de personnes qualifiées. La forme matérielle in augurée en cette occasion s'est montrée aussi utile qu'agréable et il y a lieu d'insister sur l'honneur qui rejaillit sur l'Université de Nancy, du fait qu'elle ait été choisie comme théâtre de la première réunion de cette nature." Archives départementales de Meurthe-et-Moselle, Nancy, W 1018/96, Rapport sur le colloque.

[12]The editions and reviews are listed in Beaulieu (1989), Annexe II B, Annexe III B] from 1960 to 1986. Some of the reviews will be quoted here.

which had been published earlier in the corresponding volumes of Bourbaki's *Eléments de mathématique*.

3.1 A Choice Made Early

The first historical notes appeared in 1940, appended to a booklet consisting of Chaps. 1 and 2 of the *Elements of mathematics*, Book III ("General Topology"). This in fact was the very first booklet that truly belonged to the treatise, that is, apart from the digest of mathematical results on set theory [*fascicule des résultats*], published a year earlier.[13] In the following volumes, historical notes were likewise appended to most chapters, although sometimes historical elements concerning several successive chapters were gathered in a single note. In their original form, historical notes had no titles. Chapter headings found in the *Eléments d'histoire des mathématiques* were given in 1960. These notes greatly varied in length (see Table 1).

Table 1 Table of Contents of the *Éléments d'histoire des mathématiques* (Bourbaki 1994a) with number of pages, book of the *Éléments de mathématique* and year of first publication

	Chapter title	Pages	Book	First publ.
1.	Foundations of mathematics; logic; set theory	44	I	1957
2.	Notations, combinatorial analysis	2	?	?
3.	The evolution of algebra	10	II	1942
4.	Linear algebra and multilinear algebra	12	II	1947–1948
5.	Polynomials and commutative fields	16	II	1950
6.	Divisibility; ordered fields	8	II	1952
7.	Commutative algebra. Algebraic number theory	24	VIII	1965
8.	Non commutative algebra	8	II	1958
9.	Quadratic forms; elementary geometry	14	II	1959
10.	Topological spaces	6	III	1940
11.	Uniform spaces	2	III	1940
12.	Real numbers	10	III	1942
13.	Exponentials and logarithms	2	III	1947
14.	*n* dimensional spaces	2	III	1947
15.	Complex numbers; measurement of angles	4	III	1947
16.	Metric spaces	2	III	1949
17.	Infinitesimal calculus	32	IV	1949

(continued)

[13]The list of all published volumes can be found in Beaulieu (1989, Annexe II). See also http://archives-bourbaki.ahp-numerique.fr/publications.

Table 1 (continued)

	Chapter title	Pages	Book	First publ.
18.	Asymptotic expansions	4	IV	1951
19.	The gamma function	2	IV	1951
20.	Function spaces	2	III	1949
21.	Topological vector spaces	12	V	1955
22.	Integration in locally compact spaces	12	VI	1956
23.	Haar measure, convolution	6	VI	1963
24.	Integration in non locally compact spaces	10	VI	1969
25.	Lie groups and lie algebra	22	VII	1972
26.	Groups generated by reflections root systems	6	VII	1969

In the "*Mode d'emploi du traité*", a separate leaflet of instructions inserted in each volume of the mathematical treatise, Bourbaki explained the role of these historical notes. The first paragraph specified the objectives of the treatise and discussed norms adopted throughout concerning typography, terminology, symbols, etc.:

> To the reader, [...]
>
> 1. The Elements of Mathematics Series takes up mathematics at the beginning and gives complete proofs. In principle, it requires no particular knowledge of mathematics on the reader's part, but only a certain familiarity with mathematical reasoning and a certain capacity for abstract thought.[14]

About historical notes, it was further specified:

> 12. Since in principle the text consists of the dogmatic exposition of a theory, it contains in general no references to the literature. Bibliographical references are gathered together in *Historical Notes*. The bibliography which follows each historical note contains in general only those books and original memoirs that have been of the greatest importance in the evolution of the theory under discussion. It makes no sort of pretense to completeness.[15]

[14]We quote here from the published translation, to be found in every volume of the *Elements of Mathematics*. A draft can be found in "Bourbaki—Mode d'emploi de ce traité," Archives Bourbaki, http://purl.oclc.org/net/archives-bourbaki/438. In this document, the original text is: "Le traité prend les mathématiques à leur début, et donne des démonstrations complètes. Sa lecture ne suppose donc en principle, chez le lecteur, aucune connaissance mathématique particulière, mais seulement une certaine habitude du raisonnement mathématique, et un certain pouvoir d'abstraction."

[15]The original text is a bit more specific and keeps the same content, viz. "Bourbaki—Mode d'emploi de ce traité," p. 5. Archives Bourbaki, http://purl.oclc.org/net/archives-bourbaki/438: "Le texte étant consacré, en principe, à l'exposé dogmatique d'une théorie, on n'y trouvera qu'exceptionnellement des références bibliographiques; les références seront regroupées dans *un exposé historique*, placé le plus souvent *à la fin de chaque chapitre* et où l'on trouvera, le cas échéant, des indications sur les problèmes non résolus de la théorie. On se bornera à donner les références aux livres et mémoires originaux dont l'étude peut être le plus profitable au lecteur. Les références qui servent seulement à fixer des points de priorités sont presque toujours omises; à plus forte raison, le lecteur ne doit pas s'attendre à trouver ici de bibliographie complète des sujets traités."

In a retrospective account, Henri Cartan discussed the meaning of those historical notes.

> Bourbaki often places an historical report at the end of a chapter. Some of them are quite brief, while others are detailed commentaries. Each pertains to the whole matter treated in the chapter. There are never any historical references in the text itself, for Bourbaki never allowed the slightest deviation from the logical organization of the work. It is only in the historical report that Bourbaki explains the connection between his text and traditional mathematics and such explanations often reach far back into the past. It is interesting to note that the style of the "Notes Historiques" is vastly different from that of the rigorous canon of the rest of Bourbaki's text. I can imagine that the historians of the future will be hard put to explain the reasons for these stylistic deviations. (Cartan 1980, 178)[16]

Historical notes therefore had a distinct status from the rest of the treatise. Their main function was to provide a space for bibliographical references, which were banned from the treatise due to its self-contained nature. Added to the style difference underscored by Cartan, this suggests that writing processes for historical notes and the mathematical parts of the treatise also differed. Indeed, relatively little information can be found in the Bourbaki archives about the way members of the group decided to include and actually write historical notes.

3.2 The Conception of Bourbaki's Historical Notes

The accounts of the meetings of the "Committee for a treatise of analysis" show that in 1934–1935 the history of mathematics was a very minor concern of the participants. Indeed, one could go as far as saying that history was what they would be writing *against*. At the very first meeting, on December 10, 1934, a consensus was quickly reached on the principle of adopting a general abstract point of view. The scope of the preliminary "abstract package," on the other hand, was debated, and then "discussions became," according to the meeting report by Jean Delsarte (1903–1968), "confused—historical and philosophical."[17] Clearly, historical considerations here were opposed to the "modern" (Delsarte's term), systematic approach that was desired.

Traces of discussions can nevertheless be found in the accounts of some of Bourbaki's early "congresses." Thus, at Besse-en-Chandesse in July 1935, we find the following requirement among the long list of desiderata spelt out for the final form of the treatise: "Dictionary of used terminology (history and references)" [*Dictionnaire des termes usités (historique et références)*].[18] Already at this time,

---

[16]This talk was delivered in Düsseldorf on January 8, 1958 at the 76th Meeting of the *Arbeitsgemeinschaft für Forschung des Landes Nordrhein-Westfalen*.

[17]"La discussion devient confuse—historique et philosophique." Jean Delsarte, "Réunion du 10/12/1934," p. 3. Archives Bourbaki, http://purl.oclc.org/net/archives-bourbaki/1.

[18]"Desiderata," p. 5. Archives Bourbaki, http://purl.oclc.org/net/archives-bourbaki/27. An example of such a dictionary for topology is discussed in "Compte rendu du Congrès de Nancy (9–13 avril 1948)," p. 5. Archives Bourbaki, http://purl.oclc.org/net/archives-bourbaki/93.

history was thus envisioned by Bourbaki as being inseparable from the need to refer to the existing literature, as well as from emphasis on terminology.

The various types of texts that would figure in Bourbaki's treatise are specified in more details at the next annual congress, called *Congrès de l'Escorial*, in September 1936. A document was drafted with editorial decisions reached by the Bourbakis. Among decisions concerning names to be given to theorems, mathematical notations, size of fonts, etc., the Bourbakis listed three points of interest to us. Before we explain them, let us first give the original French text:

(x) *Laïus scurrile, toute latitude, en caractères normaux.*
(y) *Laïus historique, en fin de chapitre, quand ce sera utile.*
(z) *Laïus excitateur, en fin de chapitre, avec références. (Comme bon exemple voir Severi, Traité de géométrie algébrique).*[19]

As is often the case with Bourbaki's historical archives, a lot of unpacking may be necessary to understand the above. The term *laïus*, to start with, was frequently used by Bourbaki to refer to a written piece that was not purely mathematical, in the sense that it involved some rhetorical elements that could not be easily translated into formal language. According to the authoritative Littré dictionary, the term was slang at the Ecole Polytechnique for a long speech. The word *scurrile* is even more pedantic and of Latin origin. It can be translated in English as scurrilous, that is vulgar and clownesque, but it has no obscene connotation in French. A document from the 1935 Besse-en-Chandesse Congress explains that, at Cartan's insistance, this term (together with *futile*) "was adjoined to the mathematical vocabulary" together with their superlatives.[20] It soon became widely used in Bourbaki internal communications. The Bourbaki archives preserves an example of such "vulgar speech," which should have been printed in normal characters in the treatise: a document titled "*Ensembles—Décisions escoriales—Projet de laïus scurrile*," from 1936, opening with the following sentence as an introduction to the theory of sets:

> The object of a mathematical theory is a structure organizing a set of element: the words "structure," "set," "éléments," being not subject to a definition but constituting primordial notions [that are] common to all mathematicians, will become self-evident as soon as structures will be defined, as this will be done as early as within this very chapter.[21]

[19]"Décisions Escorial (typographie et rédaction)," Archives Bourbaki, http://purl.oclc.org/net/archives-bourbaki/36.

[20]"Les termes *scurrile* et *futile* ainsi que leurs superlatifs, dont la nécessité est mise hors de doute par les remarques de Cartan, sont adjoints au vocabulaire mathématique." "Brève histoire des travaux de Bourbaki," p. 3. Archives Bourbaki, http://purl.oclc.org/net/archives-bourbaki/19.

[21]"L'objet d'une théorie mathématique est une structure organisant un ensemble d'éléments: les mots 'structure', 'ensemble', 'éléments' n'étant pas susceptibles de définition mais constituant des notions premières communes à tous les mathématiciens, ils s'éclaireront d'eux-mêmes dès qu'on aura eu l'occasion de définir des structures, comme il va être fait dès ce chapitre même." "Ensembles—Décisions Escoriales [sic]," Archives Bourbaki, http://purl.oclc.org/net/archives-bourbaki/29.

Another example can be found in the detailed outline titled "Topologica Bourbachica," which tries to set out general directions for Bourbaki's view on topology where two "scurrilous speeches" are planned in the main text to argue for the necessity of adopting a general point of view in topology.[22] Although this was generally not the case, a scurrilous speech could involve historical references. The introduction to Integration written by Claude Chevalley (1909–1984) in 1936 apparently mentioned Ancient Egyptians ["*son laïus scurrile sur l'intégration muni des égyptiens*"].[23] Discussions about this cropped up again in 1937 when historical examples had been excluded in favor of reflections by Dieudonné that were deemed to be "of a great philosophical weakness." In order to introduce the axioms of a σ-algebra [*tribu* in Bourbaki's terminology], Delsarte argued that "scurrilities" should be as short as possible and Chevalley agreed to write a "scurrilous counter-speech starting with the weight of thin plates."[24]

In other words, the early Bourbakis talked about "scurrilous speeches" when they referred to remarks that required common parlance and, as such, lay outside of Bourbaki's ambition to produce, not a mathematical text in formal language, but at least one that could be formalized almost automatically. Generally, such speeches figured in the introduction of each book of the treatise: in Weil's outline for Topology in 1936, the first paragraph was thus called "*introduction et scurrilités*."[25] Nevertheless, scurrilous speeches were to be printed in standard characters, like most of the mathematical developments. As the above shows, scurrilous speeches were thoroughly involved in the original, collective thinking at the basis of each book of the treatise.

Very different was the understanding of the kinds of "speech" Bourbaki called "historical" and "*excitatoire*" ["excitatory," or perhaps even "arousing"]. From the outset, they were thought as appendices that were not logically needed to understand the argument of the treatise, nor even part of its original conception. As such, they could be gathered at the end of a chapter without any damage to the edifice. They could be written belatedly, after the final text of the chapter was unanimously adopted, sometimes even using the work of hired staff (something wholly unthinkable for other parts of the treatise). Indeed the 1936 list of desiderata included the following statement about outside sources:

> *References external to Bourkaki*. Fundamental references will be carefully given in the historic or excitatory speeches; possibly in the text (this therefore concerns references [that have been] checked, [and are] complete and correct).

[22] "Topologia Bourbachica," p. 4. Archives Bourbaki, http://purl.oclc.org/net/archives-bourbaki/492.

[23] "Intégration escoriale," Archives Bourbaki, http://purl.oclc.org/net/archives-bourbaki/33.

[24] "Journal de Bourbaki No. 9; 16/03/1937," Archives Bourbaki, http://purl.oclc.org/net/archives-bourbaki/73.

[25] André Weil, "Topologie 1 (Weil) (Exemplaire archétype) Plan général de topologie," p. 12, Archives Bourbaki, http://purl.oclc.org/net/archives-bourbaki/399.

Concerning technical references, a mere mention of the presumed author. Concerning other [references], as it can be done, without conferring too much importance to them (use of staff for screening).[26]

As a self-contained organic whole, Bourbaki's *Elements of Mathematics* contained a complex system of cross-references, but outside sources had to be kept at bay. Bourbaki therefore distinguished three levels of external references in his treatise. Only fundamental ones were properly considered. Others were to be only casually mentioned. In effect, all the bibliography would be placed in historical notes.

Examples and exercises were other parts of the text that had a special status. At the L'Escorial Congress, it was decided that exercises would be placed at the end of a paragraph and that examples and counterexamples would figure in the main text. However, to indicate their lower status with regards to the logical architecture of the treatise, both would be printed in smaller fonts. The "*Mode d'emploi*" had been very explicit about this lower status: exercises were "results which have no place in the text but which are nonetheless of interest." Here again, however, there was a significant difference with the process that led to historical notes, since both examples and exercises often figured in outlines and drafts examined by the Bourbaki group, at all stages of production.

3.3 A Lower Status for Historical Notes?

Unfortunately, no example of an early draft of an historical note seems to have survived.[27] We may take this hole in the records as an indication of the fact that historical notes were never extensively discussed by the group, nor taken into consideration when designing general arguments.

The quote from the L'Escorial desirata list above at least shows that the Bourbakis had an explicit model in mind as far as historical notes were concerned. Severi's Treatise of Algebraic Geometry was deemed a "good example." In his *Trattato di geometria algebrica*, he had indeed included a few "*notizie storico-bibliografiche*" (Severi 1926, 350), which clearly were a stylistic guide for Bourbaki. Although he also had bibliographical footnotes, Francesco Severi (1879–1961) placed his historical notes at the end of corresponding numbered paragraphs. Varying greatly in length from a one-line sentence to two densely printed pages, they appeared in a smaller font and included many bibliographical references.

[26]"*Références extérieures à Bourbaki.* Pour les références fondamentales, elles seront données avec soin dans les laïus historiques ou excitateurs; éventuellement dans le texte; (il s'agit donc de références vérifiées, correctes et complète)."Pour les références de technique, simple mention de l'auteur présumé. Pour les autres, comme on pourra, sans y attacher trop d'importance (emploi de négres pour dépistage)" "Décisions Escorial (typographie et rédaction)," p. 3. Archives Bourbaki, 36

[27]A document is mentioned in the Bourbaki Archives with the title "Note historique: topologie Chap. I," http://purl.oclc.org/net/archives-bourbaki/425, but when we last checked, it was unavailable.

It seems that the writers of Bourbaki's treatise took dearly the recommendation of following Severi's model not only in the physical layout, but in style as well. Compared to Bourbaki's however, Severi's historical notes appear much more closely related to the mathematical material just introduced, more factual, narrower in scope, and mostly concerned with recent developments.[28]

In a review of the *Elements of the History of Mathematics*, the French historian of mathematics René Taton (1915–2004) suggested another model for Bourbaki's historical notes. Praising the originality of the work, Taton recalled the French-German edition of the *Encyclopédie des sciences mathématiques*, originally edited by Felix Klein (1849–1925). More elementary in mathematical terms, Taton thought, the Encyclopedia was more erudite in the historical sense:

> Very schematically, one may characterize the different spirit by saying that the notes of Bourbaki's *Eléments de mathématique* have been written by mathematicians interested in the history of their discipline for the purpose of other mathematicians who are equally curious of the origins of their science, while those of the *Encyclopédie* were written with the active participation of professional historians of mathematics who, in part at least, wrote for historians' purpose.[29]

In 1984, Saunders Mac Lane's (1909–2005) account of the book for the *Mathematical Reviews* was harsher. To him, historical notes were little more than Bourbaki's mathematics retroprojected in the past:

> In virtue of its origin as appendices to separate texts on topics of current interest, these elements of history are just that: Former mathematics as it seems now to Bourbaki, and not as it seemed to its practitioners then. In the terminology of historiography, it is "Whig history" (MR0782480)

Many elements (typography, place in the treatise, style, lack of discussion at Bourbaki meeting, etc.) therefore converge to give the strong impression that historical notes held a low status in the project as a whole. Archives, where only a few mentions of historical notes can be found, partly confirm this impression. Admittedly, some consideration was given to the mathematical literature, for example what should be on avail on the location of early Congresses, which often took place in isolated spots.[30] When they are mentioned, historical notes seemed to have been discussed rather quickly, and late in the writing process. In July 1945, for instance, a new outline was adopted for Chaps. 5 and 6 of *Topology*, but the final text remained unchanged: "As usual, historical notes will be discussed later," the report

[28]Earlier books by the same author also included short historical discussions, e.g., Severi (1921).

[29]"Pour caractériser d'une façon très schématique cette différence d'esprit, on peut dire que les notices des *Eléments de mathématique* de N. Bourbaki ont été rédigées par des mathématicien s'intéressant à l'histoire de leur discipline, à l'intention d'autres mathématiciens également curieux des origines de leur science, tandis que celles de l'*Encyclopédie* l'ont été avec la participation active d'historiens des mathématiques professionnels qui écrivent, en partie du moins, à l'intention d'un public d'historiens." (Taton 1961, 158–159).

[30]Two instances in the archives: Jean Delsarte, "Sous-commission bibliographique 13/04/1935" Archives Bourbaki, http://purl.oclc.org/net/archives-bourbaki/10; and "Bourbaki's Dicktat [sic]—Congrès du 18-28/09/1936" Archives Bourbaki, http://purl.oclc.org/net/archives-bourbaki/28.

said, "at the small Congress that will take place in Nancy in December 1945; Weil and Chevalley will send their remarks to this congress by correspondence."[31] Unfortunately, no record of this meeting, if indeed it took place, subsists.

Historical notes could sometimes be used to address an issue that was deemed unworthy of figuring in the main structure of the treatise:

> The resolution of equations by radicals is abandoned (even in an annex), but the historical note will have to make briefly the link between this question and modern field theory.[32]

To emphasize the lower status of historical notes, let us finally quote from one of the famous ironic comments adorning much of Bourbaki's internal documents where the notes are presented as if they were mere social entertainment:

> Read as the opening speech, the historical note to chap. II–III of Algebra put the Congress "*in the right mood*" [in English in the original] toward Algebra: it glorified Fermat, dutifully followed the meanderings of the linear, and examined the influence of Mallarmé on Bourbaki.[33]

3.4 The Role of History in Collective Thinking

Contrary to what was just said, however, we want to claim that historical notes also came out, at least in part, from the necessity of working collectively through the mathematical material. In some rather rare occasions, collective thinking on mathematical topics involved historical considerations. Not always "dead" mathematics, historical notes sometimes were reflections of very lively debates among the writers of the treatise. Often, remnants of collective reflections about introductions to each book of the treatise found a place in historical notes. This is hardly surprising since historical considerations were often raised in the conception of these introductions. At Dieulefit, in 1938, it was for example specified that: "*Weil fait*

[31]"Comme d'habitude, les Notes historiques seront discutées ultérieurement, lors du petit Congrès qui se tiendra en décembre 1945 à Nancy; Weil et Chevalley enverront leurs observations à ce Congrès par correspondance." "No. 11[sic]—15 juillet 1945 CR du Congrès de Paris," p. 5. Archives Bourbaki, http://purl.oclc.org/net/archives-bourbaki/88.

[32]"On renonce à faire (même en Appendice) la résolution des équations par radicaux, mais il faudra que la Note historique expose succinctement le lien entre cette question et la théorie moderne des corps." "Compte rendu du Congrès de Nancy (9–13 avril 1948)," p. 2. Archives Bourbaki, http://purl.oclc.org/net/archives-bourbaki/93.

[33]"La Note historique des Chap. II–III d'Algèbre, lue en guise de discours inaugural, mit le Congrès 'in the right mood' [sic] quant à l'Algèbre: il glorifia Fermat, suivit docilement les méandres du linéaire et scruta l'influence de Mallarmé sur Bourbaki." "Observations du Congrès de Paris (18–20 Janvier 1947)," p. 1. Archives Bourbaki, http://purl.oclc.org/net/archives-bourbaki/92. On the humorous tone of Bourbaki's documents, see Beaulieu (1998).

l'introduction et laïus historico-bibliographico-existants" (or in more sober language, he promised to write, for October 15, the introduction and the history of topology).[34]

As seen above, integration was one of the topics for which historical considerations were discussed in early outlines and drafts. On February 11, 1935, the Committee approvingly discussed Chevalley's historically-minded project for measure and integration theory. Chevalley's outline, which was given orally, "seemed rather intuitive; he reaches the notion of measure by following the historical order Egyptians → Archimedes → Lebesgue → de Possel."[35] On the next meeting, on February 25, Chevalley presented his report which began with a definition of measure as a "number reported to *some* sets" [*nombre rapporté à certains ensemble*], which was to be illustrated by "historical examples."[36] From what we find in some later draft, however, we may suspect that mentions of historical considerations remained cursory:

> the notions of *quality*, *quantity*, *magnitude*, and the notion of *measure*, which established the link between them all to the notion of number, have appeared very early in the history of the human thought. They are at the basis of civilized life and of present-day experimental science; […] mathematics owes its origin and its most essential tools (such as the *real number* or the *integral*) to problems raised by the measure of magnitudes. The goal of this introductions to show how the abstract mathematical problems, which will be studied in the next chapters, can be, by means of the analysis of these *concrete* notions, disentangled from them.[37]

In the end, it was the lengthy note devoted to the history of logic and set theory that accommodated some of these reflections rather than the introduction. Archival materials allows us partly to follow the collective decision process that led to this result. The introduction to *Set Theory* occupied various Bourbakis over the years and consensus about what it should be was especially hard to reach.

At the Congress of Nancy in April 1948, for example, this introduction, which was to serve as the general introduction to the treatise, was discussed at length. A pragmatic viewpoint on foundational issues was adopted, according to which the

[34]"[version ronéo de l'ensemble des textes]," Archives Bourbaki, http://purl.oclc.org/net/archives-bourbaki/63. A preliminary document rather mentions "*laïus historico-bibliographico-excitants*," which is probably what was truly meant. See "Engagements de Dieulefit," Archives Bourbaki, http://purl.oclc.org/net/archives-bourbaki/56.

[35]Jean Delsarte, "Comité du Traité d'analyse—Réunion du 11/02/1935," Archives Bourbaki, http://purl.oclc.org/net/archives-bourbaki/5.

[36]Jean Delsarte, "Traité d' Analyse—Réunion du 25/02/1935," Archives Bourbaki, http://purl.oclc.org/net/archives-bourbaki/6.

[37]"les notions de *qualité*, de *quantité*, de *grandeur*, et la notion de *mesure*, qui les relie à celle de nombre, sont apparues très tôt dans l'histoire de la pensée humaine. Elles sont à la base de la vie civilisée et de la science expérimentale actuelle; quant aux mathématiques, elles doivent leur origine et leurs outils les plus essentiels (comme le *nombre réel* ou *l'intégrale*) aux problèmes que pose la mesure des grandeurs. Le but de cette introduction et de montrer comment, par une analyse de ces notions *concrètes*, on peut en dégager les problèmes mathématiques abstraits qui seront étudiés dans les chapitres suivants." "Théorie de la mesure et de l'intégration: Introduction (Etat 2)," Archives Bourbaki, http://purl.oclc.org/net/archives-bourbaki/414.

thorniest issues regarding the philosophy of language or the metaphysics of mathematical objects were simply to be set aside. In a separate article soon to be published, they deemed them irrelevant to the "working mathematician" (Bourbaki 1949). Concerning the history of such debates, the report of the 1948 Congress bluntly declared: "State explicitly that we are not interested in the quarrels from the beginning of the century about sets and their paradoxes."[38] Although Weil was asked to come up with a first draft of this introduction, it was Chevalley's text that the Bourbaki Congress at Royaumont unhappily examined in April 1950. "For the explication of the axiomatic viewpoint," it was decided that "the best place is the historical note, which will have to be thorough [*chiadée*]."[39] At the next Congress in Pelvoux, this decision was confirmed: "Introduction. It is decided to make it very short, and to leave as many things as possible to the historical note."[40] Once again, Weil was assigned the task of writing the historical note on sets, with the help of the logician John Barkley Rosser, Sr. In October 1951, a reference to the 25 centuries of experience with elementary geometry and arithmetic that increased mathematicians' confidence in set theory (see the quote about the unicity of *mathématique* above, p. 3), was discarded.[41] The writing of this historical notice was assigned to a younger recruit, Pierre Samuel, who was to be "tutored" by Rosser.[42] The last mention of historical notes for logic and set theory occurred in October 1952 when Samuel was still in charge, but no due date was given for the manuscript.[43] We shall come back in the next section to these extended historical notes, first published in 1957.

In some cases, historical notes can therefore be read as the graveyard where earlier debates among Bourbakis were put to rest. Such debates were not limited to the various introductions, which always held a special status. They sometimes concerned core issues such as the role of Hilbert spaces and the emergence of Book V on *Topological Vector Spaces*. Considered a masterpiece (Dhombres 2006) or criticized as not the best of Bourbaki's books (Ferrier, nd), Book V was acknowledged by all as a major restructuring of the field, more specifically with

[38]"Dire explicitement que nous nous désintéressons des querelles du début du siècle sur les ensembles et leurs paradoxes." "Compte rendu du Congrès de Nancy (9–13 avril 1948)," p. 6. Archives Bourbaki, http://purl.oclc.org/net/archives-bourbaki/93.

[39]"No. 22 Compte rendu du Congrès de la revanche" Archives Bourbaki, http://purl.oclc.org/net/archives-bourbaki/100.

[40]"No. 25 Compte rendu du Congrès oecuménique de Pelvoux (1951)," Archives Bourbaki, http://purl.oclc.org/net/archives-bourbaki/102.

[41]"No. 26 Compte rendu du Congrés croupion (oct 1951)," Archives Bourbaki, http://purl.oclc.org/net/archives-bourbaki/103.

[42]In the report of the Congress in the summer 1952, one reads that: "*Note historique: Samuel se fera tapirer par Rosser à Ithaca.*" See "Pelvoux 25 juin–8 juillet 1952," Archives Bourbaki, http://purl.oclc.org/net/archives-bourbaki/105. The verb *tapirer* is ENS slang for privately tutoring. Samuel was introduced to Bourbaki in 1945. See "No. 11 [sic]–15 juillet 1945 CR du Congrès de Paris," p. 2. Archives Bourbaki, http://purl.oclc.org/net/archives-bourbaki/88.

[43]"No. 29 Celles-sur-Plaine, 19–26 oct. 1952," Archives Bourbaki, http://purl.oclc.org/net/archives-bourbaki/106.

respect to its treatment of duality.[44] Although Book V was quite remote from integral equations, we think it is enlightening to follow original discussions in this way. Not only does this case help understand how the consideration of the historical literature was involved in the collective writing processes, it also provides clues as to what kind of historical reflections seemed useful to the writers of the treatise.

After an exchange of ideas on the theory of integral equations at the Committee meeting of March 25, 1935, Weil, Delsarte, Cartan, and Dieudonné reached the conclusion that the theory could be presented from three different "points of view". The first of these was that of Hilbert spaces "which gives a complete, perfectly esthetic theory." Then, there was Fredholm's "old" point of view. A final viewpoint was called "more modern, à la Rusz [sic, i.e. Frigyes Riesz] or à la Leray," and was placed in normed vector spaces. Acknowledging their poor command of the question, present members felt that, if Riesz's viewpoint covered Fredholm's and although it was more general, it nevertheless seemed not to "go as far in the results as Hilbert's point of view." But in Leray's absence, no decision was reached.[45] On May 6, 1935, Leray presented his own views on the matter before the Committee. He distinguished two parts: non symmetrical integral equations in Banach spaces and symmetric integral equations in Hilbert spaces. Although this raised some objections by Delsarte and Chevalley, Leray believed that there was no need to mention Fredholm's method.[46] In discussing Fredholm's theory, it is to be noticed that contrary to most manuscripts, participants to Bourbaki congresses often gave an exact reference, that of Riesz's article (published in *Acta mathematica* in 1918).[47] Lest we hastily conclude that this had nothing to do with proper historical work and that it was no more than mere bibliographical work, we should remember how history and bibliography were mixed in Bourbaki's practice. This clearly was the type of work Bourbaki considered history.

Although at the meeting at Besse-en-Chandesse in 1935, the report on integral equations was assigned to Mandelbrojt, a complete consideration of Hilbert spaces was only published as a part of Book V ("Topological Vector Spaces") in 1955.[48] At Besse-en-Chandesse, a section on "Fredholm determinant function" was planned as part of the theory of linear functional equations, for which the global existence

[44]An otherwise critical reviewer thus wrote in 1956: "Chap. IV, entitled *Duality in topological vector spaces*, is […] the most useful of all five chapters. Here is a complete and readable account of the various topologies for the space of continuous linear functionals on a topological vector space "(Hewitt 1956, 508).

[45]Jean Delsarte, "Traité d'Analyse—Réunion du 25 mars 1935," Archives Bourbaki, http://purl.oclc.org/net/archives-bourbaki/8.

[46]Jean Delsarte, "Traité d'Analyse—Comité de rédaction 06/05/1935," Archives Bourbaki, http://purl.oclc.org/net/archives-bourbaki/11.

[47]For example in Jean Leray, "Théorie des systèmes de *n* équations à *n* inconnues. Théorie des équations fonctionnelles. A titre documentaire: Projet d'exposé des théorèmes d'existence topologiques par J. Leray (1935)," Archives Bourbaki, http://purl.oclc.org/net/archives-bourbaki/619; or "Avant-projet—équations intégrales (Delsarte)," (dated September 18, 1936), Archives Bourbaki, http://purl.oclc.org/net/archives-bourbaki/38.

[48]"Serment," Archives Bourbaki, http://purl.oclc.org/net/archives-bourbaki/18.

theorem was assumed.[49] Later, Jean Coulomb (1904–1999) was assigned the task of drafting this chapter, but declined in 1937.[50] In March, 1937, Delsarte presented a short draft concerning integral equations where considerations of Riesz's work was judged faulty and he was asked to rewrite the project.[51] The complete history of Book V's origins is fascinating and remains to be written, but it lies outside of the scope of this paper.[52]

Consideration about Fredholm theory however cropped out in the historical note long before Book V was finished that were appended to Chap. 1 of *General Topology*, published in 1940. In this note, Fredholm theory is presented as a way mathematicians got used to consider functions as elements of general topological spaces. Hilbert's work on Fredholm integrals and Erhard Schmidt's generalization (1905) are praised as "memorable." In 1907, Riesz and Fréchet both developed general approaches to function spaces. But in Bourbaki's presentation of their work, the latter is said to have only succeeded in producing a cumbersome system of axioms, while the former's theory was judged to be "still incomplete, and remain [ing] besides as only a sketch" (Bourbaki 1994b, 143). Not discussed here, Riesz' 1918 paper was celebrated in the historical note that came with Book V and called "a chef-d'œuvre of axiomatic analysis [through which] the whole of Fredholm's theory (in its qualitative aspect) is reduced to a single fundamental theorem, namely that every normed locally compact space is finite dimensional" (Bourbaki 1994b, 214). It is interesting to note that a French reviewer of the 1940 booklet (A. Appert) regretted that Fréchet's contribution to general topology was minimized in the historical note, underscoring that Fréchet had done this while Riesz' work was unavailable to him due to war [JFM 66.1357.01].

Be that as it may, although our analysis of the path from Fredholm theory to Book V is sketchy and no more than provisional, it shows that, in the writing process of Bourbaki's treatise, reviews of recent literature—if not full historical considerations of the questions involved—were an essential part of the collective conception of the *Elements of Mathematics*. Moreover, our study underscores that traces of the process by which Bourbaki collectively restructured whole mathematical fields appeared in the historical notes. More precisely, elements taken from the historical notes, when reread in the light of discussions extracted from the archives, are witness to various stages in the organization of the treatise at various levels: general outline of the treatise, the order in which various topics were

[49]"Équations fonctionnelles linéaires," Archives Bourbaki, http://purl.oclc.org/net/archives-bourbaki/26.

[50]"Journal de Bourbaki No. 8; 16/02/1937," Archives Bourbaki, http://purl.oclc.org/net/archives-bourbaki/72.

[51]"Journal de Bourbaki No. 9; 16/03/1937," Archives Bourbaki, http://purl.oclc.org/net/archives-bourbaki/73.

[52]This is the current research topic of a working group organized by Christian Houzel. See an abstract in Beaulieu et al. "Bourbaki et les espaces vectoriels topologiques: histoire et création mathématique," http://hal.archives-ouvertes.fr/hal-00593455.

introduces, choices of viewpoints, etc. Let us now see whether this type of collective mathematical practices were reflected in the final product.

4 Collective Practices in Bourbaki's Historiography

When the *Elements of History of Mathematics* were published in 1960, Bourbaki made no pretense of giving a complete history of mathematics. In an "*avertissement*" placed at the start of the volume, authors explained that separate studies written for another purpose had merely been gathered here without major revisions. As a consequence, many portions of the history of mathematics like differential geometry, algebraic geometry, and the calculus of variations were absent from the volume, because the corresponding parts in the *Element of Mathematics* had not been published yet. Bourbaki kept silent about branches of mathematics, such as probability and statistics or more generally all applied mathematics, that seemed of little relevance to the scheme the group was in the process of diffusing. Mostly they warned their readers that they would find in this book:

> no bibliographic [sic] or anecdotal information about the mathematicians in question; what has been attempted above all is for each theory to bring out as clearly as possible what were the guiding ideas, and how these ideas developed and interacted on the others. (Bourbaki 1994b, v)[53]

Although these notes reflected a historiography that was sometimes more than 20 years old already in 1960, historians of mathematics received the book with enthusiasm and lauded its originality. While being critical on many relatively minor points, Itard for example emphasized that this was "a nice and good work" [Itard, 1965, 123]. He also mentioned that one would do well to read Bourbaki's book in parallel with another collective project, the *Histoire générale des sciences*, directed by Taton, whose chapter on the history of mathematics was partly written by the Bourbaki Dieudonné. While underscoring more important inadequacies in Bourbaki's book (like the provisional character of the endeavor and the lack of reference to contemporary research by historians of mathematics), Taton repeated without comment Bourbaki's point of view quoted above. Pointing out that the history of "the working mathematician" was not to be found in Bourbaki's book which he deemed "factual rather than interpretative," the historian Ivor Grattan-Guinness (1941–2014) nonetheless called it "the most important history yet produced of the mathematics of recent times" (Grattan-Guinness 1970). Mathematicians likewise repeated approvingly Bourbaki's words about the lack of biographical information.

[53]Note that the French version mentions *biographical* rather than *bibliographical* information (Bourbaki 1994a, v).

As Alexander Craig Aitken (1895–1967) noted: "the work has no concern whatever with anecdote, legend or personality of the authors concerned: authors are related solely to theorems or contributions to theory which mathematics owes them" (Aitken 1961).

Anecdotal evidence concerning Weil ironically exist to illustrate the point that this historiographical assumption hardly entailed a lack of attention to the collective aspects of mathematics:

> For a 1968–69 guest lecture in topology, the audience was packed into the lecture room of Old Fine Hall in Princeton and included Weil and many other notables. At one point someone in the audience rose to object that the lecturer was not giving proper credit for a particular theorem. The questioner went on in impassioned tones for what seemed an eternity. Finally Weil rose, turned to the questioner, and said in a loud voice, "I am not interested in priorities!" The discussion was over, and the lecturer resumed without further interruption. This was the quintessential Weil. Mathematics to him was a collective enterprise. (Knapp 1999)

In this sense, the assertion was quite trivial. In a strong rebuttal piece, Serge Lang wrote: "I object. In the sense that mathematics progresses by using results of others, Knapp's assertion is tautologically true, and mathematics is a collective enterprise not only to Weil but to every mathematician" (Lang 2001, 46). Polemically contending that Weil may have been disingenuous—or worse mischievous—on more than one occasion, and faulting him for not only neglecting other mathematicians' contributions but also deliberately misrepresenting their work, Lang at least showed that collective aspects in mathematics went much beyond the trivial conception put forward by Knapp.

4.1 Collective Practices Emerging

The table of contents (Table 1) greatly reinforced the view that Bourbaki's book was above all concerned with mathematical notions, and modern ones more especially. The focus on ideas erased much of the social dynamics at play in the historical development of mathematics. The collective result of a group of mathematicians who had embarked 25 years earlier on a highly original project of collectively and anonymously rewriting vast portions of mathematics, Bourbaki's *Elements of the History of Mathematics* thus appeared, at face value, as a paradoxical product: a collectively written history of mathematics whose content eschewed any serious consideration of the collective social dimensions of mathematics. Of course, one may say that by keeping silent about collective social dynamics of the past, Bourbaki was merely mirroring its own practice as a group, which remained discrete, and even secretive, about its internal workings. We argue, however, that a deeper examination can nevertheless dig up crucial concerns for collective aspects of mathematical practice. Faint as it is, partial discussions on collective and social dynamics in Bourbakis' historiography is nonetheless a reflection of their practical experience as mathematicians.

Institutions

Unsurprisingly, there were few institutions in Bourbaki's account of the history of mathematics. Universities were barely mentioned twice and in passing [pp. 136 and 170]; so is the École polytechnique [pp. 59 and 133].[54] Most of the time, this was to recall that formal training played a crucial part in the flow of ideas from one generation to the next.[55] Journals, academies, and learned societies scarcely appear, except in titles of cited literature. In a rare instance, Bourbaki recalled the role played by personal discussions in the formal setting of the London Mathematical Society between Benjamin Peirce and William Clifford [p. 118n].

True to their belief in international exchanges, however, the Bourbakis paid a bit more attention to the International Congresses of Mathematicians. In their book, ICMs mostly served as vehicle for quickly diffusing ideas expressed by individual mathematicians. In Zurich (dated once erroneously 1896 and once correctly 1897), Hadamard and Hurwitz drew attention to the applications of set theory to analysis; in Paris in 1900, Hilbert included the problem on the noncontradiction of arithmetic among his famous list; in Rome in 1904, the same Hilbert attacked the same problem [pp. 30, 142, 39 and 40]. Little more can be said about the now standard sociological foci of the historiography of mathematics. To have any chance of catching other ways in which social or collective aspects of mathematical practice may nonetheless surface in Bourbaki's historiography, our net needs a finer mesh.

Individual in Tension

While mathematical notions indeed structured Bourbaki's text, we note that very many individuals mathematicians were explicitly named. According to the index, which was added to the second edition of the work, 456 mathematicians were mentioned by name. As Fig. 1 exhibits, among the 26 most cited mathematicians geographical and chronological distributions were far from uniform. The first 8 in this list came from Germany and Switzerland (out of a total of 10 who do). 10 of these 26 mathematicians were from France, 3 from Great Britain, 2 from the Ancient Greek world, and 1 one from Norway. Some of the names served as labels for mathematical ideas, notions or theorems. Sophus Lie's name for example arose frequently mostly due to the emphasis put on Lie groups and algebras in the treatise. One is also struck by the chronological imbalance of this list, where barely 9 mathematicians had died before the start of nineteenth century. Such statistics confirm common views about Bourbaki's image of mathematics. In total, German mathematicians received roughly 100 more citations than French ones. Recent developments in the German cultural sphere indeed were what Bourbaki valued

[54]In the following, all pages numbers in square brackets without any other indication will be understood as taken from the English edition of the *Elements of the History of Mathematics* (Bourbaki 1994b).

[55]The teachings of Cauchy at the École polytechnique, of Kronecker and Weierstrass at the university of Berlin where they introduced the "axiomatics" of determinants, and Gauss' courses followed by Riemann at Göttingen in 1846–1847 were thus mentioned (resp. pp. 59, 64, and 129n).

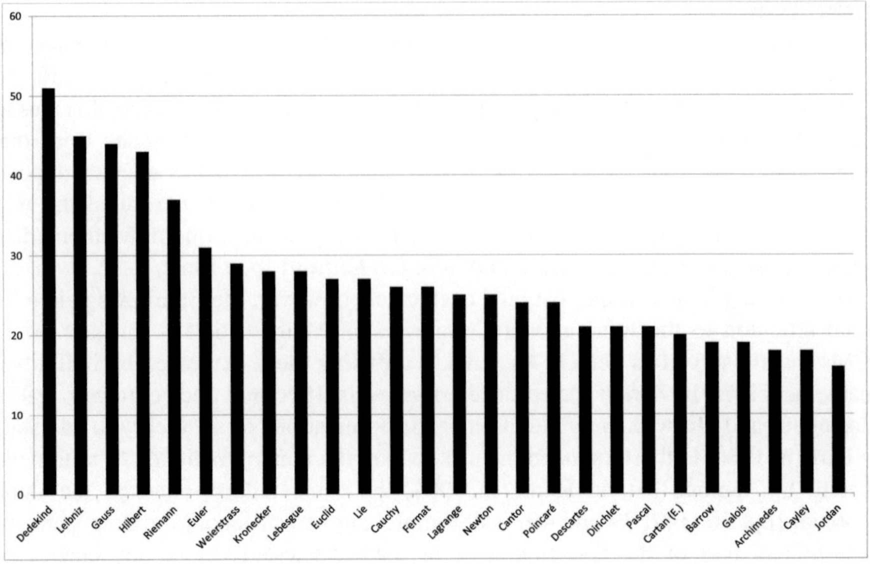

Fig. 1 The 26 mathematicians who are mentioned the most in Bourbaki's *Elements of the history of mathematics* and the number of pages where their name occurs

most. At the same time, our numbers indicate that discussions of mathematical influences on Bourbaki might have downplayed the importance of French mathematicians, including twentieth-century ones, like Henri Poincaré (1854–1912), Henri Lebesgue (1875–1941), or Élie Cartan (1869–1951).

A mentioned, these names in Bourbaki's text represented mathematical ideas and seldom flesh-and-blood people. This deliberate viewpoint announced in the introduction was expressed in the text itself. In the chapter on "Polynomials and Commutative Fields," Bourbaki only reluctantly refrained from rehearsing the convoluted story of the Cardano formula for solving third-degree algebraic equations:

> We cannot describe here the picturesque side of this sensational discovery — the quarrels that it provoked between Tartaglia on the one hand, Cardano and his school on the other — nor the figures, often appealing, of the scholars who were its protagonists [p. 72].[56]

That this is a history of mathematics emphasizing ideas at the expanse of social practice and institutions is of course well known: what we would like to point out here is that, by doing so, Bourbaki inadvertently produced a very odd result. Indeed, in this text—as, for that matter, it is often the case in the history of ideas—much agency is placed in the hands (or minds) of (a selected set of) individuals. In the

[56]"Nous ne pouvons décrire ici le côté pittoresque de cette sensationnelle découverte—les querelles qu'elle provoqua entre Tartaglia, d'une part, Cardan et son école de l'autre—ni les figures, souvent attachantes, des savants qui en furent les protagonistes" (Bourbaki 1994a, 96).

historical note on "Topological Sets," published in 1940, Bourbaki thus wrote: "It is Riemann who must be considered as the creator of topology; as of so many other branches of modern mathematics" [p. 139].[57] In another place, the authors underscored the sole agency of the inventors of the calculus in overcoming epistemological obstacles:

> it must be realised that this way was not open for modern analysis until Newton and Leibniz, turning their back on the past, accept that they must seek provisionally the justification for their new methods, not in rigorous proofs, but in the fruitfulness and the coherence of the results [p. 175].[58]

This view—according to which crucial innovations towards modernity "must await" the intervention of a chosen individual—was typical of Bourbaki's historiography. The authors for example wrote that we "must await" Cauchy [p. 129], Chasles [p. 131], or Möbius [p. 126] for the emergence of various mathematical concepts. Obviously, Bourbaki's historiography was filled with value judgments that emphasized the worth of great mathematicians and sometimes their "genius" [Poincaré on p. 35; Cantor's on p. 27]. For such luminaries, Bourbaki often preferred to talk of "mathematicians of the first rank" [pp. 6, 11, and 61].

As has often been emphasized, this conception of history certainly contained elements useful to Bourbaki's self-promotion and self-aggrandizing. Often starting with the Greeks or even the Babylonians, Bourbaki systematically saw the notions emphasized in its treatise and, at times, the work of Bourbakis themselves as the rightful culmination of a continuous, progressive history of mathematics. This has been discussed as the "royal road to me" historiography of mathematics, where mathematicians "confound the question, 'How did we get here?', with the different question, 'What happened in the past?'" [Grattan-Guinness 1990, 157].[59] Chevalley's initial plan for measure theory, whereby the line from the Egyptians to Bourbaki was almost direct (quoted above), was a caricature of this pattern. Deservedly or not, the work of individual members of the Bourbaki group was often given prominence. In Book V, as a reviewer noted at the time, Bourbaki "trace[d] the history of the subject from the contributions of D. Bernoulli to those of L. Schwartz" (a postwar recruit of Bourbaki's) (Hewitt 1956, 508). Most striking perhaps, in the published record, was a development where work by Weil and Jean-Pierre Serre (*1926), another postwar recruit, was identified as having suc-

[57]Early in the book, Bourbaki similarly underscored that Boole "must be considered to be the real creator of modern symbolic logic" [p. 8]. Galois was considered as the "real initiator" of the theory of substitution [p. 51]. The notion of tensor product of two algebras "must be attributed" to Benjamin Pierce [p. 118].

[58]One may note that Bourbaki is using the present tense to talk about the past and try to infer something from this unusual practice in English. In our view, this would be mistaken since this was then already a common practice in French historical writing to use the present tense.

[59]On Bourbaki's implicit place in the historical notes, see also Beaulieu (1998, 114).

cessfully dispelled mistrust regarding the theory of normed algebras developed by Israel Gelfand (1913–2009) [114–115].[60]

In Bourbaki's hand, the history of ideas therefore became a teleological account where agency was placed in the hands of individuals. This of course was paradoxical since nowhere was it explained how these special individuals could have been able to discern ahead of time the direction that history would take. From the text emerged a tension between individual contributions and collective aspects. This tension, we argue, was resolved in various ways that were not fully thought out.

Fluid Metaphors, the Practice of the Mathematicians, and *Zeitgeist*

Intent on capturing the flow of mathematical ideas, Bourbaki used a mixed bag of fluid metaphors to characterize the "stream of ideas" they wished to capture [pp. 19, 228, 240, and 270]. The notion of existence at the beginning of the 20th century was to be at the center of a "philosophico mathematics maelstrom" [p. 24]. Ideas were said to be "bubbling" [*bouillonnement*] in algebra at the start of the 19th century [p. 52].

But, in some case, it was impossible to ignore that the flow of ideas had met some serious obstacles in the history of mathematics. The first chapter of the book, dealing with the foundations of mathematics and set theory, contained remnants of the discussions sketched above about the introduction to Book I on Set Theory. There, Bourbaki uncharacteristically acknowledged that one needed to pay attention to "problem[s] which visibly ha[ve] nothing anymore to do with Mathematics" [p. 15], such as deciding whether geometry corresponds to experimental reality. Experience, intuition, and the "practice of the mathematicians" [pp. 10, 13, 14, and 35] entered the discussion, and, significantly, issues were debated in reference to conflicting collective understanding of the nature of mathematics. Discussing the *Grundlagen* crisis, Bourbaki identified several groups of mathematicians who held different views—rather than focusing, as usual, on various ideas for the foundation of mathematics. "Idealists" and "Formalists" looked for an axiomatic basis of mathematics [p. 31]; "empiricists," "realists," and "intuitionists" [p. 35] clung on to the need for inner certainty about the "existence" of mathematical objects.

> The intuitionist school, of which the memory is no doubt destined to remain only as a historical curiosity, would at least have been of service by having forced its adversaries, that is to say the immense majority of mathematicians, to make their position precise and to take more clearly notice of the reasons (the ones of a logical kind, the others of a sentimental kind) for their confidence in mathematics. [p. 38]

Our goal is not to discuss the validity, or not, of Bourbaki's views on the foundational crisis here. We merely want to point out that, when pressed to provide an account for a diversity of opinions on a topic that was related to mathematics, Bourbaki decided to frame the question in collective terms.

---

[60]In Bourbaki's history of mathematics, one also finds mentions of Bourbaki's *Elements of Mathematics*, as well as Henri Cartan's notion of filters [160 and 180]. Chevalley is mentioned eight times in the text.

A similar issue sprang up in the chapter on the birth of differential calculus where Bourbaki struggled with famous priority disputes between Newton and Leibniz, which were difficult to ignore. Interestingly, the authors here emphasized, as being characteristic of the period, the fact that "mathematical creations, the arithmetic of Fermat, the dynamic of Newton, carried a strong *individual* cachet" [p. 173, our emphasis]. But Bourbaki wished to undermine individual idiosyncrasies in order to stress the unstoppable motion of history:

> it is very much the gradual and inevitable development of a symphony, where the '*Zeitgeist*,' at the same time composer and conductor hold the baton, that we are reminded by the development of the infinitesimal calculus of the XVIIth century: each [individual mathematician] undertakes his part with characteristic timbre, but no one is master of themes that he is creating for the listener, themes that a scholarly counterpoint has almost inextricably entwined. It is thus under the form of a thematic analysis that the history of this must be written. [p. 173]

In this remarkable excerpt, history is likened to a symphony where individual performers are allowed to express their individuality within limits. An invisible director and composer (the spirit of the time, or *Zeitgeist*) is invoked to ensure that the individual priority claims at stake in the Newton-Leibniz debate remained as irrelevant to the history of mathematics as are quibbles between music performers to the rise of a symphonic theme. Deliberately confusing two different senses of the term "theme" (a short melodic subject and a subject of discourse), Bourbaki concluded that history of mathematics needed to be "thematic," that is, to follow the *Zeitgeist*'s lead rather than individual idiosyncrasy. In a letter to his sister, dated February 29, 1940, Weil developed a strikingly similar idea:

> As for speaking to nonspecialists about my research or any other mathematical research, it seems it would be better to try and explain a symphony to a deaf person. This can be done: one uses images, speaks of themes that run after each other, that intermingle [...]: but what have we at the end? Sentences, or at most a problem, good or bad, but without relation to what it was meant to describe. [61]

Significantly, this discussion about the Newton-Leibniz debate was also where Bourbaki was the most explicit about his method as a historian. In a rare acknowledgment of the need for historians to pay attention to context, Bourbaki underscored that quarrels about the invention of the calculus have a lot to do with organizational "deficiencies" in 17th-century mathematics:

> The historian must take account also of the organisation of the scientific world of the time, very defective still at the beginning of the XVIIth century, whereas at the end of the end of the same century, by means of the creation of scholarly societies and scientific periodicals, by means of consolidation and development of the universities, it ends up by resembling

[61]"Quant à parler à des non-spécialistes de mes recherches ou de toute autre recherche mathématique, autant vaudrait, il me semble, expliquer une symphonie à un sourd. Cela peut se faire; on emploie des images, on parle des thèmes qui se poursuivent, qui s'entrelacent, qui se marient ou divorcent; d'harmonies tristes ou de dissonances triomphantes: mais qu'a-t-on fait quand on a fini? Des phrases, ou tout au plus un poème, bon ou mauvais, sans rapport avec ce qu'il prétendait décrire" (Weil 1979, 255).

> strongly what we know today. Deprived of all periodicals until 1665, mathematicians did not have the choice in order to make their work known, of anything other than by way of letters, and the printing of a book, most often at their own cost, or at the cost of a patron if one could be found. The editors and printers capable of work of this sort were rare [...]. After the long delays and the innumerable troubles that a publication of this kind implied, the author had most often to face up to interminable controversies, provoked by adversaries who were not always in good faith, and carried on sometimes in a surprising bitterness of tone. [pp. 170–171]

Clearly, obstacles to the smooth flow of ideas came from social inadequacies. In the absence of proper scientific institutions, some "science amateurs, such as Mersenne in Paris, and later Collins in London" filled the void with a vast correspondence network, "not without mixing in with these extract stupidities of their own vintage" [ibid.]. Other types of social dynamics however might have had a more positive effect. In a thinly veiled allusion to their youthful travels sponsored by the Rockefeller Foundation, which had shaped their image of mathematics, the Bourbakis noted: "The studious youth journeyed, and more perhaps than today; and the ideas of such a scholar were spread sometimes better as a result of the journeys of his pupils than by his own publication" [p. 171]. From all these considerations, Bourbaki concluded that: "It is therefore in the letters and private papers of the scholars of the time, as much or even more than in their publications proper, that the historian must seek his documents" [p. 171].

While Bourbaki refrained from going in this direction, this implicitly recalls the division of tasks famously suggested by Weil at the Helsinki ICM in 1978: "The historian can help [since we mathematicians] all know by experience how much is to be gained through personal acquaintance when we wish to study contemporary work; our meetings and congresses have hardly any other purpose." (Weil 1980, 229). As can be seen from the above, Weil's notorious article was a suitable development from Bourbaki's historiography. But a crucial reversal had occurred. At Helsinki, Weil indeed went on:

> It is also necessary not to yield to the temptation (a natural one to the mathematician) of concentrating upon the greatest among past mathematicians and neglecting work of only subsidiary value. Even from the point of view of esthetic enjoyment one stands to lose a great deal by such an attitude, as every art-lover knows; historically it can be fatal, since genius seldom thrives in the absence of a suitable environment, and some familiarity with the latter is an essential prerequisite for a proper understanding and appreciation of the former. Even the textbooks in use at every stage of mathematical development should be carefully examined in order to find out, whenever possible, what was and what was not common knowledge at a given time. (Weil 1980, 335)

In Weil's view, social circumstances appeared no longer as blocks to the natural flow of ideas; they were the ground on which the seed of genius was allowed to blossom. We would like to argue that the notion of school which is very prominent in Bourbaki's historiography prefigured this reversal.

4.2 The Notion of School

Schools in Bourbaki's Historiography

The notion of "school" was by far the most commonly used by Bourbaki in its *Elements of the History of Mathematics* to refer to social and collective aspects.[62] It occurred on 40 pages of the book (out of 274). On the very first page of the first chapter, a mention of the "Vienna School" appeared, alongside "the Sophists," reinforcing long-term resonances between Greek Antiquity and the modern period in the history of mathematics.[63] Significantly, also mentioned in the same sentence as these schools, were "controversies [...] which have never stopped dividing philosophers" [p. 1]. Like all extra-mathematical entities, schools of thought were thus associated with a lack of certainty undermining the mathematical enterprise as understood by Bourbaki.

In the *Element of the History of Mathematics*, the concept of mathematical schools was used to refer to rather specific entities, but its general signification was loosely defined. It is of course impossible to analyze, for each and every case, the criteria that were implicitly used to identify mathematical schools and whether this identification holds up to historical scrutiny. In the following, we merely want to exhibit the many occurrences of the term in Bourbaki's historiography. This will enable us more precisely to characterize the understanding of collective practices that was put forward.

First, it is to be noted that Bourbaki often linked schools with prominent names: Brouwer [p. 37], Riemann [p. 54], Banach [p. 66], Cardano [p. 72], Clebsch and M. Noether [p. 106], Gelfand [p. 114], Monge [p. 132], etc. If this list in itself was not enough to show that the school concept had a positive value in Bourbaki's eyes, the fact that Hilbert and his school appeared prominently confirmed this impression: "a whole school of young mathematicians take part (Ackermann, Bernays, Herbrand, von Neumann)" in his work on proof theory [p. 40].[64]

Second, schools in the *Elements* could also be associated with cities or countries as we have seen apropos the "Vienna School," but also "the school of Moscow" in general topology [p. 143]. Mostly, schools were identified by one or several countries, the most prominent being of course the German school(s). Let us give a few examples:

---

[62]To examine the relationship between Bourbaki's notion of "school" and attempts by historians of science to give some substance to the notion of "research schools" (Geison and Holmes 1993) (and esp. Servos' paper therein) is a suggestive idea, but a path not taken here. For a discussion of the "modern mathematical research schools," see Ferreirós (1999) and Rowe (2003).

[63]There also were several mentions of the Pythagorean "school" [pp. 2, 69, 147], whose heirs, the Peripathetics, opposed the "Megaric and Stoic schools" [p. 5].

[64]There were many more instances where schools are associated with individual mathematicians: Boole's system as the basis for an active school of logicians [p. 9]; "the Peano school" suffering a "heavy blow" from Poincaré's "unjustified" criticism that "became an obstacle to the diffusion of his [Peano's] doctrine in the world" [p. 10]; Zariski and his school of algebraic geometry [p. 52]; a school working on Lie algebra in Leipzig [pp. 119 and 254]; a school whose main representative was von Staudt [p. 134].

- "the work of the modern German school: begun by Dedekind and Hilbert in the last years of the XIXth century, the work of axiomatisation of Algebra was vigorously pursued by E. Steinitz, then, from 1920, under the impulsion of E. Artin, E. Noether and the algebraists of their schools (Hasse, Krull, O. Schreier, van der Waerden)" [p. 55].
- "the German school of the XIXth century (Dirichlet, Kummer, Kronecker, Dedekind, Hilbert) of the theory of algebraic numbers, coming out of the work of Gauss" [p. 53];
- "the German school around E. Noether and E. Artin, in the period 1921–1931 which sees the creation of modern algebra" [p. 122].[65]

Other national schools were mentioned, often in relation with German schools. The development of abstract algebra was attributed to both the American (Wedderburn and Dickson) and the German (E. Noether and Artin) schools [p. 67]. In analysis, Bourbaki lumped together "the French and German schools of the theory of functions (Jordan, Poincaré, Klein, Mittag-Leffler, then Hadamard, Borel, Baire, ...)" [p. 142]. There are many other examples.[66] Finally, as seen in passing in the above, some schools were on occasions identified by mathematical criteria. One can for instance find: the formalist and intuitionist school [pp. 32 and 38]; a school of "fanatical 'quaternionists'" [p. 62]; a school studying quadratic forms [p. 61]; or the "school of 'synthetic geometry'" [p. 131].

From this survey, we conclude that the notion of school in Bourbaki's historiography was indeed very vague. Above all, it was a catchall, never defined nor discussed generally. Bourbaki had no wish to develop the notion of mathematical school nor to explain its meaning. But its use was systematic: this was the principal tool with which Bourbaki dealt with collective aspects of mathematical research. The only place in the whole book where the term is used, not to refer to one or several specific "schools," but more generally, shows that schools were the repository of some mathematical values. In this instance, Bourbaki saw in Euclid's Book VIII "the rigidity of the rather pedantic reasoning that does not fail to appear in all mathematical schools where 'rigour' is discovered or believed to have been discovered" [p. 13].

Where did this use of the term come from is not too clear. The historian José Ferreirós claims that it was already common in the 19th century to speak of mathematical "schools" (Ferreirós 1999, xviii). In the French language, the term was commonly used to refer to "a sect or doctrine of a few individuals," especially

---

[65]And also: the "German school of number Theory" [p. 98]; and "the German school of Geometry in the years 1870–1880" [p. 104].

[66]One can find mentions of: the "American school, around E.H. Moore and L.E. Dickson" for the study of finite fields [p. 120]; an "the anglo-American school" in algebra [p. 118] (partly overlapping with the English school of algebraists, "most notably Morgan and Cayley" [pp. 52, 117]); an Italian school (Dini and Arzelà) as well as a German school (Hankel, du Bois-Reymond) on uniform convergence [p. 205]; the Russian and Polish schools in topology [p. 156]; and, from a different time period, the "Italian school" at the beginning of the 16th century, solving algebraic equations by radicals [pp. 50 and 73].

in philosophy and art and in reference to actual schools like Plato's Academy or Raphael's workshop.[67] In the 19th century, one found occurrences of the phrase "*l'école mathématique française*" (Fourier 1825, xvi), and, at the start of the 20th century, frequent mentions of other national schools in mathematics. Although the expression "school of thought" obviously is classic in English as well, it seemed that its more systematic usage among historians of science may have been related to the mathematicians' usage (Rowe 2003, 121).

A tantalizing possibility would be that the Bourbakis themselves felt that they formed a "school." In historiographical terms, one may wonder how their own understanding of the collective work they had undertaken informed their discussion of the importance of research schools for the development of mathematics. We may at least say that, in a positive or in a negative light, they were indeed regarded as forming a school at the time of Bourbaki's greatest fame. As we have mentioned, the "*École Bourbaki*" was the term used in early characterizations of the collective enterprise (Delachet 1949, 13–116); one may even find earlier mentions of it (Bouligand 1947, 318). Even abroad, this seemed clear, as witnessed by the Hungarian mathematician Béla Szökefalvi-Nagy (1913–1998): "Without doubt, Bourbaki will form a school [*fera son école*] and will have a considerable influence on the development of mathematics" (Szökefalvi-Nagy et al. 1950, 258). In a more critical way, older mathematics professors at the Sorbonne, like Arnaud Denjoy (1884–1974), harshly criticized the Bourbakis' clanic behavior: "I fear your absolutism, your certainty of holding the true faith in mathematics, your mechanical move to take out the sword to exterminate the infidel to the Bourbakist Coran […]. We are many to think that you are despotic, capricious, and sectarian."[68]

Schools in Bourbakist Historiography and Beyond

Albeit loosely defined, mathematical schools played a crucial role in some of the Bourbakis' understanding of the dynamical development of mathematics. In 1940, while imprisoned in Finland for having refused to be drafted in the army, Weil wrote to his sister Simone Weil (1909–1943):

> The current organization of science does not take into account […] the fact that very few persons are capable of grasping the entire forefront of science, of seizing not only the weak points of resistance, but also the part that is most important to take on, the art of massing the troops, of making each sector work toward the success of the others, etc. Of course, when I speak of troops the term (for the mathematician, at least) is essentially metaphoric, each mathematician being himself his own troops. If, under the leadership given by certain teachers, certain "schools" have notable success, the role of the individual in mathematics remains preponderant. (Weil 2005, 341)

[67] *Dictionnaire de l'Académie française*, 4th ed. (1762); online edition (accessed October 28, 2013): http://artfl-project.uchicago.edu/.

[68] "Je redoute votre absolutisme, votre certitude de détenir la vraie foi en mathématiques, votre geste mécanique de tirer le glaive pour exterminer l'infidèle au Coran bourbakiste. […] Nous sommes nombreux à vous juger despotique, capricieux, sectaire." Arnaud Denjoy to Henri Cartan (May 22, 1954). Archives de l'Académie des sciences, fonds Montel, carton 1.

In other words, Weil though of mathematical schools as extensions of the powers of the individual mathematician, like armies extended the power of army generals, with the crucial difference being that while a general without his army was powerless, a single mathematician was able to accomplish much. At a time when "it is not possible to have someone who can master enough of both mathematics and physics at the same time to control their development alternatively or simultaneously" [ibid.], schools were individuals writ large and often relied on charismatic leaders. Schools were natural extensions of individual agency in mathematics and, indeed, another way to address the anxiety caused by the unfettered growth of science.

Among Bourbaki's founding generation, Weil and Dieudonné were, as we know, the most prolific producers of historical texts under their own names. They paid distinct attention to historical contextualization, which generally aroused much less interest from Dieudonné's part, while Weil always remained critical of "extra-mathematical" asides in the history of mathematics (Weil 1984). Both Dieudonné and Weil used the notion of school in their individual work in ways we shall not study here, besides suggesting that Weil might have been more careful in doing so.[69] Other mathematicians also reflected on the term, e.g. Mordell (1959, 41).

In the wake of Bourbaki, historians of mathematics widely adopted this term and, perhaps independently, "research schools" entered the vocabulary of the critical historian of science as well (Morrell 1972, Geison and Holmes 1993). But many felt uneasy about the loose use of the term:

> That the word "school," which has often been invoked in the history of mathematics, has been understood in a loose sense is indicated by the pervasive usage of the word in quotation marks. (Parshall 2004, 271)

And some historians of mathematics tried better to circumscribe historically and conceptually the meaning of a "mathematical research school."

> [C]ollaborative research presupposes suitable working conditions and, in particular, a critical mass of researchers with similar backgrounds and shared interests. A work group may be composed of peers, but often one of the individuals assumes a leadership role, most typically as the academic mentor to the junior members of the group. This type of arrangement—the modern mathematical research school—has persisted in various forms throughout the nineteenth and twentieth centuries. (Rowe 2003, 120)[70]

As a group, Bourbaki certainly fitted Rowe's description, albeit without a clear leader, hence perhaps, the necessity of inventing a fictitious one. In any case, a term casually used by mathematicians has become an inescapable descriptor for some social dynamics in mathematics. Even if Bourbaki cannot be held accountable for originating the wide use of the term, we suggest that the group's writings may have

[69]See, e.g., Dieudonné (1981, 39, 81, 83, and 212), Dieudonné (1989, 19n, 39n, 52, 68, 198 and 288), and Weil (1992, 50).

[70]For other attempts at restricting the notion of school in the historiography of mathematics, see Ferreirós (1999, xviii–xx) and Parshall (2004, 271–274).

helped to diffuse it widely. Mostly, we venture that "schools," as a loose concept, had become a useful way to resolve a historiographical tensions between individual and collective agencies in the history of mathematics precisely because "schools," in the more restricted sense quoted above, now corresponded to a social situation that was experienced more commonly than ever.

5 Conclusion

Historical notes held an ambiguous status in Bourbaki's *Elements of Mathematics*. The original emphasis on history in a mathematical treatise clearly played a part in shaping the "image of mathematics" the group wanted to project (Corry 1996). In a self-contained whole from which all reference to the literature, to historical development, or even to the mathematicians themselves had all but vanished, historical notes allowed the Bourbakis to re-humanize mathematics somewhat. We have moreover established here that this historical work sometimes impacted the architecture of the mathematical enterprise. In these cases, historical notes can be read as vestiges of mathematical discussions from which the treatise is a result. Entering through the backdoor, historical notes were however never allowed to take precedence over real Bourbakist mathematics. This relatively lower status was reflected in the different treatment historical notes received in the writing process as opposed to the other parts of the treatise. Although historical notes, like the rest of the treatise, certainly emerged through collective writing practices, this was achieved through much less back-and-forth motion among the various authors.

Despite being collectively conceived, the historical notices that were assembled in the *Elements of the History of Mathematics* exhibited great unity and belonged to a well-defined historiographical genre that stressed the stream of ideas from the remotest Antiquity to the Bourbakist present while emphasizing the contributions of a selected set of individuals. As we have shown, while collective aspects of mathematical work hardly surfaced in this book, the notion of "school" was used extensively for the purpose of capturing some of these aspects. For Bourbaki, the consideration of loosely-defined "mathematical schools," while often insisting on charismatic leaders, was a way to resolve the historiographical tension between streams of idea and individual agency. Its success in Bourbakist historiography also stemmed from the term's appropriateness as a reflection of the group's self-understanding in terms of social dynamics.

Recalling that collective research practices were also increasingly experienced by historians of mathematics, too, in the same period, it comes as no surprise that this usage of the term "schools" was widely adopted rather uncritically at first, with more subtlety later. In 1948, the "*Séminaire d'histoire des mathématiques*" was indeed launched by Taton, among others, at the Institut Henri Poincaré, where the Bourbaki Seminar also took place. Taton would soon be called to direct the ambitious collective project of the *Histoire générale des sciences* in four thick volumes (Taton 1964), to which he "devoted so many hours" (Huard 1959, 74).[71] Itard with whose remarks about Euclid we have opened this paper was part of both of these collective undertakings. By then, it seemed not only that Bourbaki had replaced the old Euclidean approach to mathematics based on intuition and experience, but also that the mere appearance of Bourbaki as a "polycephalic mathematician" was enough to cast doubt on the old master's very existence. Euclid's metamorphosis into a "school" with no identified leader was the culmination of both Bourbakist mathematics and Bourbaki's historiography. Perhaps this was the crime of *lèse-majesté* Dieudonné had in the back of his mind when he famously exclaimed at a European conference on the teaching of geometry in secondary schools: "Down with Euclid!" (Dugac 1995, 15)?

References

Acerbi, F. (2010). Euclide. *Images des Mathématiques*. CNRS. http://images.math.cnrs.fr/Euclide.html

Aitken, A. C. (1961). Review of Bourbaki, *Éléments d'histoire des mathématiques. Proceedings of the Edinburgh Mathematical Society, 12*, 217.

Aubin, D. (1997). The withering immortality of Nicolas Bourbaki: A cultural connector at the confluence of mathematics. *Science in Context, 10*, 297–342.

Beaulieu, L. (1989). *Bourbaki: une histoire du groupe de mathématiciens français et de ses travaux (1934–1944)* (PhD thesis). Université de Montréal.

Beaulieu, L. (1993). A Parisian Café and Ten Proto-Bourbaki Meetings (1934–1935). *The Mathematical Intelligencer, 15*(1), 27–35.

Beaulieu, L. (1994). Questions and answers about Bourbaki's early work (1934–1944). In S. Chikara, S. Mitsuo, J.W. Dauben (Eds.), *The intersection of history and mathematics* (pp. 241–252), Basel: Birkhäuser.

Beaulieu, L. (1998). Jeux d'esprit et jeux de mémoire chez N. Bourbaki. In P. Abir-Am (Eds.), *La mise en mémoire de la science: Pour une ethnographie historique des rites commémoratifs* (pp. 75–123). Amsterdam: Éd. des archives contemporaines.

Bouligand, G. (1947). Le mathématicien au seuil de la métaphysique. *Revue de Métaphysique et de Morale, 52*, 308–321.

Bourbaki, N. (1949). Foundations of mathematics for the working mathematician. *Journal of Symbolic Logic, 14*, 1–8.

Bourbaki, N. (1950). The architecture of mathematics. *American Mathematical Monthly, 57*, 221–232.

Bourbaki, N. (1994a). *Éléments d'histoire des mathématiques*. Berlin: Springer, repr. of Paris: Masson, 1984.

Bourbaki, N. (1994b). *Elements of the history of mathematics* (J. Meldrum, Trans.). Berlin: Springer.

Cartan, H. (1980). Nicolas Bourbaki and contemporary mathematics. *The Mathematical Intelligencer, 2*, 175–181.

Chabert, J.-L., & Gilain, C. (Forthcoming). *Sans amour et sans haine: le difficile principe de Jacques Hadamard.*

Corry, L. (1996). *Modern algebra and the rise of mathematical structures*. Basel: Birkhäuser.

[71]For a list of talks delivered at the Seminar since 1948, see http://www.ihp.fr/seminaire/SHM-histoire.

Dauben, J.W., & Scriba, C.J. (Eds.). (2002). *Writing the history of mathematics: Its historical development*. Basel: Birkhäuser.
Delachet, A. (1949). *L'Analyse mathématique*. "Que sais-je?" Paris: Presses Universitaires de France.
Dhombres, J. (2006). Analyse fonctionelle. In D. Lecourt (Ed.), *Dictionnaire d'histoire et philosophie des sciences*. Paris: Presses universitaires de France.
Dieudonné, J. (1978). *Abrégé d'histoire des mathématiques* (2 vols.). Paris: Hermann.
Dieudonné, J. (1981). *History of functional analysis* (2 vols.). Amsterdam: North-Holland Publishing.
Dieudonné, J. (1989). *A history of algebraic and differential topology, 1900–1960* (2 vols.). Basel: Birkhäuser.
Dugac, P. (Ed.). (1995). *Jean Dieudonné, mathématicien complet*. Paris: Jacques Gabay.
Ferreirós, J. (1999). *Labyrinth of thought: A history of set theory and its role in modern mathematics*. Basel: Birkhäuser.
Ferrier, J.-P. (n.d.). La dualité dans les EVT de Bourbaki: entre exposition et création. Online unpublished manuscript http://www.iecn.u-nancy.fr/ferrier/EVT/DualEVT.pdf
Fourier, J. (1825). Analyse des travaux de l'Académie royale des sciences pendant l'année 1824: partie mathématiques. *Mémoires de l'Académie des Sciences de l'Institut Impérial de France, 5*, i–xci.
Galison, P. (1992). *Big science: The growth of large-scale research*. Stanford: Stanford University Press.
Geison, G.L., & Holmes, F.L. (Eds.). (1993). *Research schools: A reappraisal*. Chicago: University of Chicago Press. Special issue of *Osiris, 8*.
Goldstein, C. (2010). Des passés utiles: mathématiques, mathématiciens et histoires des mathématiques. *Noesis, 17*, 137–152.
Grattan-Guinness, I. (1970). Review of Bourbaki, *Éléments d'histoire des mathématiques*. *British Journal for the History of Science, 5*, 190–191.
Grattan-Guinness, I. (1990). Does the history of science treat of the history of science? The case of mathematics. *History of Science, 28*, 149–173.
Hewitt, E. (1956). Review of Bourbaki's *Espaces vectoriels topologiques* (Actualités scientifiques et industrielles, nos. 1189, 1229, 1230). *Bulletin of the American Mathematical Society, 62*, 507–508.
Huard, P. (1959). Review of René Taton, *Histoire générale des sciences*. *Revue d'histoire des sciences, 12*, 71–74.
Itard, J. (1962). *Les Livres arithmétiques d'Euclide*. Paris: Hermann.
Itard, J. (1965). Review of Bourbaki, *Éléments d'histoire des mathématiques* (1964 ed.). *Revue d'histoire des sciences et de leurs applications, 18*, 120–123
Knapp, A. W. (1999). Editorial. *Notices of the American Mathematical Society, 46*, 420.
Lang, S. (2001). Comments on non-references in Weil's works. *Gazette des Mathématiciens, 90*, 46–52.
Mashaal, M. (2006). *Bourbaki: A secret society of mathematicians* (A. Pierrehumbert, Trans.). Providence, R.I.: American Mathematical Society.
Mordell, L. J. (1959). *Reflections of a mathematician*. Montreal: École polytechnique.
Morrell, J. B. (1972). The chemist breeders: The research schools of Liebig and Thomas Thomson. *Ambix, 19*, 1–46.
Parshall, K. H. (2004). Defining a mathematical research school: The case of algebra at the University of Chicago, 1892–1945. *Historia Mathematica, 31*, 263–278.
Paumier, A.-S. (2014). *Laurent Schwartz et la vie collective des mathématiques* (PhD thesis). Université Pierre et Marie Curie, Paris.
Rowe, D.E. (2003). Mathematical schools, communities, and networks. In M.J. Nye (Ed.), *Cambridge history of science* (Vol. 5, pp. 113–132). Cambridge: Cambridge University Press.
Severi, F. (1921). *Vorlesungen über algebraische Geometrie*. Leipzig: Teubner.
Severi, F. (1926). *Trattato di geometria algebrica* (Vol. 1). Bologne: Nicola Zanichelli.

Szőkefalvi-Nagy, Béla, et al. (1950). Review of Bourbaki, *Éléments de mathématique*. *Acta Scientiarum Mathematicarum, 13*, 258–263.

Taton, R. (Ed.). (1957–1964). *Histoire générale des sciences* (4 vols). Paris: Presses Universitaires de France.

Taton, R. (1961). Review of Bourbaki, *Éléments d'histoire des mathématiques*. *Bulletin des Sciences Mathématiques, 85*, 150–155.

Weil, A. (1979). *Œuvres scientifiques* (Vol. 1). New York: Springer.

Weil, A. (1980). History of mathematics: Why and how. In O. Letho (Ed.), *Proceedings of the International Congress of Mathematicians, Helsinki 1978* (Vol. 1, pp. 227–236). Helsinki: Academia Scientiarum Fennica.

Weil, A. (1984). *Number theory. An approach through history: From Hammurapi to Legendre*. Basel: Birkhäuser.

Weil, A. (1992). *The apprenticeship of a mathematician* (J. Gage, Trans.). Basel: Birkhäuser.

Weil, A. (2005). A letter of André Weil on analogy in mathematics. *Notices of the American Mathematical Society, 52*, 334–341.

Zallen, D.T. (1989). The Rockfeller foundation and French research. *Cahiers pour l'histoire du CNRS, 5*, 24 pp.

Pulling Harriot Out of Newton's Shadow: How the Norwegian Outsider Johannes Lohne Came to Contribute to Mainstream History of Mathematics

Reinhard Siegmund-Schultze

Abstract

The focus of this paper is on the peculiarities of Lohne's "outsider approach" to the historiography of mathematics and physics. The main thesis is that the circumstances of Lohne's outsider position both narrowed down and sharpened the focus of his research, but that he succeeded because he joined the then rising tide of archival based and content oriented internalist research in the history of physics and mathematics and because he was supported by scholars such as D.T. Whiteside and J.E. Hofmann, whose connections to the international community were better than his own. The main conclusions are based on Lohne's Nachlass in Oslo and on some other archival sources. In addition Lohne's publications are used, which were mostly in English and German, and partly in Norwegian. We do not, however, discuss Lohne's contributions to the historiography of mathematics and physics in any detail, because they are published and accessible. We merely sum up two outstanding results of Lohne's research on Thomas Harriot (1560–1621), concerning the discovery of the sine law of light refraction and the calculation of the Mercator map, in order to stress the quality of his work and his critical mind and to give some idea about his historical method. Concluding the paper we reflect once again, on the basis of our biographical evidence in the case of Lohne, and with some emphasis on Scandinavia, on the relative notions of the "outsider" and the "main stream"

R. Siegmund-Schultze (✉)
Faculty of Engineering and Science, University of Agder, Gimlemoen 25a,
Postboks 422, 4604 Kristiansand S, Norway
e-mail: reinhard.siegmund-schultze@uia.no

V.R. Remmert et al. (eds.), *Historiography of Mathematics in the 19th and 20th Centuries*, Trends in the History of Science,
DOI 10.1007/978-3-319-39649-1_11

within the process of professionalization of the historiography of mathematics and sciences in the second half of the 20th century.

Keywords
Norwegian school teacher Lohne as outsider · Harriot and sine law of light refraction · Calculation of Mercator map · Language problems · Political isolation

1 The Origin of Lohne's Interest in the History of Science and Mathematics

Johannes Lohne (1908–1993, pronounced Lu:nə) came to the historiography of physics and mathematics in unusual ways.[1] In a Norwegian popular journal we read the following editorial note referring to Lohne, the author of an article on "The dream of the moon: Newton's apple" (Lohne 1963b) in the same issue, partly quoting the author directly (Fig. 1):

> Lector Johannes August Lohne, Cand. Real. Address: Flekkefjord.
>
> Born in Flekkefjord 1908. Studied science at the University of Oslo, is now lector at a grammar school. Has in addition done work on the sources of the history of physics at the British Museum and in Cambridge. 'Over the years my interests have gone through the following stages: history, language, science, detective novels, and in the last ten years I have combined these interests in a study of the history of natural science, where as a rule, one finds, just like in detective novels, that everything is different from what the layman thinks.'[2]

Lohne had a master's degree in physics, specialising in optics, from Oslo University in 1932, but he never took a Ph.D. He became a school teacher in small provincial towns (first Odda, then Flekkefjord after WWII) and remained in this profession during his entire working life. He came to historiographical research late, and published his first historical article when he was 50 years old, in 1959. Lohne's interest in history grew after having served three and a half years (1945–1948) in jail for volunteering as a soldier in a unit of the Waffen–SS which was established by the German occupiers of Norway and deployed at the Eastern Front. As will be argued more fully below, Lohne's political conservatism and a certain personal

[1]More details about Lohne's life, as well as about his political actions before he became a historian, one finds in Siegmund-Schultze (2010) and in a forthcoming scientific biography of Lohne in Norwegian, written by Rolf Nossum and the present author. It will be published by Novus Publishers Oslo in 2016.

[2]"Lektor Johannes August Lohne, Cand. Real, Adresse: Flekkefjord. Født i Flekkefjord 1908. Studerte realfag ved Universitetet i Oslo, er nå lektor i Den høgre skolen. Har dessuten drevet kildestudier over fysikkens historie ved British Museum og i Cambridge. 'Mine interesser har gjennom årene gjennomløpet følgende stadier: historie, språk, realfag, detektivromaner, og i det siste tiår har jeg forent disse interesser i studiet av realvitenskapenes historie, hvor det som regel viser seg, likesom i detektivromanene, at alt forholder seg annerledes enn godtfolk tror.'" (*Naturen* 87 (1963), 3: 192).

Fig. 1 Portrait of Johannes Lohne (1908–1993) around 1980. Courtesy of Bernt Øksendal (Oslo)

stubbornness contributed to his outsider position in the scientific community. His personality influenced the choice of his research topics but it did not invalidate his historical results. Neither was Lohne's political stance during the German occupation of Norway, in the long run, held against him by the authorities, as permanent support by the Norwegian Research Council for his travels to England shows. On the contrary, it seems likely that assumed or alleged marginalization of Lohne in Norway prompted historians such as Hofmann and Whiteside to actions in Lohne's favor, which will be documented below.

Nevertheless, to a certain degree, Lohne remained isolated from the international community of historians of science and mathematics throughout his life. This was not least due to his heavy teaching load at school, which basically prevented him from attending international conferences, with the notable exception of the Oberwolfach seminars on the history of mathematics organized by J.E. Hofmann (cf. below). Lohne's native language was Norwegian, and this certainly restricted his international impact too, although he was talented with languages. He found support by German and English colleagues; but unlike his rival in matters pertaining to Thomas Harriot, John Shirley, Lohne never published a book.

Being more of a physics than a mathematics teacher, Lohne's historical interests were originally in physics, and he remained self-taught in general historical and in mathematical respects, as partly indicated in the editorial note quoted above. His sources of learning were apparently not those which a young student of the history of science would have regularly used at the time. Lohne describes his first encounter with the history of physics in a letter to Joseph Ehrenfried Hofmann (1900–1973) of 15 August 1961: "I came to the history of science 12 years ago, at the age of 40, through Mach's books." (See a picture of the German letter below.) Clifford Truesdell (1919–2000), who at the time helped Lohne with the language in his first publications, would certainly not have approved of Lohne's reliance on the physicist and philosopher Ernst Mach (1838–1916), of whom as a historian of science Truesdell was very critical. Indeed, Lohne remained ignorant about parts of the relevant literature in the historiography of physics and mathematics although he soon became widely read on literature specialized on the 17th century. Commenting, in 1992, in a letter to one of his former high school pupils, on geometrical lectures by the topologist Paul Heegaard (1871–1948) which he had attended in Oslo around 1930, Lohne said:

> Although Heegaard encouraged his students to read Euclid, it was only during my trips to England [in the 1960s!] that I obtained closer familiarity with Euclid's Elements. In a secondhand bookshop I came across an English edition, translated and commented upon by Sir Thomas Heath.[3]

As we know, Heath's edition is originally from 1908, the year of Lohne's birth. In a Norwegian review of biographical books published in 1972 Lohne calls E.T. Bell's notoriously unreliable "Men of mathematics" an example of "books of high quality" (Lohne 1972, 836: "bøker av høy kvalitet"), which gives another impression of his rudimentary education in and layman's approach to the history of science and mathematics.

2 Restrictions for Lohne Due to His Origin from a Small Country at the Periphery

Johannes Lohne came from a Scandinavian country at the periphery of Europe with a small language and a very limited tradition for history of science, at least in the older periods. (We will comment on the language question in some detail further below.) Beside language help from his German, English, and American colleagues Lohne was profiting from the rise of the new Danish journal for the history of science *Centaurus*, founded in 1950. This journal offered opportunities to publish

[3]"Skjønt Heegaard oppmuntret sine studenter til å lese i Euklid, er det først under mine Englandreiser at jeg fikk nøyere kjennskap til Euklids 'Elementer.' I et antikvariat kom jeg over en engelsk utgave, oversatt og kommentert av Sir Thomas Heath." The former high school pupil is Bernt Øksendal, today a prominent Norwegian mathematician in Oslo. He keeps the Nachlass of Lohne, to which I will refer repeatedly in this article. The quote is from a letter by Lohne to Øksendal, 12 February 1992.

also to other unconventional Scandinavian authors of the time such as Aage Gerhard Drachmann (1891–1980), the noted historian of ancient technology (Anon. 1972).[4] Lohne had regular contact to the editor of *Centaurus* since 1957, Mogens Pihl (1907–1986). Pihl recommended Lohne to the leading Norwegian physicist Harald Wergeland (1912–1987) and, indirectly, to the Norwegian research council (Siegmund-Schultze 2010, 586). A biography of Thomas Harriot which Lohne planned for the Danish *Acta Historica Scientiarum Naturalium et Medicinalium*, which was edited by Pihl as well, did not appear for unknown reasons. (For the table of contents of the unpublished book cf. Siegmund-Schultze 2010, 594–596).

Lohne not only contributed to the methodology of the discipline, as will be described below. He was also critical, in various reviews and letters, of the state of the history of mathematics and the sciences in Norway and Scandinavia, echoing others such as the Norwegian number theorist Viggo Brun (1885–1978). On 8 August 1970 the latter wrote a critical reply in the Swedish daily *Dagens Nyheter* to a contribution by the leading Swedish mathematician Lars Gårding (1919–2014) in the same newspaper the month before, on 4 July. Gårding had criticized the decision of the Swedish government to establish mandatory lectures in the history of mathematics for mathematics teacher students for one week, including exam. Gårding felt this would only produce superficial knowledge. Brun replied under the Swedish title "Of course we do need teaching in the history of mathematics." ("Visst behövs undervisning i matematikens historia"). Brun said in particular in Norwegian:

> In my opinion there should be a professorship in the history of mathematics at all bigger universities. In this context it should be interesting to mention that recently at the Technical University in West Berlin a chair has been created for history of the exact sciences and technology. Both the professor and one of his assistants are historians of mathematics.[5]

The professorship at Berlin had been given to Christoph J. Scriba (1929–2013), the student of Hofmann's, who by this step progressed further in his academic career than his teacher. Viggo Brun had sent copies of the two articles (by Gårding and by himself) to Scriba, while Lohne, who had been cooperating with Scriba for several years, provided German translations of the Swedish and Norwegian texts.[6]

Early in his late career as a historian of science Lohne had similarly demanded conspicuous steps in the institutionalization of that field in Norway. In a report to the Norwegian Research Council he wrote 4 April 1960:

---

[4]A detailed study of Drachmann and what he had in common with Lohne cannot be pursued here. While both stressed the importance of instruments and experiments in the history of science and technology, their style in writing on theoretical science seems to have been very different, with Lohne being much more mathematical in his approach (Cf. Drachmann 1967).

[5]"I denne forbindelse kan det være av interesse å nevne at man nylig ved det tekniske universitet i Vest-Berlin har opprettet en lærestol for de eksakte videnskapers og teknikkens historie. Både professoren og en av assistentene er matematikk-historikere."

[6]About ten years ago the late professor Scriba kindly sent me copies of these documents, including an interesting autograph by Brun, who published occasionally in the history of mathematics, in particular pertaining to his field of number theory. From Brun's autograph we will quote below.

> I should remark that Denmark and Sweden both have chairs in the history of natural sciences and publish international journals in this field, namely Centaurus and Lychnos. Scientific and technical progress and new ideas have contributed more to our civilization that the kings' wars or the statesmen's wit. So Norway must finally also give some support to the history of the exact sciences.[7]

On another occasion, in a positive review of a 1961 book by Hofmann, Lohne criticized the lack of any reliable information on the history of mathematics, given to Norwegian school students: "The history of mathematics is lamentably neglected at Norwegian schools."[8]

10 years later, Lohne pleaded for a detailed presentation of the more historical and traditional parts of physics within secondary schools. He thus saw history as a regulatory force and source of reflection against uncontrolled modernist research, a point which often has been made for mathematics teaching as well:

> Very much of traditional physics will with high probability maintain its value. Therefore one should not force-feed students in secondary school with subjects that - in the form they are presented now - have few chances to survive. The fusion of optics and atomic physics, accomplished in our century, forces parts of optical theories out of teaching and text books.[9] (Lohne 1971, 4)

Both the politically and scientifically conservative Lohne and the at least politically more liberal Brun[10] had their reasons to demand more teaching and research in the history of mathematics and physics in Norway, in the well-understood interest of science and mathematics and their teaching.

3 Lohne's Criticism of Newton and of Newton Scholars and His Admiration for Neugebauer and Hofmann

Johannes Kepler had been the first and foremost interest of Johannes Lohne in the 1950s. Beyond the accidental likeness of their first names, Lohne's admiration for Germany and for Kepler as an upright and stubborn man and as a religious thinker

[7]"Det bemerkes at både Danmark og Sverige har lærestoler i naturvitenskapenes historie og at de utgir internasjonale tidsskrift på dette område, nemlig publikasjonene Centaurus og Lychnos. Vitenskapelige og tekniske framskritt og nye idéer dér har vel bidratt mer til vår sivilisasjon enn kongers kriger og statsmenns kløkt. Så Norge må vel omsider satse noe også på realvitenskapenes historie." (From Lohne's Nachlass with Bernt Øksendal in Oslo).

[8]"Matematikkens historie er sørgelig forsømt i norsk skole." (Lohne 1961b, 591).

[9]"Så meget av den tradisjonelle fysikk vil med stor sannsynlighet bevare sin verdi at man burde unngå å tvangsfore gymnasiaster med stoff som i den form det nå presenteres, har liten sjanse å overleve. Vårt århundres fusjon av optikk og atomfysikk skyver en del av lyslæren utover i leseplan og lærebøker." (Lohne 1971, 4).

[10]Brun's autograph, which is dated 4 October 1970, also contains a congratulation for Scriba to the election of Willy Brandt as chancellor, whom Brun called "one of the few great politicians of our time." ("In meinen Augen ist er einer der wenigen grossen Politiker unserer Zeit"). It seems unlikely to this author that Scriba or Lohne would have approved of this political statement.

seems to have played a role.[11] However, given Lohne's personal history, his country's difficult relationship with Germany, and Norway's embrace of its war ally Britain, it was certainly a reasonable move for Lohne to focus—for all his proven nonconformity—more on the work of English than of German scholars. This enabled him to receive regular support by the Norwegian Research Council to travel to England during summer school vacations. Lohne would soon focus on Newton and, as we will see, somewhat accidentally, on Harriot. From the outset Lohne's approach was twofold critical: with respect to the work of the scientists to be investigated and with respect to the legends which some historians constructed around them. In his first application to the Norwegian Research Council, not having been yet to England, Lohne wrote in November 1957:

> Newton and the theory of colours: …Newton does not seem to have been an outstanding observer, because the experiments which he carefully performed should have taught him other things too than those which he saw. But he wished to see only what his theory told him.[12]

After his first work in English archives in the summer 1958, Lohne sent a 3-page "Short overview of the research into Newton until today"[13] to the Norwegian Research council, which was very critical of the existing literature on Newton's life and work, basically exempting only the German Hofmann from his criticism:

> Of the now living it is probably only professor J.E. Hofmann who has such a broad view over the mathematics of the 17th century that he is able to give a full and reliable appreciation of Newton's contributions.[14]

Indeed, it was the personal example of historians of mathematics such as Otto Neugebauer (1899–1990) and Hofmann which guided Lohne, even at a time when his interest in the history of physics still prevailed over mathematics. This is what Lohne wrote to the Norwegian Research Council in December 1959, partly reporting on his first findings in England:

> Analysis of Newton's work with prismatic colours. … I have tried to analyze tables, instruments and methods of measurement. It has become clear that a superficial report and analysis is not sufficient. Control calculation and control experiments have to be performed. Such exact methods of investigation have long been used in the history of mathematics (Jos. E. Hofmann and O. Neugebauer) but not until now in the history of physics.[15]

---

[11]In Lohne's Nachlass there is for example the draft of a talk in Norwegian on "Kepler and religion" from about 1952.

[12]"*Newton og fargelæren.* …. Newton synes ikke å ha vært noen fremragende observator, for hans nitid utførte eksperimenter skulle også ha lært ham andre ting enn det han så. Men han ønsket kun å se det han, etter sin forut oppstilte teori, ventet å få se" (Lohne to the Norwegian Research Council, November 29, 1957, Riksarkiv Oslo.).

[13]"Kort oversikt over Newton forskningen hittil." Published in English translation with commentary by Niccolò Guicciardini in Siegmund-Schultze (2010, 589–593). The following quote is reproduced there p. 589.

[14]"Av nålevende er det antagelig bare professor J.E. Hofmann som har slik oversikt over det 17. århundres matematikk at han kan gi en fullstendig og pålitelig vurdering av Newtons innsats her."

[15]"Analyse av Newtons arbeide med prismefargene … Her har jeg forsøkt å analysere tabeller, apparater og målingsmetoder. Det har vist seg at en mer overfladisk beretning og resonnering ikke

Lohne's immediate next step was to replicate historical experiments in the physics classes at his grammar school in Flekkefjord and to draw some anti-Whig conclusions from them about Newton's original experiments, although he repeatedly expressed admiration for Newton the experimenter as well. Lohne's results on Newton, particularly on his optics, such as (Lohne 1961a), have found critical acclaim, for instance in the biography of Newton in the *Dictionary of Scientific Biography*, where I.B. Cohen writes in 1980:

> Lohne finds great difficulty in repeating Newton's experimentum crucis, but more important, he has traced the influence of Descartes, Hooke, and Boyle on Newton's work in optics. He has further found that Newton used a prism in optical experiments much earlier than hitherto suspected – certainly before 1666 … and has shown that very early in his optical research Newton was explaining his experiments by the 'corpuscular hypothesis'. … Newton must, according to Lohne, have 'found it opportune to let his theory of colours appear as a Baconian induction from experiments, although it primarily was deduced from speculations.'[16]

When Lohne went to England for archival study in the late 1950s, the much younger D.T. Whiteside (1932–2009) had just begun his deep studies of Newton's mathematical papers, which he would publish in the 1960s and 1970s in his ground-breaking eight volume edition.

Lohne expressed repeatedly his admiration for Whiteside's thoroughness as an editor, but remained somewhat critical even of this new level of contemporary research on Newton, and he sometimes stressed the differences in his and Whiteside's views of Newton (see below).

4 Lohne's Turn to Harriot's Physics by Chance and with Some Anti-whig Tendency

It was certainly a lucky coincidence for Lohne when in 1958 he found in the library of Oslo University Harriot's former private copy of Alhazen-Vitelo's optics (F. Risner 1572 ed.) with handwritten notes in the margins. It was this book which got him started on Harriot (Lohne 1959).[17] This was the same year when he undertook his first Newton studies in England. Lohne's gradual turn to Thomas Harriot can partly also be seen as an outsider decision in order to restore the "fair fame" of a scientist so far neglected by historians (Lohne 1963a). Such restoration seemed warranted since Harriot did not publish a single line during his lifetime about his many results in physics and mathematics, the well-known *Ars analyticae Praxis*

(Footnote 15 continued)
strekker til. Der må kontrollregning og kontrollforsøk til. Slik nøyaktig analyse er for lengst tatt i bruk i matematikkens historie (Jos. E. Hofmann og O. Neugebauer), men hittil ikke i fysikkens historie" (Lohne to the Norwegian Research Council, December 1958, Riksarkiv Oslo.).

[16]Cohen (1980, 55). Cohen refers here to the paper (Lohne 1968). In the last sentence, Cohen quotes the concluding words in Lohne (1965b, 138).

[17]Harriot's book was temporarily lost, as reported in Siegmund-Schultze (2010). It has now been retraced by a Norwegian journalist, maybe partly influenced by our publications on Lohne.

being published posthumously in 1631 by his student Walter Warner (1563–1643), unfortunately with some distortions of Harriot's original manuscripts. This is not to say that Harriot himself—for instance compared to Newton—was an outsider to the science of his time. One must not forget that scholars published less back then and that the Royal Society had yet to be founded.

In his first publication in the history of science and based on Harriot's unpublished manuscripts Lohne showed in *Centaurus* in 1959 what had been occasionally conjectured but not proven earlier (by J. Shirley, see below), namely that Harriot was in possession of the sine law of light refraction decades before Snellius and Descartes, to whom the law is usually attributed even today in less well informed publications.

5 Excursus 1: Sketch of Harriot's Proof of the Sine Law of Light Refraction, Based on Lohne's Publications

Lohne found in Harriot's copy of Alhazen-Vitelo a handwritten table with coefficients for light refraction. In addition he found in Harriot's handwritten papers further tables but, above all, some pictures with concentric circles as reproduced by him in the figure (Fig. 3) below in Lohne (1959). In the figure with the two concentric circles *n* is the eye, *k* below is the real, observed point, *h* on the vertical above the virtual image of it.[18] I have marked the angles α and β in addition to Lohne's drawing. In careful experiments with astrolabe-like instruments Harriot measured β in dependence of α. When he found for many measurements that the points of intersection between the vertical line and the observed ray always lie on the same inner circle he was justified to conclude the sine law. The sine law says that with varying α the light-refraction index remains constant, depending only on the medium. In the given case (for air and water) the value is about 1.3 (Figs. 2 and 3):

$$\frac{\sin \alpha}{\sin \beta} = const$$

It has to be said that the American John Shirley (1908–1988), in his Harriot biography (Shirley 1983), does not give Lohne full credit for his historical proof that Harriot had the sine law of refraction. Shirley, who long before Lohne had seen and copied some of Harriot's manuscripts in England, published 1951 in the *American Journal of Physics* the very short note "An early experimental determination of Snell's law" (Shirley 1951). He quoted in it from a manuscript by the English mathematician John Pell (1611–1685). Pell could not have known Harriot personally, but he convincingly reported in the manuscript about his conversations around 1640 with Harriot's student, the editor of Harriot's *Praxis*, Walter Warner. The sine law of light refraction is mentioned in the conversations and it is claimed that Harriot knew it. But no proof is given in (Shirley 1951) for Harriot's possession of the sine law, in particular not Harriot's convincing diagram with the two concentric circles.

[18]Note that the position of the image is not quite according to modern optical theory, but this does not affect the direction (angles) of the light rays.

THOMAS HARRIOTT (1560–1621)

The Tycho Brahe of Optics

Preliminary notice by

Johs. Lohne*

Doctissimus vir et Geometra
Thomas Harriotus longe peritissimus ...
(Briggs to Kepler 1625)

The world has hitherto known Thomas Harriott only as an eminent mathematician. On his four hundredth anniversary he should be remembered also for his optical investigations where he anticipated discoveries of Snell, Descartes and Newton.

Let us first recall what was known about optical refraction before Harriott's time. Vitelo (1225?–1280?) had written ten books on optics which were printed three times in the sixteenth century. In the best known edition (Risner's, Basel 1572) we find on page 412 the following table of refractions:

Tabula quantitatis angulorum incidentiæ	Anguli refracti	
	ab aere ad aquam	ab aere ad vitrum
10°	7°45′	7°
20°	15°30′	13°30′
30°	22°30′	19°30′
40°	29°	25°
50°	35°	30°
60°	40°30′	34°30′
70°	45°30′	38°30′
80°	50°	42°

* Flekkefjord, Norway.

Centaurus 1959: vol. 6: no. 2: pp. 113–121

Fig. 2 First page of Lohne (1959), Lohne's first and most influential publication in the history of science

6 Lohne's "Anxious" Turn to the History of Mathematics

While Harriot's sine law is basically a simple law of mathematical physics, but no mathematics in itself, Lohne had caught fire with Harriot as a mathematician too, after studying his manuscripts. He wrote to Hofmann in an undated New Year's postcard (probably 1962/63 or a year later) with "a certain anxiety":

> Dear Professor Hofmann, I am deeply moved by your willingness to help. After the optics I am now turning to Harriott's mathematics: not without a certain anxiety. 'Argumentationes

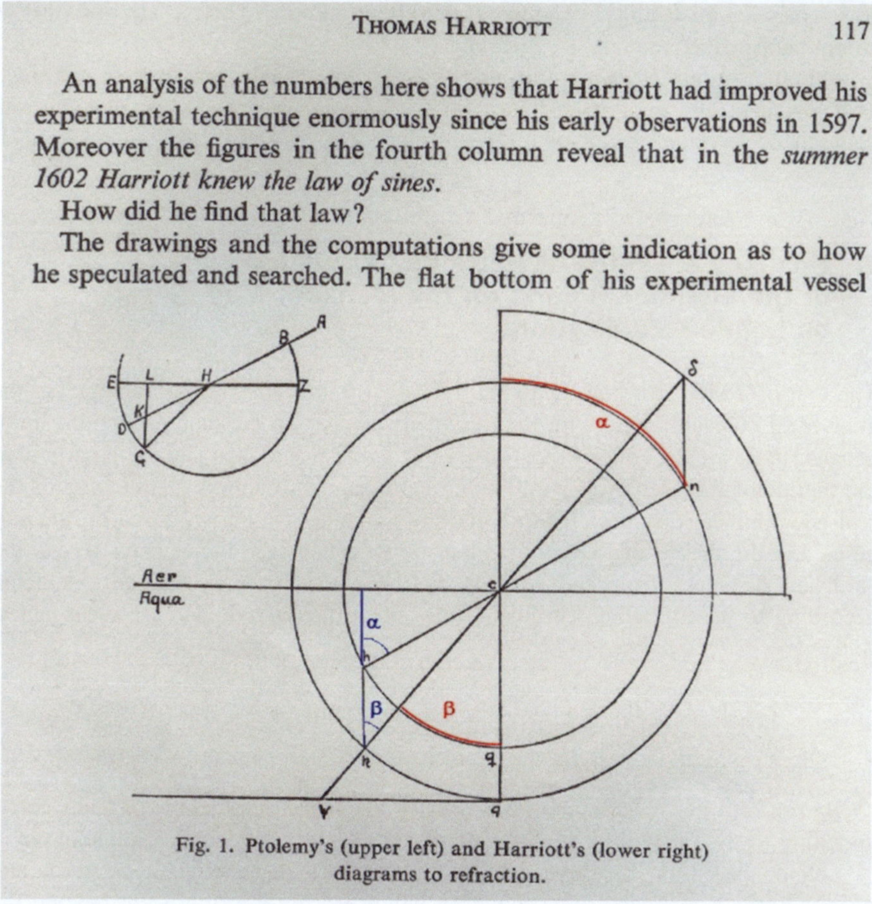
THOMAS HARRIOTT 117

An analysis of the numbers here shows that Harriott had improved his experimental technique enormously since his early observations in 1597. Moreover the figures in the fourth column reveal that in the *summer 1602 Harriott knew the law of sines.*

How did he find that law?

The drawings and the computations give some indication as to how he speculated and searched. The flat bottom of his experimental vessel

Fig. 1. Ptolemy’s (upper left) and Harriott’s (lower right) diagrams to refraction.

Fig. 3 From Lohne (1959, p. 117), the symbols α and β for the angles are added twice by the present author in order to show the geometric argument used by Harriot

has tu ridebis, conscius Thesauri tui scientifici in quem ego non inscipiens, misere forte erro.’[19]

The Latin is a quote from a letter written by Kepler to Harriot in 1609, where Kepler describes his lack of familiarity with Harriot’s mathematical work. Meanwhile, in 1962, Lohne had apparently realized that the introductory sentence of his article on Harriot as the “Tycho Brahe of optics” (as reproduced above in Fig. 2)

[19]The German part of the original is: “Lieber Prof. Hofmann. Ihre Hilfsbereitschaft hat mich tief gerührt. Nach der Optik wende ich mich zu der Mathematik Harriotts, doch mit einer gewissen Angst:” The correspondence between Lohne and Hofmann is kept at the Leopoldina Academy in Halle (Germany), see below.

was somewhat misleading: "The world has hitherto known Thomas Harriott only as an eminent mathematician."

Although rightly stressing with this sentence that Harriot was largely unknown as a physicist and geographer, it gradually became clear to Lohne that the real mathematical accomplishments of Harriot were not yet known to the world either.

7 Excursus 2: Sketch of Harriot's Calculation of the Meridional Parts on the Mercator Map, Based on Lohne's Publications

The aim of a Mercator map is that all so-called rhumb lines (loxodromes), i.e. the courses of constant angle with the meridians, appear on the map as straight lines, because these are the courses most easily navigable at sea. This makes, in particular, the meridians themselves to parallel vertical lines, and requires a stretching of the real geographical distances towards the poles. Today we know, based on the calculus and the theory of logarithms, that the so-called meridional parts M, i.e. the stretched distances from the equator on the Mercator map, can be calculated according to the following formula, with λ being the latitude and k a scale factor:

$$(*) \int_0^\lambda \sec\phi \, d\phi = \int_0^\lambda \frac{d\phi}{\cos\phi} = k\,M = \ln \tan\left(\frac{\pi}{4} + \frac{\lambda}{2}\right), \quad 0 \leq \lambda \leq \frac{\pi}{2}$$

Harriot and his contemporaries basically knew the summation formula corresponding to the integral at the left side of (*) which was needed to calculate M.

Lohne—based on his investigation of Harriot's manuscripts located in and nearby London—showed three things in which Harriot's results went beyond the knowledge of his contemporaries:

1. Harriot proved the conformality of the stereographic projection and the "logarithmic property" of the projection of the loxodromes into the equatorial plane, later to be called the logarithmic spiral.
2. Harriot used the latter property and subsequently numerical interpolation in order to find the "meridional parts" M of a reasonably dense net (grid) of latitudes (needed for the Mercator map) out of given singular values for M, which he calculated first and separately by a summation formula equivalent to the left hand side of (*).
3. Harriot's work could have influenced both the history of the theory of logarithms and of the calculus if this work had been known to his contemporaries. This is, because Harriot showed the logarithmic nature of the anti-derivative, and it only remained the relatively easier task to show that derivation leads back to the integrand on the left hand side.

Lohne's article (1965a), which contains his results, is placed at the centre of a group of historical contributions on Harriot's work on the Mercator map in terms of its timing and its contents. The other contributions were all written in English, namely by Sadler (1953), George (1956), and Pepper (1968). Lohne was the only non-English scholar in this group, and his article was published in German in the Danish journal *Centaurus*, the German resulting from the fact that the article was based on Lohne's talk in Oberwolfach in September 1964. This and the fact that Lohne elaborated much less on the details than later Jon Pepper, may have diminished the effect of his paper also in the eyes of Shirley (1983), when he wrote his standard Harriot biography. Generally speaking, Lohne's style of writing sometimes lacks lucidity, and reference to sources is often vague. However, the fact that Lohne does not mention the remarkable article by Frank George (1956) which conjectures (!) all main results of Lohne's analysis without being based on an investigation of Harriot's papers, cannot be held against Lohne. The present author himself discovered George's paper accidentally when browsing the *Journal of Navigation*. In fact, even the Englishman Pepper did not see George's paper although he himself published in that journal a decade later and although Sadler (1953) had in the very same journal alerted the public to the logarithmic tables which are in Harriot's papers (Fig. 4).[20]

8 Remarks on Lohne's Overall Contribution to Harriot Research

Lohne's research has contributed to the now widespread acknowledgement that Harriot was the most important, at least the most versatile English mathematician before Newton.[21] To be sure, Thomas Harriot, the very influential scholar, at least in English scientific circles in his time around 1600, had never been forgotten. But, as Lohne showed, he was partly remembered for the wrong things. As much as Lohne was justified in fighting for the "fair fame of Harriot," who had been underestimated by historians even in his own country,[22] he exaggerated Harriot's influence when calling him, in the title of his important paper quoted above, the "Tycho Brahe of Optics". Here, as well as in recent accounts which celebrate

[20]J. Pepper wrote to me 3 March 2012: "This is not as strange as it might seem, as it is not mentioned in David Waters' monumental Art of Navigation & c & c" Cf. (Waters 1958).

[21]As for instance expressed by Whiteside in a seminar in 1983, see below. Qualifications seem to be needed though, for instance in comparison between Harriot's and John Wallis's contribution.

[22]There was a controversial discussion around 1830 between the German-Austrian astronomer Franz von Zach (1754–1832) and the professor at Oxford Stephen Peter Rigaud (1774–1839) on a possible edition of Harriot's papers, in which, paradoxically, the Englishman advised against publication. Lohne published a paper on that discussion and may have felt in a similar position as once the other foreigner Zach: "The feud was paradoxical. A German blazoned abroad outstanding discoveries of the English Harriott. The English professors from Harriott's university did their utmost to conjure the inopportune Harriott back again into relative obscurity." (Lohne 1963a, 69).

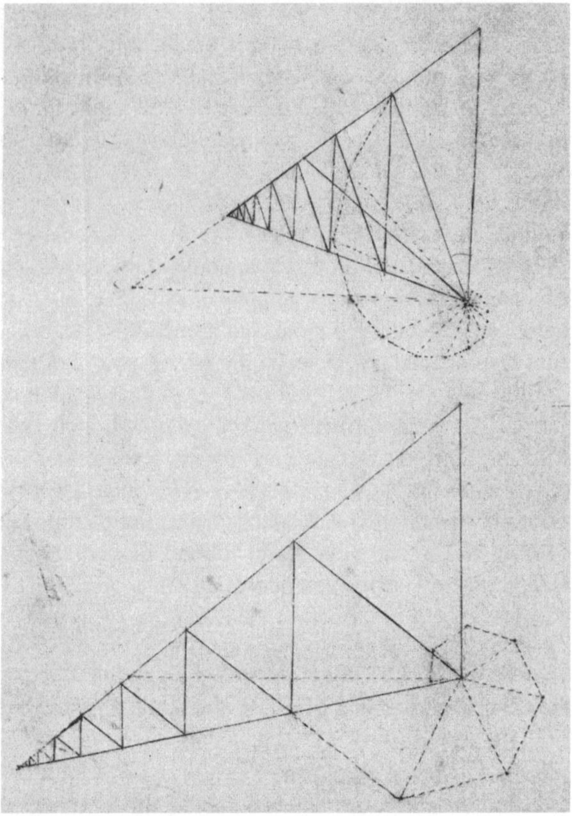

Fig. 4 A photograph from Harriot's manuscripts reproduced in Lohne (1979, p. 273), and showing the use of logarithmic spirals for the calculation of the meridional parts on the Mercator map

Harriot as the "English Galileo" (Schemmel 2008), Harriot's comparatively lesser actual influence on his contemporaries and on later generations is neglected. The lesser influence of Harriot compared to Tycho Brahe and Galileo is well described by Lohne himself for instance by pointing to the private interests of Harriot's sponsors which prevented publication (Lohne 1965a, 20). How little was known in detail about Harriot's manuscripts on mathematics and physics around 1960 is perhaps visible in Whiteside's dissertation "Patterns of Mathematical Thought in the Seventeenth Century" (Whiteside 1961), although admittedly the latter paper was devoted primarily to the second part of the 17th century. Whiteside, in his 1961 paper, mentions Harriot only once, in a footnote related to the posthumously printed *Praxis* of 1631. While it is not surprising that Newton's immediate predecessors such as J. Gregory, Wallis and I. Barrow are discussed in much more detail, it is nevertheless striking to compare Whiteside's work of 1961 with a remark which Whiteside made at a Seminar on "Harriot's geometrical papers" on 28 April of 1983:

> Thomas Harriot (?1560-1621) was Britain's most learned and brilliant mathematician and exact scientist before Newton; of all his contemporaries, indeed, only Kepler had claim to be his peer.[23]

Whiteside in this abstract to his talk recommends Lohne (1979) and Lohne's article on Harriot in the *Dictionary of Scientific Biography* (1972) as "preliminary reading." Whiteside indicates in the same abstract that he considered going fully into Harriot research after he had finished the edition of Newton's mathematical manuscripts in 1981, something, however, which would not materialize:

> Recently, however, I have been organising and studying his geometrical papers: a mere 1500 folios or so which (so far as I know) no one has even bothered to put into any semblance of proper order, let alone examine critically.

Lohne's most important publications on Harriot's mathematics are the following three Lohne (1965a), Lohne (1966), and Lohne (1979). The last of the three is written in a period of declining strength (eyesight etc.) at an age of about 70, and Lohne therefore confines himself in Lohne (1979) to listing and compiling manuscripts and giving suggestions for further research.

Research on Harriot's mathematics in recent decades has been extended above all by the late Jacqueline Stedall in Oxford. She showed among other things that Lohne had underestimated the influence of François Viète (1540–1603) on Harriot and that John Wallis's (1616–1703) historical judgment on Harriot, apparently based on the latter's manuscript and criticized by Lohne as partly too euphoric, and as partly belittling, was basically sound.[24] The "rediscovery" of Harriot has recently culminated in the effort to put his extensive unpublished manuscripts on various scientific and non-scientific matters on the Internet.[25]

9 Lohne's Relationship with Established Historians of Science of His Time, in Particular Hofmann and Whiteside

Among Lohne's international contacts, the best documented is Lohne's relationship to Joseph Ehrenfried Hofmann in Germany, whom he admired from the outset, as we have seen. Their correspondence is kept at the Leopoldina Academy in Halle (Germany).[26] The following excerpt from the beginning of their exchange in August 1961 may serve as an example, where—in addition to his remark about Mach quoted

[23]Norwegian Research Council, file Johannes Lohne, copy of Whiteside's abstract from 28 April 1983.

[24]"Wallis's treatment still stands as the most thorough and detailed analysis of Harriot's algebra to date" (Stedall 2000, 489).

[25]Gradually to be extended under the auspices of ECHO (European Cultural Heritage Online) at http://echo.mpiwg-berlin.mpg.de/home.

[26]Thanks to M. Folkerts (Munich) who deposited the letters there and gave me access to them in advance.

above—Lohne expresses his gratitude for the first invitation to Oberwolfach (of which many would follow) and his astonishment that he as a school teacher and "amateur" would have the chance to attend such an important meeting (Fig. 5).

Hofmann, on his part, seems to have harboured special sympathies for the eight years younger Norwegian. The historian of mathematics, Menso Folkerts, wrote to me on 1 February 2008:

> I knew Mr. Lohne from frequent meetings in Oberwolfach in the late 1960s and early 1970s. He mainly talked on Harriot. He always came with his wife... He was kind of a friend to Hofmann, in case Hofmann had friends at all. Lohne's talks were clear, he spoke excellent German and he was very friendly. He appeared to me somewhat like a solitary person.[27]

As to Hofmann's motives to support Lohne—beyond being impressed by his work—one may speculatively consider the shared past of close Norwegian-German relationships and one may assume a certain feeling of obligation for a Norwegian who had come into political problems due to the German invasion of the country in 1940. (The present author dares say that a similar motive stimulated his interest in Lohne as well.) Hofmann himself had profited in several respects from the political regime in the 1930s and 1940s, in particular through his protector in Berlin, Ludwig Bieberbach (1886–1982) (Siegmund-Schultze 2012).

In the letter from August 1961, quoted above, Lohne thanked Hofmann also for mediating his somewhat iconoclastic[28] article on Newton's theory of colors to the first volume of Truesdell's *Archive for the History of Exact Sciences*. Clifford Truesdell, who published the article (Whiteside 1961) in the same first volume of his new journal, was—as we will see below—another role model as an historian of science in the eyes of the eleven years older late-bloomer Lohne.

Lohne's and Whiteside's shared interest in Newton and Lohne's regular visits to England during school vacations in the summer raise the question of their mutual relationship. Although scarcely documented,[29] it is of particular interest and probably influenced the work of both men during the following two decades when they had found their final research topics, in Lohne's case the physical and mathematical work of Thomas Harriot. It seems to me that Whiteside, who at times was less than polished in his social manners, was particularly sensitive towards the outsider Lohne,[30] maybe partly because Whiteside himself came to the field from

[27]"Ich kenne Herrn Lohne von zahlreichen Oberwolfach-Tagungen Ende der 1960er und Anfang der 1970er Jahre. Er hat … vor allem über Harriot vorgetragen. Er kam immer mit Frau …. Mit Hofmann verband ihn eine Art Freundschaft (sofern Hofmann überhaupt Freunde hatte). Lohnes Vorträge waren klar; er sprach vorzüglich deutsch, und er war sehr freundlich. Mir kam er wie ein isolierter Einzelgänger vor".

[28]Lohne (1961a). Lohne puts the word "proof" in the title into ironic quotation marks. The word 'iconoclastic' is ours here. It is not in the letter.

[29]Maybe in the future some correspondence may be found in Whiteside's Nachlass.

[30]Lohne too had some deficits in communicating with people on a private basis. Personal information by M. Folkerts, and I. Grattan-Guinness who met him in Oberwolfach and described him as shy and reclusive.

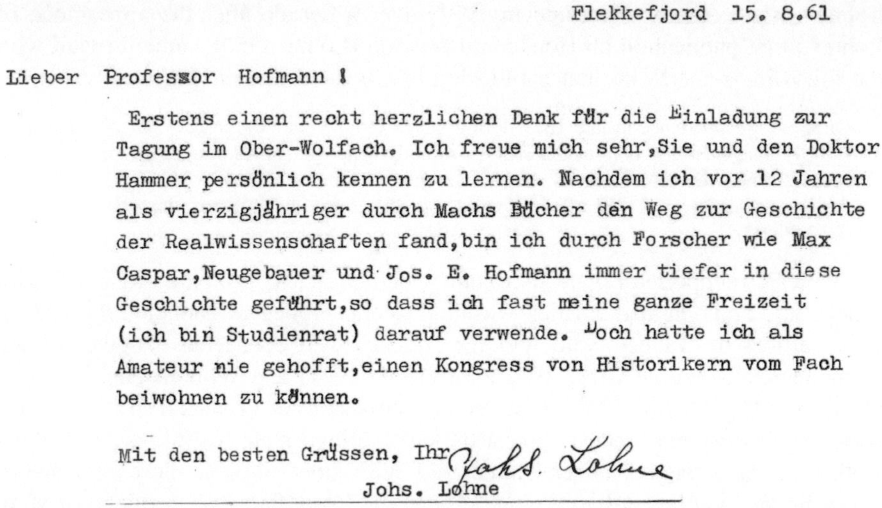

Flekkefjord 15. 8.61

Lieber Professor Hofmann !

Erstens einen recht herzlichen Dank für die Einladung zur Tagung im Ober-Wolfach. Ich freue mich sehr,Sie und den Doktor Hammer persönlich kennen zu lernen. Nachdem ich vor 12 Jahren als vierzigjähriger durch Machs Bücher den Weg zur Geschichte der Realwissenschaften fand,bin ich durch Forscher wie Max Caspar,Neugebauer und Jos. E. Hofmann immer tiefer in diese Geschichte geführt,so dass ich fast meine ganze Freizeit (ich bin Studienrat) darauf verwende. Doch hatte ich als Amateur nie gehofft,einen Kongress von Historikern vom Fach beiwohnen zu können.

Mit den besten Grüssen, Ihr Johs. Lohne
Johs. Lohne

Fig. 5 Excerpt from a letter by J. Lohne to J.E. Hofmann. Courtesy of the Archives Leopoldina Halle (Germany)

an underprivileged, nonscientific background.[31] The very first review I am aware of, which Whiteside wrote on Lohne's work, is from 1964, published in *Zentralblatt für Mathematik*. It is about Lohne's article on Newton's optical work, just mentioned, and strikes me as somewhat condescending, and maybe a bit jealous:

> The present paper is more in the nature of a progress report than a definite statement of viewpoint and much of the author's fertile, suggestive account must await a more closely reasoned and adequately documented sequel. Inevitably, the scholar familiar with Newton's unpublished papers will find a good deal to disagree with in interpretation of detail, and the author would seem to have an insufficient grasp of the wealth and thoughtfulness of allied mathematical papers. … But the author's intrinsic honesty and thoughtfulness outshine his clumsiness, and his paper is a most welcome addition to and symptom of the recent new-found interest in Newton's scientific work. (Zbl 104.00404)

This verdict on the much older Lohne alludes, not unexpectedly, to Whiteside's superior knowledge about Newton's mathematical manuscripts. Whiteside's judgment about Lohne became much more enthusiastic during the years, although both historians insisted on some differences of opinion regarding Newton, as we will see. Footnotes in Whiteside's Newton edition refer quite frequently to Lohne's research not only on Harriot but also on Hooke's relationship with Newton.[32] It seems very likely—given Lohne's yearly visits to England—that Whiteside profited from personal discussions with Lohne particularly on the physical and experimental aspects of Newton's work. Moreover, it seems like a deliberate action to promote

[31]For a biographical account cf. (Guicciardini 2009).

[32]Whiteside indicates e.g. that he used Lohne (1960) with benefit for his edition of Newton's manuscript "On Motion" (Whiteside 1974, 80, 151).

Lohne's name, when Whiteside in 1971, over a decade after the appearance of Lohne's first publication on Hooke and Newton (Lohne 1960), came forward with the following remarks on that publication in a belated *Zentralblatt* review:

> On the correspondence in late 1679 during which Hooke (as Newton later complained to Halley) 'magisterially' corrected Newton's initially somewhat crude notions on the path of free fall under terrestrial gravity, the author gives a lengthy critique which cuts sharply through a century of muddled thought on the topic by Newton historians. (ZB 0213.00402)[33]

Whiteside supported Lohne also against jealous attacks from the English Harriot scholar and Ph.D. in mathematics, Rosalind Cecilia Hildegard Tanner (1900–1992). In an article by Tanner, who was the socially well connected daughter of the mathematicians couple Grace Chisholm (1868–1944) and William Henry Young (1863–1942) published 1967 in the Italian journal *Physis* (Tanner 1967), one feels some condescension directed against the outsider and mere school teacher Lohne. In the correspondence between Tanner and Whiteside[34] there is more evidence of her jealousy with regard to Lohne. There one also finds Whiteside protesting against one passage of her article. After announcing his intent to Tanner, Whiteside sent a letter to the editor of *Physis* (Whiteside 1968), from which the following excerpt is taken:

> I would insist that Dr. Lohne has *at no time* either asked for or received my help in 'piecing together' his many insights into Harriot's achievements in mathematics and exact science. We have been, it is true, in regular correspondence since 1962 and in his letters he has revealed to me much of the detail of his important discoveries regarding Harriot (as I have told him, in return, of my own discoveries regarding Newton), but at all times my rôle has been that of a friendly, interested critic: I have never helped him in any more active rôle, either as technical collaborator, assistant or editor of his discoveries, with regard to any of his Harriot researches.

Whiteside would support Lohne on many other occasions as well, and he seems to have had a special agenda to help the disadvantaged school teacher. In 2000 Jackie Stedall kindly shared with me the following impression from her talks with Whiteside:

> Tom Whiteside spoke to me about Lohne on more than one occasion and clearly held him in high regard. His personal opinion was that Lohne had been badly treated in Norway for events he had been caught up in many years before. However, Whiteside was not always accurate in such political matters.

Lohne on his part kept his independence of judgment. His review of the Proceedings of a 1967 Texas conference which was the "Tricentennial Celebration of the The Annus Mirabilis of Sir Isaac Newton" begins with the following "qualifying" remarks:

---

[33]Some results of this work by Lohne of 1960 on Hooke and Newton are even today appreciated by the Newton specialist N. Guicciardini, cf. (Siegmund-Schultze 2010, 571).

[34]Cf. the Tanner Papers at the Liverpool University Archives.

> The hero of the Texas conference is introduced as follows: 'By general consent [of the English speaking people] Sir Isaac Newton is the greatest scientist [in mathematics and physics] who ever lived.' The qualifying brackets are ours. (Lohne 1970)

The contributions by the leading Newton scholars D.T. Whiteside and I.B. Cohen to the Texas Proceedings are mentioned with praise in the review. Even more enthusiastic are Lohne's remarks on Clifford Truesdell, whom Lohne without any doubt placed in the category of Hofmann, Neugebauer, Whiteside and others with their preference for internalist historiography:

> C. Truesdell (pp. 238–258) is again refreshingly unconventional and not a little ironical. When he sees a spade he calls it a spade, even if it be Newton's. According to Truesdell, *Principia*'s second book is 'a failure as an essay toward a unified, mathematical mechanics' and he concludes: 'It is from Newton's *Principia* that our forebears learned how to use the concept of force given a priori. The structure which that concept binds together we owe to many great mathematicians; among the founders, Newton is joined by Huygens, Leibniz, and James Bernoulli, and the great architect is Euler.'

It is well known, as summarized in Niccolò Guicciardini's obituary of Whiteside, that Truesdell's and Whiteside's views on Newton differed in several points:

> In general, Whiteside showed little interest in or respect for the achievements of eighteenth century mathematicians: he was inclined to say (if not to write) that all that Pierre Varignon or Johann Bernoulli had done on central force and resisted motion was to reformulate Newtonian mathematical results in a different language, the language of ordinary differential equations. Whiteside passionately defended a view of the long seventeenth century that is opposite to the one endorsed by another, equally great and idiosyncratic, scholar, Clifford Truesdell, who rather sided with Euler in casting a critical eye on Newton's geometrical style. (Guicciardini 2009, 8).

Anyway, Whiteside's disclaimer in his letter to *Physis* and their occasional disagreements on Newton notwithstanding, Lohne wrote in 1979 at the end of one of his last articles, which contains his legacy for further research on Harriot:

> For eighteen years I have had the privilege of being advised and helped very generously by Dr. D. T. Whiteside in Cambridge. (Lohne 1979, 311–312).

10 The Language Problem

In a letter from February 1966 Lohne thanked Hofmann for support with language editing, which was given to him as a "member of a smaller nation", thus enabling him to publish in the "world language" German.[35] From his grammar school in

[35]Lohne to Hofmann, 4 February, 1966, Hofmann Papers, Leopoldina Academy in Halle "Angehörigen kleiner Nationen". In the same letter Lohne mentions in passing that he did not know what a "Habilitation" was when he heard about this second German academic degree, which had been recently conferred to Hofmann's student C. Scriba.

Stavanger in the 1920s, where Lohne attended the "Latin track",[36] Lohne was apparently well educated in languages. The diploma of 1926 gives him a "very satisfactory" (meget tilfredsstillende) in all subjects, including the languages Norwegian, German, Latin, English and French.[37] Thus Lohne was well prepared to study Harriot's manuscripts. As to German, Lohne had a long stay in Germany during the war, German being anyway the foreign language most familiar to Norwegians at the time. As to English he was helped by Truesdell, apparently by Whiteside, Adolf Prag (1906–2004), the Jewish emigrant to England, and by his younger brother Andreas (1909–1981, who happened to be a resistance fighter against the Germans and finally fled to England during the war). Tanner makes the following slightly patronizing remark on Lohne in her article on the state of Harriot research, quoted above:

> He publishes in excellent English or German in Centaurus and Sudhoff's Archive, and recently in the Notes and Records of the Royal Society, none of them very generally accessible journals. (Tanner 1967, 244).

Tanner seemed to involuntarily exhibit her own English provincialism when classifying established journals in the history of science like *Centaurus* and *Sudhoffs Archiv* (misspelled as Sudhoff's Archive) as not very "accessible." Anyhow, she was probably not in any position to pass judgment on Lohne's German.[38] But Tanner had, of course, been right when alluding to Lohne's disadvantage with respect to languages, as compared to native speakers of English or German. Referring to another contribution by Lohne on Harriot, namely his publication in 1963 of Harriot's map of the moon, Tanner said in the same article in *Physis* about an English publication of the Russian E. Strout in 1965:

> In publishing this map, Dr. Strout bad been anticipated by J. Lohne in a paper with the title *Drømmen om manen,*[39] of which, being in Norwegian and in an obscure periodical he was naturally quite unaware. (Tanner 1967, 291)

To Tanner's credit, she did alert the readers to this article by Lohne in an "obscure" periodical. Whiteside recommended in his Essay Review "In Search of Thomas Harriot" another Norwegian article by Lohne (from 1964, on ballistics and

[36]This forced Lohne to take a supplementary exam to his high school exam in Oslo in order to be admitted to study physics at the university.

[37]In mathematics Lohne received the exceptional mark "particularly satisfactory" (særdeles tilfredsstillende) which also became the overall mark of his school diploma. A copy of the school diploma is contained in Lohne's Nachlass with B. Øksendal in Oslo.

[38]Lohne's German was good, as for instance described by Folkerts above, but not perfect, evidenced not just by his letters but also by his publications, even after having received help in language editing.

[39]Lohne 1963b: "Dream of the moon", the word "månen" in the title being misspelled.

Galileo) to the readers (Whiteside 1975, 70). Whiteside obviously felt less prepared than Tanner to accept the language barrier as an excuse for ignorance.

The fact that he was chosen to write the entry on Harriot for Charles Coulston Gillispie's *Dictionary of Scientific Biography* may partly testify to Lohne's success in breaking the English language barrier. But this publication seems to point above all to Gillispie's preference for Lohne's internalist approach compared to the more context oriented approach by the American John Shirley.

11 Lohne's Methodology

It was typical for Lohne's work—and in this respect Lohne shared the approach of other rising historians of science at the time such as Whiteside, Pepper, and Truesdell—that he focused primarily on concrete scientific results and methods, as presented in the original manuscripts, and less on the biographies and the social environments of the historical figures he described. Lohne was fortunate to hit on his research topics at a time when outsiders like him with scientific but without historical training had a chance to enter the international scene. This was not long after more professional general historians without special competence in individual sciences such as Herbert Butterfield (1900–1979) and Rupert Hall (1920–2009) had dominated much of the research environment in the history of science. That generation of scholars had attacked the Whig approach to history from a quite different angle.

Lohne set it as his task to prove concrete historical conjectures on the basis of extant manuscripts. In several instances he proved conjectures which had been proposed by others before him. Harriot's priority with respect to the sine law of light refraction (claimed by J. Pell, J. Shirley and others) had been assumed by others before Lohne. This is also true of Harriot's calculation of the meridional parts of the Mercator map (claimed by D. Sadler, F. George and others) and Harriot's drawing of the moon maps (F. von Zach and others). But Lohne was the first to prove these historical conjectures.

Lohne was without any doubt successful and influential among his peers, not just because he found single facts, but also methodologically speaking. Many took to their hearts his warnings, uttered in "The increasing corruption of Newton's Diagrams" (Lohne 1967) against some historians' careless treatment or even distortion of original historical drawings.

Lohne's focus on the details of the historical methods of scientific measurement (Lohne 1968) influenced even "constructivist" approaches to the history of physics (Schaffer 1989). Lohne was one of the first historians to replicate historical experiments, which he did in his class room. This was later in the 1980s systematically used as a method by a modern school of historical research, several of them belonging to a group of scholars at the German university of Oldenburg. Both with his emphasis on drawings and on experiments Lohne went "beyond armchair study" of classical texts, comparable in these respects with the other Scandinavian historian, Drachmann, whom we mentioned before and for whom this description was coined (Anon. 1972, 438).

Lohne's deep focus on the internal history of physics and mathematics during the 17th century and his predilection for physical experiments came at a price. In some of his utterings about later periods of physics and mathematics he seemed rather uninformed, sometimes prejudiced. I was unable to find in his articles or reviews any allusion to Einstein's theory of relativity. Instead he calls Max Planck, who deeply admired Einstein, "today's Newton" ("nåtidas Newton") in an application to the Norwegian Research Council,[40] a title which in the literature is usually reserved for Einstein.

Lohne was, of course, conscious of his own merits. His 1975 review of the Proceedings of a Harriot conference at the University of Delaware in 1971 in which he had not taken part and which was organized by the American John Shirley, Lohne began with the following words, which alluded to Shirley:

> The author is an authority on the period but is less well-informed on the science of Harriot himself. In view of the neglect which has hitherto surrounded the work of Harriot, a short survey may be of value.[41]

Shirley, on his part, praised in his 1983 biography on Harriot the work of Lohne particularly on the history of optics, although, as described above, he did not give him full credit in the case of the sine law. Shirley quoted in his Harriot biography a long passage on the "secret of Harriot's accuracy" from an unpublished talk by Lohne in 1973.[42] Shirley apparently did not ask Lohne for permission to publish that passage, which he had received from Whiteside who had once read the talk in Oxford on Lohne's behalf. This shows something of the lack of communication between the two leading Harriot scholars of exactly the same age, Shirley and Lohne. It also points to different scientific cultures in which they lived. As to Shirley's and Lohne's reputation within Harriot research, the lack of a book from Lohne's pen has probably acted in Shirley's favor. Lohne undertook several efforts in this direction,[43] but he did not succeed. This was due to objective (isolated position in Norway) and subjective (unwillingness to accept advice for changes) limitations of the outsider Lohne.

[40]Lohne on 5 April 1960 to Norwegian Research Council. Riksarkiv Oslo.

[41]The whole passage containing a short description of Harriot's scientific results by Lohne is quoted as appendix E 1975 in (Siegmund-Schultze 2010).

[42]This passage has been re-published by me as an appendix (E 1973) to (Siegmund-Schultze 2010).

[43]As documented in the appendix of Siegmund-Schultze (2010), which contains drafts of tables of contents of two books which Lohne planned but which were never published.

12 Lohne's Legacy, and Conclusions

Concluding this paper I will reflect once again, on the basis of our biographical evidence in the case of Lohne and with some emphasis on Scandinavia, on the relative notions of the "outsider" and the "main stream" in the historiography of mathematics and sciences.

It seems to me that the process of professionalization of this discipline was in no small measure shaped by various outsiders and by strong, partly even idiosyncratic individuals. These came from different countries and social strata, were influenced by varying scientific traditions, and by various research standpoints, including Marxist historiography, and religious feelings.

Lohne was presented here mainly as an "outsider" to the profession of the historiography of mathematics and the sciences and I stressed the new perspectives which he brought to the field and indicated that there were other "outsiders" particular in smaller countries.

However, the "outsider-position" is not restricted to individuals, languages or countries but, in a way, it pertains to the discipline of the historiography of mathematics and sciences as a whole. The rise, and in particular the professionalization and institutionalization of that discipline during the second half of the 20th century was, and is, intimately connected with its hybrid character, its political and philosophical aspects, and with the volatility of institutional support for it. Much in its development depended on publishers and on journals such as *Centaurus, Isis*, and *Archive for History of Exact Sciences*. Societies such as the History of Science Society (HSS) in the U.S., and the influence and support of the practitioners in the contact disciplines, such as mathematics and physics, played a huge role. Not least due to political events and economic restraints the discipline experienced ups and downs in the process of its professionalization, and that process might not be fully accomplished or finished even today. Political changes such as the fall of the Iron Curtain destroyed entire national schools in the history of mathematics such as the East German and the Russian ones (Siegmund-Schultze 2012).

Looking at the question of the "outsiders" and the "main stream" from this more general standpoint allows to put into perspective the restrictions, the peculiarities and also the indubitable uniqueness of historians such as Johannes Lohne, who remains a remarkable and memorable contributor and member, even a pioneer within the history of science community half a century ago.

Acknowledgments I thank my colleague Rolf Nossum (Kristiansand) for advice with respect to content and language. Bernt Øksendal (Oslo) and Menso Folkerts (Munich) receive thanks for providing archival material related to Lohne. The editors of the present volume deserve thanks for various recommendations with respect to content and form.

References

Anon. (1972). The Leonardo Da Vinci medal. *Technology and Culture, 13*, 434–439.
Cohen, I. B. (1980). Newton, Isaac. *Dictionary of Scientific Biography, 10*, 42–103.
Drachmann, A. G. (1967). Archimedes and the science of physics. *Centaurus, 12*, 1–11.
George, F. (1956). Hariot's meridional parts. *Journal of Navigation, 9*, 66–69.
Guicciardini, N. (2009). In memoriam Derek Thomas Whiteside (1932–2008). *Historia Mathematica, 36*, 4–9.
Lohne, J. (1959). Thomas Harriott (1560–1621): The Tycho Brahe of optics. Preliminary notice. *Centaurus, 6*, 113–121.
Lohne, J. (1960). Hooke versus Newton: An analysis of the documents in the case on [sic] free fall and planetary motion. *Centaurus, 7*(1), 6–52.
Lohne, J. (1961a). Newton's "Proof" of the Sine Law and his mathematical principles of colors. *Archive for History of Exact Sciences, 1*, 389–405.
Lohne, J. (1961b). Jos. E. Hofmann: Problemgeschichte der Mathematik im Unterricht. Stuttgart: Klett; 1961. *Den Høgre Skolen, 60*(17), 591–592 [Review].
Lohne, J. (1963a). The fair fame of Thomas Harriott: Rigaud versus Baron von Zach. *Centaurus, 8*, 69–84.
Lohne, J. (1963b). Drømmen om månen: Newtons eple. *Naturen, 87*(3), 159–174.
Lohne, J. (1965a). Thomas Harriot als Mathematiker. *Centaurus, 11*, 19–45.
Lohne, J. (1965b). Isaac Newton: The rise of a scientist 1661–1671. *Notes and Records of the Royal Society of London, 20*(2), 125–139.
Lohne, J. (1966). Dokumente zur Revalidierung von Thomas Harriot als Algebraiker. *Archive for History of Exact Sciences, 3*, 185–205.
Lohne, J. (1967). The increasing corruption of Newton's diagrams. *History of Science, 6*, 69–89.
Lohne, J. (1968). Experimentum crucis. *Notes and Records of the Royal Society of London, 23*(2), 169–199.
Lohne, J. (1970). The Annus Mirabilis of Sir Isaac Newton. *The Texas Quarterly, 10* (1967), 3: no paging available. *Mathematical Reviews, 39*, # 2581 [Review].
Lohne, J. (1971). Glimt fra 700 års skole- og populærfysikk. *Fra Fysikkens Verden, 33*(3), 1–4.
Lohne, J. (1972). Biografiske leksika. Matematikere og naturforskere. *Den Høgre Skolen, 71*(18), 836–838 [Review].
Lohne, J. (1979). Essays on Thomas Harriot: I. Billiard Balls and Laws of Collision, II. Ballistic Parabolas, III. A survey of Harriot's scientific writings. *Archive for History of Exact Sciences, 20*, 189–312.
Pepper, J. V. (1968). Harriot's calculations of the meridional parts as logarithmic tangents. *Archive for History of Exact Sciences, 4*, 359–413.
Sadler, D. H. (1953). The Doctrine of Nauticall Triangles Compendious. II—Calculating the Meridional Parts. *Journal of Navigation, 6*, 141–147.
Schaffer, S. (1989). Glass works: Newton's prisms and the uses of experiment. In D. Gooding, T. Pinch, & S. Schaffer (Eds.), *The uses of experiment* (pp. 67–104). Cambridge: Cambridge University Press.
Schemmel, M. (2008). *The English Galileo: Thomas Harriot's work on motion as an example of preclassical mechanics* (Vol. 2). Berlin: Springer.
Shirley, J. W. (1951). An early experimental determination of Snell's law. *American Journal of Physics, 19*, 507–508.
Shirley, J. W. (1983). *Thomas Harriot: A biography*. Oxford: Clarendon Press.
Siegmund-Schultze, R. (2010). Johannes Lohne (1908–1993) revisited: documents for his life and work, half a century after his pioneering research on Harriot and Newton. *Archives Internationales d'histoire des sciences, 60*(165), 569–596.
Siegmund-Schultze, R. (2012). Hans Wußing (1927–2011) and the blooming of the history of mathematics and sciences in the German Democratic Republic—A biographical essay. *Historia Mathematica, 39*, 143–173.

Stedall, J. (2000). Rob'd of Glories: The posthumous misfortunes of Thomas Harriot and His Algebra. *Archive for History of Exact Sciences, 54*, 455–497.

Tanner, R. C. H. (1967). Thomas Harriot as mathematician. A legacy of hearsay. *Physis, 9*, 235–247, 257–292.

Waters, D. W. (1958). *The art of navigation in England in Elizabethan and Early Stuart Times*. New Haven: Yale University Press.

Whiteside, D. T. (1961). Patterns of mathematical thought in the later seventeenth century. *Archive for History of Exact Sciences, 1*, 179–388.

Whiteside, D. T. (1968). Lettere alla redazione. *Physis, 10*, 258.

Whiteside, D. T. (1974). *The mathematical papers of Isaac Newton* (Vol. VI). Cambridge: Cambridge University Press.

Whiteside, D. T. (1975). Essay review: In Search of Thomas Harriot. *History of Science, 13*, 61–70. [Review] Shirley, J. W. (1974). *Thomas Harriot: Renaissance Scientist*. Oxford: Clarendon Press.

Contextualizing Unguru's 1975 Attack on the Historiography of Ancient Greek Mathematics

Martina R. Schneider

Abstract

In 1975 S. Unguru published his controversial paper on the need to rewrite the history of ancient Greek mathematics. The origin of the paper is sketched according to Unguru's own story, and then the paper is contextualized in some of the historiographic and disciplinary discussions and shifts taking place during the decade before its publication. The focus is not only on the history of (Greek) mathematics (J. Klein, A. Szabó, M. S. Mahoney), but a rather broad approach is taken to capture the wider (U.S.-American, academic) discourse around questions of professionalisation of history of science/mathematics. This analysis shows the complexity of the discursive field in which Unguru's paper was written and received.

Keywords

Historiography of mathematics · Historiography of science · 20th century · Sabetai Unguru · Jacob Klein · Árpád Szabó · Michael S. Mahoney · André Weil · Robert N. Bellah

M.R. Schneider (✉)
Institut für Mathematik, AG Geschichte der Mathematik, Johannes Gutenberg Universität Mainz, Staudinger Weg 9, 55099 Mainz, Germany
e-mail: mschneider@mathematik.uni-mainz.de

M.R. Schneider
ERC-Project "Mathematical Sciences in the Ancient World" (SAW), CNRS, SPHERE, University Paris 7 (Diderot), Paris, France

V.R. Remmert et al. (eds.), *Historiography of Mathematics in the 19th and 20th Centuries*, Trends in the History of Science,
DOI 10.1007/978-3-319-39649-1_12

1 Introduction

It is well known that Unguru's 1975 article (Unguru 1975a) on the need to rewrite the history of Greek mathematics set off a rather fierce controversy (van der Waerden 1976; Freudenthal 1977; Weil 1978a, b; Unguru 1979). The controversy —as well as the methodological issues raised in Unguru's article—went beyond the historiography of the small field of ancient Greek mathematics, and touched upon the history of mathematics as a whole. The debate has been analyzed in detail (see for example Berggren 1984; Knorr 1990; Artmann 1991; Rowe 1996; Hoyrup 2011). However, what is often missing is a closer contextualization of the discourse (an exception is Kastanis and Thomaidis 1993).

In this paper, I will contextualize the catalyst of this debate. For this purpose I use the term contextualization in a very broad sense to include the prehistory of Unguru's article as well as some concurrent events in the academic world that are not directly connected to it, but are in my opinion nevertheless relevant for a more comprehensive understanding of its genesis. In this way I hope to capture some features of the on-going methodological discussions and of the North-American academic milieu at the time.

I will start by giving Unguru's personal recollection of the paper's genesis. Then I will examine cross-currents in the standard historiography of Greek mathematics during the decade before Unguru's paper. Despite having different agendas, scholars like Mahoney, Klein, and Szabó also raised methodological issues many of which were centered around the concept of geometric algebra. Finally I will focus on academic controversies on a wider disciplinary level in the USA during that period. They concern discussions of the role of scientists in the history of science, and the status of mathematics and mathematicians in the academic world. Thus, they are related to broader issues of professionalisation and to tensions between humanists and scientists.

2 The Story as Told by Unguru Himself

Sabetai Unguru, born in Romania in 1931, started studying philosophy at the University Alexandru Ioan Cuza in Jassy after the Second World War.[1] Founded in 1860, it was the oldest university in Romania. One year after Unguru enrolled, the philosophy faculty in Jassy had to close down. Unguru then changed to German philology which also ceased to exist soon after. However, Unguru continued studying both philosophy and philology on his own. He then took up history and wrote a (Master's) thesis on the independence of the Romanian principalities as reflected in the diplomatic reports by French representatives in Romania.

[1]The biographical information provided here is, unless stated otherwise, based on an interview with Unguru conducted by D. E. Rowe and the author in Wiesbaden in May 2013.

In the course of his studies, Unguru became interested in ancient history, especially in Greek history and in the history of Greek mathematics. Therefore he decided to study mathematics as well, and attended lectures on the history of mathematics given by the mathematician Ilie Popa (1907–1983). Popa, who did research in geometry and analysis, had a strong interest in the history of mathematics, especially that of Romanian mathematics (Poggendorff 2004, 790 ff.). Unguru became his teaching assistant, and he began reading editions of ancient Greek mathematical texts and studying secondary literature on Greek mathematics.

Like many Jews in Romania after World War II (Glass 2002, esp. Chap. 5 and Table 8; Glass 2005), Unguru's family asked for permission to emigrate to Israel. When it was finally granted to his parents and his sister, it turned out that Unguru was not allowed to leave Romania and that they had to leave without him. It later became clear to Unguru that this was due to the fact that Unguru had refused to collaborate with the Romanian secret service Securitate. After the incident with the Securitate Unguru was sacked as a teaching assistant, and he would make ends meet mainly by giving private lessons. However, due to the help of former colleagues, he was able to continue studying privately at the university. The director of the library gave Unguru full access to library books, even those in the restricted section. For Unguru these three and a half years of studying turned out to be intellectually his most fruitful ones. In 1961 he was allowed to leave Romania and joined his family in Tel Aviv in the same year.

In Israel Unguru taught mathematics in a school. He wanted to do a Ph.D. in the history of science, but at the time there was no possibility to do so in Israel. So he applied at various US-American universities. As he was not able to obtain the required letter of recommendation from his Romanian professors he turned to the physicist and historian of science Shmuel Sambursky (1900–1990). In 1959 Sambursky was appointed professor of the history and philosophy of science at the newly created department at the Hebrew University of Jerusalem. He specialized among other things in Greek science (Elkana and Segre 1985). The letter of recommendation Sambursky wrote for Unguru after careful examination was so good that 4–5 universities were willing to take him on. However, the University of Wisconsin-Madison was the only one to offer Unguru a stipend from the very beginning. So Unguru decided to go there, and left Israel in 1966.

Wisconsin was the first university in the USA to establish a separate department of history of science in 1941. When Unguru studied there, the department was thriving. Between 1968 and 1973 one sixth of all the Ph.D.s in the history of science in the USA were completed at Wisconsin (Hilts 1984, 88). Unguru wrote his thesis on the mathematical part of Witelo's *Perspectiva* on optics (see Unguru 1972). Unguru continued studying ancient Greek mathematics and the relevant secondary literature. It was then that he arrived at the conclusion of "the historical unsoundness of the notion of 'geometrical algebra' independently" (Unguru 1975a, 81, fn.26). The on-going discourse about ahistoricism—Unguru mentioned texts by Herbert Butterfield (1900–1979) and Thomas Kuhn (1922–1996) as part of the reading list for graduate students in history of science at Wisconsin—supported Unguru's way of thinking.

After finishing his Ph.D. in 1970, Unguru obtained a position as assistant professor at the University of Oklahoma. There he taught a graduate seminar on Euclid's *Elements* in the fall of 1972. It was then that he came to his "final conviction about the necessity to discard and repudiate 'geometric algebra' as an explanatory device in the study of the history of Greek mathematics and about the need, growing out of this rejection, to rewrite that history on a sound basis" (Unguru 1975a, 81, fn.26).

During his time at the University of Oklahoma, Unguru started to write his article and to present his theory. In the fall of 1972 he gave a talk at the University of Jerusalem. Unguru wrote that he got a "mixed reception" to this talk: "historians and the (very few) historically-minded mathematicians present seemed to like its conclusions, while the mathematicians (to put it mildly) remained unconvinced" (Unguru 1975a, 81, fn.26). He also gave a paper at the 14th International Congress for the History of Science in Japan in August 1974. Its title was "On the Need to Rewrite the History of Greek Mathematics" and a shortened version of it was published in the conference proceedings (Unguru 1975b). It is clear that by that time Unguru had everything worked out.

Unguru remembered that he had shown his findings to a friend while he was at Oklahoma. This friend did not work in the history of science field. He told Unguru nevertheless that his article was insulting and that as a result he would be cut into pieces. So Unguru decided not to show his manuscript to anyone any more.

When the paper was finished and Unguru tried to publish it, Unguru recalled the following: He sent it to *Isis*. The editor rejected it on the basis of it being too lengthy. Then Unguru thought about publishing it with the *Archive for History of Exact Sciences*. Willy Hartner (1905–1981) was willing to be his referee (communicator). He told Unguru that he had raised important issues and fundamental questions, and that his paper should be published. But he also remarked that Unguru was saying things in a harsh way, and warned him of possible consequences without however explicitly asking him to change the paper's style.

The founder and editor of the journal, Clifford Truesdell (1919–2000), however was—according to Unguru—against its publication. Usually, if one member of the editorial board was willing to communicate a paper, the paper was printed in the *Archive* without further ado. Only by threatening to resign from the editorial board of the *Archive* did Hartner manage to get Unguru's article published (Unguru 1975a). Later on, however, Unguru was denied the opportunity to write a reply to the critical articles launched by Hans Freudenthal (1905–1990), Bartel van der Waerden (1903–1996) and André Weil (1906–1998) in the *Archive* (van der Waerden 1976; Freudenthal 1977; Weil 1978b). Truesdell argued then that the journal was not intended for the publication of controversies. In the end, Unguru was able to publish his response to the criticism in *Isis* in 1979 (Unguru 1979).

In retrospect, Unguru thinks that it was the style of his paper—one might call it polemical—that helped its reception and contributed to its notoriety. The provocation required a response. The issue could not be just dropped or passed over in silence. Unguru also claims that this was not his intention at the time. The style was rather his very personal style of writing. As a young man Unguru had wanted to become a literary critic. He had read many Romanian and international literary critiques and loved the work of the Danish critic Georg Brandes (1847–1931). He even had tried to imitate him when he was a student of German philology. In the 1975 article Unguru quoted the German Jewish writer and critic Heinrich Heine (1797–1856) (Unguru 1975a, 68, fn. 4). This critical way of writing and thinking became his habit. When reading texts—be it fiction or non-fiction—Unguru focuses on the weak points of the narrative. It is exactly this critical attitude with which he scrutinized the secondary literature on Greek mathematics. Unguru believes that he would not have been able to change his style even if he had been asked to do so.

The story on the origin of his controversial paper as told by Unguru himself in the paper and the interview has basically two main messages: Firstly, Unguru had been working on the topic for a long time before the article was published. He had already started while he was studying in Romania. Secondly, Unguru had arrived at his conclusions independently. He was not influenced by others, even though he was aware of the critical papers of other scientists and cited them in support of his cause. He portrayed himself basically as a lone wolf. As fascinating as Unguru's autobiographical account is, historically one is tempted to search for further contextualization of the article.

3 Cross-Currents to the Standard Historiography

A central point of Unguru's criticism was the concept of geometric algebra. It had been introduced by the Danish mathematician and historian of mathematics Hieronymus Georg Zeuthen (1839–1920) in his study of conic sections in 1885 (Zeuthen 1886).[2] It was then taken up (and modified) by the French historian of science Paul Tannery (1843–1904), and later regularly used by the British philologist and historian of science Thomas Heath (1861–1940) in his edition of Euclid's *Elements*.[3] In the 1930s the concept was an integral part of Neugebauer's studies on the history of algebra in ancient civilisations. Otto Neugebauer (1899–1990) used it firstly to explain how the Greek mathematicians got around the problem of incommensurability, and secondly to argue for a line of transmission of mathematical knowledge from Mesopotamia to Greece (Neugebauer 1936). According to Neugebauer, the Greek method of application of areas became "immediately comprehensible with respect to both its orientation and its methods of solution if

[2]The German edition was published one year later. One can also argue that the concept was implicitly introduced much earlier: e.g. Mahoney (1971b, 25) traced it back to the 16th century French humanist Petrus Ramus.

[3]On Heath see Wardhaugh's contribution "Greek Mathematics in English: The Work of Sir Thomas L. Heath (1861–1940)".

one takes it as nothing but the translation of Babylonian methods in the language of geometric algebra".[4]

In the 1950s the concept entered into two widely popular introductions into the history of ancient science: Neugebauer's *The exact sciences in Antiquity* (1952), which was based on six lectures given by Neugebauer at Cornell university in the autumn 1949, and B.L. van der Waerden's *Science Awakening* (1954 English edition; first edition in Dutch published 1950). Neugebauer remarked that the explanation of the method of application of areas by geometrical algebra was only one among others—without giving any references to those alternative interpretations. He claimed that it was "by far the most simple and direct explanation", then, however, pointed out that "simplicity is by no means equivalent with historical proof" (Neugebauer 1957, 150). This kind of historiographical cautiousness or warning is missing in van der Waerden's presentation. He argued that "there is no danger of misrepresentation, if we reconvert the derivations into algebraic language and use modern notations" (van der Waerden 1954, 119). Thus, by the time Unguru had entered the field, geometric algebra was already part of the standard narrative on Greek mathematics.

3.1 First Criticism and Critical Remarks

In the standard interpretation the crisis following the discovery of irrational numbers played a central role. It was said to be the key motivation for Greek scholars to create geometric algebra. The concept of a foundational crisis in Greek mathematics was introduced by the mathematician Helmut Hasse (1898–1979) and the philosopher Heinrich Scholz (1884–1956) in 1928 (Hasse and Scholz 1928)—precisely at the time when Luitzen Egbertus Jan Brouwer (1881–1966) and David Hilbert (1862–1943) were fighting over the right foundations of mathematics (Weyl 1921). Discussions followed about dating the crisis, but otherwise it was widely accepted until the 1960s. In 1966 Freudenthal published severe doubts whether there had actually been a foundational crisis in Greek mathematics (Freudenthal 1966).

Apart from this harsh criticism of a central element in the standard interpretation, minor critical remarks were also made. In his paper "Another look at Greek geometrical analysis" Michael S. Mahoney (1939–2008) called for a historisation of Greek mathematics (Mahoney 1968). Mahoney had just finished his Ph.D. with Thomas Kuhn on Pierre de Fermat (1607–1665) in the 'Program in History and Philosophy of Science', founded by Charles C. Gillispie (1918–2015), at Princeton University. Mahoney (1968, 319; emphasis in the original) claimed in the introduction that

[4]Neugebauer (1936, 252): "die ganze Flächenanlegungsaufgabe wird sowohl hinsichtlich Fragestellung wie hinsichtlich Lösungsmethode unmittelbar verständlich, wenn man sie nur als die sinngemäße Übersetzung der babylonischen Methoden in die Sprache der geometrischen Algebra auffaßt". For Neugebauer, see also Rowe's contribution to the present volume "Otto Neugebauer's Vision for Rewriting the History of Ancient Mathematics".

> [p]ast discussions have failed to take proper account of Greek geometrical analysis as an historical phenomenon. Rather, they have concentrated on the logical and epistemological problems raised by the accounts of analysis by *Pappus* and a scholist on the *Elements* of *Euclid*. [...] But examinations of the logical and epistemological status of analysis do not lead, or they have not led, to the central historical problem that Greek geometrical analysis poses: what was its role in the continuing tradition of Greek mathematical research?

In his analysis of that role, Mahoney took the concept of geometric algebra for granted (see for example Mahoney 1968, 331). He concluded that "[i]n the realm of geometrical analysis in particular, *Tannery's* remark holds true: the Greeks did not so much lack methods of mathematics as means to express them" (Mahoney 1968, 348)—thereby stressing the importance of symbolism in the development of mathematics.

3.2 Criticism of the Concept of Geometric Algebra

However, as Neugebauer had hinted, even before Unguru's attack of the concept and its contemporary representatives, the concept had already been criticized by various scholars—some of whom are also mentioned in Unguru's article (Unguru 1975a). I will not give a complete analysis of this critical line, but will focus on criticism raised by Unguru's contemporaries Jacob Klein (1899–1978) and Árpád Szabó (1913–2001) when Unguru was in the USA.

Jacob Klein—Criticism from a Philosophical Point of View

In 1936 Jacob Klein published an article on the genesis of algebra that gave a different picture from that in Neugebauer's article in the very same issue of *Quellen und Studien* (Klein 1936a, b). It is quite remarkable that the editors of the journal, in particular Neugebauer as one of them, published simultaneously two conflicting views.

Klein, who was born in Liepaja, then part of Russia, and educated in Russia, Belgium and Germany, studied mathematics and physics in Berlin for three years from 1917 onwards. He had intended to study philosophy with Edmund Husserl (1859–1938) in Freiburg, but on Husserl's advice he began to study philosophy in Marburg. His 1922 Marburg Ph.D., supervised by Nicolai Hartmann (1882–1950), was on the relation between the logical and the historical element in Hegel's philosophy. In the following years, he studied with Martin Heidegger (1889–1976), worked for a year at the Institute of Theoretical Physics in Berlin, and became a lecturer of philosophy in Marburg. Hopkins (2011a) argues in detail that Klein's 1936 article presents a critical study of Husserl's conception of science, in particular of his concept of number. Due to his Jewish origin, Klein's life became difficult when the Nazis came into power in 1933. Klein spent time in Prague, Berlin and England, and finally emigrated to the USA in 1937/38. There he obtained a position at St. John's College in Annapolis, Maryland, where he taught for the rest of his life.[5]

It was at that time that St. John's College changed its curriculum to a thorough study of classical texts known today as the Great Books Program. Klein shaped this interdisciplinary program while he was dean in the late 1940s and throughout the 1950s. It was one of his colleagues, Eva Brann (*1929), also a Jewish emigrant from Germany, who decided to translate Klein's 1936 article into English, first without Klein's knowledge (Brann in Hopkins 2011a, xxvii). In 1968, a translation of the stylistically and content-wise unaltered article was published as a book under the title "Greek mathematical thought and the origin of algebra" (Klein 1968). While Klein's original article received only little attention, the book came out at the right time (see below).

Klein saw a strong interaction between the mathematics and the philosophy of the ancient Greeks. He believed that without a knowledge of Greek philosophy there was no chance of understanding Greek mathematics, and vice versa. In the article he studied the development of the conceptual structure which he perceived as being at the heart of the scientific revolution of the 16th and 17th centuries. He argued that the conceptual changes that had taken place during the reception of Greek mathematics at that time were vital for the development of modern symbolism, but until then had been neglected by research.

It is in this context that Klein (1968, 5, emphasis in the original) severely criticized the use of modern symbolism when dealing with ancient Greek mathematical texts:

> But they [the conventional presentations] always take for granted, and far too much as a matter of course, the *fact* of symbolic mathematics. They are not sufficiently aware of the *character* of the conceptual transformation which occurred in the course of this assimilation and which constitutes the indispensable condition of modern mathematical symbolism. Moreover, most of the standard histories attempt to grasp Greek mathematics itself with the aid of modern symbolism, as if the latter were an altogether external "form" which may be tailored to any desirable content. And even in the case of investigations intent upon a genuine understanding of Greek science, one finds that the inquiry starts out from a conceptual level which is, from the very beginning, and precisely with respect to fundamental concepts, determined by modern modes of thought. To disengage ourselves as far as possible from these modes must be the first concern of our enterprise.

Klein also attempted to show "how little this concept [of geometrical algebra of Zeuthen] does justice to the Greek procedure" (Klein 1968, 62). In Klein's view the concept of geometrical algebra does not distinguish sufficiently between

> the *generality of the method* and the *generality of the object* of investigation. Thus Zeuthen himself immediately relates his concept of "geometric algebra" to that of "general magnitude"; "geometric algebra," being a general method, has for its proper object precisely this "general magnitude." Thereby the whole complex of problems which presented itself to the ancients because their "scientific" interest was centered on questions concerning the mode of being of mathematical objects is obviated at one stroke; ancient mathematics is characterized precisely by a tension between method and object. The objects in question

[5]Cf. Hopkins (2011b) for biographical information on Klein.

> (geometric figures and curves, their relations, proportions of commensurable and incommensurable geometric magnitudes, numbers, ratios) give the inquiry its direction, for they are both its point of departure and its end. [...] modern mathematics, and thereby also the modern interpretation of ancient mathematics, turns its attention first and last to *method as such.* It determines its objects *by reflecting on the way in which these objects become accessible through a general method.* Thus, arguing from the "generality" of the linear presentation in Euclid's "arithmetical" books, namely from the fact that with their aid *all arithmoi*, in all their possible relations to one another, can be grasped, its proponents reach the conclusion that this "generality" of presentation is intent upon "general magnitude." Now what is characteristic of this "general magnitude" is its indeterminateness, of which, as such, a concept can be formed only within the realm of symbolic procedure. But the Euclidean presentation is *not* symbolic. It always intends *determinate* numbers of units of measurement, and it does this *without any detour through a "general notion" or a concept of a "general magnitude."* (Klein 1968, 122, emphasis in the original)

These are some of the main reasons why Klein rejected the use of the concept of geometric algebra in order to understand Greek mathematics. Instead, he approached Greek mathematics via Greek philosophy. In particular, Klein analysed the concept of arithmos in Greek mathematics as well as its ontological status, and its developments up to the 17th century. He found a shift in the interpretation of the concept from an ensemble of definite objects to a symbol—"number"—that signified magnitude in general and was made to be used in equations. According to Klein, it was precisely these changes that allowed the development of symbolical algebra.

Árpád Szabó—Criticism from a Philological Point of View

The Hungarian philologist Árpád Szabó took a different methodological approach to Greek mathematics. He tried to gain a better understanding of the development of Greek mathematics by looking at the philological history of central mathematical concepts. In Szabó's view, the peculiarity of ancient Greek thought could be best captured by a linguistic analysis since everything in Greek mathematical texts was put down in words alone without the use of symbols.

Szabó studied philosophy, history and philology, and was appointed professor of classical philology at the Hungarian University of Debrecen in the late 1930s.[6] In 1946, after the Second World War, he was also appointed fellow of the Eötvös Collegium in Budapest maintaining his professorship in Debrecen until 1948. At Debrecen he got to know both Imre Lakatos (1922–1974) and Alfréd Rényi (1921–1970). Lakatos and Szabó planned to work together on the history of dialectics—Lakatos in the realm of modern mathematics, Szabó of ancient mathematics. In a certain way Lakatos' *Proofs and Refutations* and Szabó's *Anfänge der griechischen Mathematik* have their common roots in this joint project (Szabó 1969; Lakatos 1976). Due to some critical remarks about Stalin's publications on linguistics, and his involvement in the anti-Soviet risings in 1956, Szabó got into trouble, and lost his job as professor in 1957. In 1958, with the help of Rényi, Szabó got a job as

[6]See Máté (2006) for biographical information of Szabó and his relation to Lakatos.

historian of mathematics at the Institute of Applied Mathematics at the Budapest Academy of Sciences, which Rényi had founded and ran.

Szabó started publishing papers on the history of ancient mathematics in Hungarian journals from the mid 1950s onwards. His first publication in a non-Hungarian journal was in German on the foundations of early Greek mathematics in *Studi italiani de filologia classica* (Szabó 1958). From the 1960s onwards Szabó published his research in various international journals—mostly in German. In English he published only two articles in *Scripta mathematica* in 1964 (Szabó 1964a, b).

Szabó's book *Anfänge der griechischen Mathematik*, published in 1969 (and translated into English in 1978), is a revised and extended resumé of all his papers on the topic.[7] It begins with the defense of Szabó's linguistico-historical method. Szabó made it clear that it is not enough to study translations and to consider the original only in cases of doubt—as van der Waerden had suggested in his *Science Awakening*. What Szabó called for and tried to engage in was a systematic linguistic study, a linguistic history of (mathematical) concepts:

> As far as I am concerned, a special significance is attached to language in this analysis. I intend to investigate Euclid's mathematical language from a historical point of view, for this technical language provides living proof of the developmental process which started long before the texts themselves came into being. (Szabó 1978, 23)

Szabó pointed out that his approach rested on "methodological novelties" and that it contributed to a "definite new picture of the early Greek science yet to be painted" (Szabó 1978, 24).

A critical examination of the concept of geometric algebra can be found in an appendix (Szabó 1969, 455–488 or Szabó 1978, 332–353). The appendix, which Szabó saw as an essential extension of the book, originated from a paper that Szabó gave at the Greek Academy of Sciences in Athens in May 1968 after the book was completed (Szabó 1969, 455). Szabó tried to show that those propositions in Euclid's *Elements* that are typically linked to geometric algebra could be understood as truly geometrical propositions in a Greek geometrical mathematical research context going back to the Pythagoreans. Szabó reconstructed the existence of a Pythagorean geometry of areas that could be used to avoid the (arithmetical) problem of incommensurability and the use of proportionals. Szabó deployed this theory to explaining the gnomon and Euclid's propositions in *Elements* II.5, II.6, II.10 and II.14. He tried to make plausible that all the other propositions linked to geometric algebra could also be interpreted as purely geometric propositions. The interpretation in terms of solutions of algebraic equations was according to Szabó "misleading because it obscures the true geometric meaning of the proposition [Euclid's *Elements* II.5] and suggests the false historical idea that the Greeks actually operated with algebraic equations in pre-Euclidean times. […] no traces of genuine algebraic ideas have yet been discovered in the mathematical tradition which culminated in Euclid's *Elements*" (Szabó 1978, 352f.). At that point he also explicitly criticized

[7]Szabó (1969). When I quote from it, the English translation (Szabó 1978) is used.

van der Waerden's approach in *Science Awakening* as a modern interpretation that is "foreign to the spirit of ancient mathematics" (Szabó 1978, 353).[8]

In addition to the topics of geometric algebra and a Greek foundational crisis, Szabó also touched upon Neugebauer's hypothesis concerning a transmission of Babylonian algebra to the Greeks (Szabó 1969, 34–36, 240–242, 488).[9] By giving an alternative Greek geometric interpretation of Euclid's propositions there was no need (and no possibility) to establish a link between Greek and Babylonian mathematics. This also meant that there was no need to question the Greek supremacy in science.

By the beginning of the 1970s, Szabó's approach was well known and Szabó was invited as a speaker at the 14th International Congress of the History of Science in Japan in 1974. His invited paper was called: The Origin of the Pythagorean "Application of Areas" (Szabó 1975). Szabó took up the theory as developed in his 1969 appendix. As we have seen already, Unguru also delivered a paper that was critical of geometric algebra at the congress. However, he could not remember whether he had met Szabó there.[10]

Reception

Both Klein's and Szabó's books were reviewed.[11] However, to my knowledge only one review was published in an American journal dedicated to the history of science (Scriba (1970) on Klein in *Isis*). Although both Szabó and Klein had raised substantial criticism concerning the concept of geometric algebra from different perspectives, the criticism was hardly taken up. It was only Michael S. Mahoney who raised this issue time and again in the early 1970s. In his review of Szabó's book for *The British Journal for the Philosophy of Science* in 1970, Mahoney recommended the book to "anyone who deals with ancient mathematics and philosophy" and pointed out Szabó's philological method. The fact that Szabó gave an alternative interpretation of what is usually described as geometric algebra is referred to only in one sentence: "Where other scholars have found in Books II and VI of Euclid's *Elements* a "geometric algebra" of Babylonian origins, Szabó finds (in part) a geometric attempt to solve problems posed by a purely Greek theory of music (or to solve their insolubility in numbers)" (Mahoney 1970, 306).

In his essay review of the second unrevised edition of the first volume on pre-Greek mathematics of Neugebauer's *Vorlesungen über Geschichte der antiken mathematischen Wissenschaften*, Mahoney questioned the existence or nature of

---

[8]In the original German edition Szabó was a bit milder: "Natürlich ist auch der moderne Beweis von B.L. van der Waerden *tadellos*. Aber mit dem antiken Gedankengang hat er in Wirklichkeit *gar nichts zu tun*." Szabó (1969, 487; emphasis in the original).

[9]Szabo saw no basis for the claim that the Greeks took over ideas from the Babylonians. As far as I know, Szabó did not mention Freudenthal's critical analysis of the idea of a Greek foundational crisis (Freudenthal 1966).

[10]According to the interview with Unguru conducted by D. E. Rowe and the author in Wiesbaden in May 2013.

[11]On Klein see Boyer (1968), Scriba (1970), Estes (1971); on Szabó see Mahoney (1970).

Babylonian algebra and its influence on Greek mathematics. In this context he outlined what van der Waerden had written on geometric algebra in *Science awakening* and pointed to Szabó's "heavy attack" at the time, but still characterized van der Waerden's interpretation as a "good theory" (Mahoney 1971a, 371). Defending more or less the Greek miracle thesis, Mahoney talked about the "dangers of reading pre-Greek mathematics with an eye trained in a mathematics that is, partly at least, of Greek origin and that reflects Greek concern" (Mahoney 1971a, 377).

These lines of thought reappeared in Mahoney's Fermat biography published in 1973. Mahoney made it clear that he aimed at a contextualization of Fermat and his work in his times.[12] He also acknowledged that there was a risk of misinterpreting Fermat's achievements by applying modern mathematical language:

> [...] the translation of past mathematics into modern symbolism and terminology represents the greatest danger of all. The symbols and terms of modern mathematics are the bearers of its concepts and methods. Their application to historical material always involves the risk of imposing on that material a content it does not in fact possess.

However after this warning, Mahoney continued by praising a cautiously applied use of symbolism:

> By the same token, the purpose of mathematical symbolism – and its foremost virtue – is to lay bare the basic structure of concepts and methods. Hence, judiciously applied, it should serve historical analysis by enabling one to cut through to the core of past mathematics without introducing anachronisms. (Mahoney 1994, xiif).

Maybe the little attention that Szabó and Klein received was due to the fact that Szabó and Klein were outsiders in the field of history of mathematics. They followed a kind of humanistic approach to the history of mathematics: Klein's drew heavily on Greek philosophy, Szabó's on philology. Klein had published nothing in that direction since the 1930s; Szabó's method of analysis was new and required a very high level of knowledge of the Greek language which was not common among historians of mathematics at the time. In addition to that, Szabó's book was written in German.

Unguru, of course, knew both of these books and cited them several times in his 1975 article. He was also aware of Mahoney's critical remarks. In contrast to Mahoney, Klein and Szabó, Unguru put methodological issues first as his title clearly shows. Geometric algebra embodied for him the prime example of wrong methodology. Unguru did not replace the standard narrative around geometric algebra by an alternative one, as Klein and Szabó tried to do. This approach together with his polemical style provoked a discussion about methodology in the history of mathematics at large—which was taken up by prominent figures. Their published replies, especially Weil's, did the rest to further the reception and

[12] "Fermat was not any mathematician anywhere at any time. He was a French mathematician of the first two-thirds of the seventeenth century. His thought, however original or novel, operated within a range of possibilities limited by that time and place. His odyssey had its boundaries; his drummer beat to a tune of the times" (Mahoney 1973, x).

notoriety of Unguru's paper. It is a different question, however, to determine to what extent Unguru's criticism was really taken up in later research in the history of (ancient Greek) mathematics.

4 Academic and Disciplinary Issues

The wider context of Unguru's 1975 article encompass the academic and disciplinary developments of the 1960s and 70s at large. Not only did the history of science flourish in the USA with lively debates around Thomas Kuhn's concept of a scientific revolution.[13] In one 1965 issue of *History of Science*, one reviewer spoke of "a revolution in historiography of science" that had taken place in recent years (Buchdahl 1965, 55). A revival of the history of mathematics could also be observed (Dauben and Scriba 2002). In the USA two new disciplinary journals were launched: the aforementioned *Archive for History of Exact Sciences* in 1960, and *Historia mathematica* in 1974. The latter was the result of an international initiative started by Adolf-Andrej Pavlovich Youshkevich (1906–1993) and Réné Taton (1915–2004) at the International Congress on the History of Science in Paris in 1968. It led to the formation of the International Commission on the History of Mathematics (ICHM) which met for the first time in Nice in 1970. Among its initial goals was the creation of a new journal on history of mathematics. The launching of *Historia mathematica* was seen by the editors as a "milestone in the development of history of mathematics as a scholarly discipline."[14]

The issues raised here deal mainly with questions of professionalisation of the history of mathematics and science. One issue explicitly touched upon by Unguru in his article is the role of mathematicians in the history of mathematics

> [...] those who have been writing the history of mathematics in general, and that of ancient mathematics (including Greek) in particular, have typically been mathematicians, abreast of the modern developments of their discipline, who have been largely unable to relinquish and discard their laboriously acquired mathematical competence when dealing with periods in history during which such competence is historically irrelevant and (I dare say) outright anachronistic. Such an approach, furthermore, stems from the unstated assumption that mathematics is a *scientia universalis*, an algebra of thought containing universal ways of inference, everlasting structures, and timeless, ideal patterns of investigation which can be identified throughout the history of civilized man and which are *completely independent of the form in which they happen to appear at a particular juncture in time.* (Unguru 1975a, 73; emphasis in the original)

The mathematicians' work in the history of ancient mathematics was characterized as largely anachronistic and their approach to mathematics as Platonist, and thus downright ahistorical. The mathematicians' competence in modern mathematics was seen as irrelevant for the history of ancient mathematics. Unguru was

[13]For the impact of Kuhn's concept in the history of mathematics, see Gillies (1992).

[14]See the editorial of the first issue of *Historia Mathematica*.

not alone in raising these issues. Time and again, this controversy with respect to the history of science, and thus to the discipline at large, appeared in print in journals.

4.1 The Debate Around Bernal's Science in History

One incident linked to the Wisconsin faculty of history of science is a debate about the third and new edition of *Science in History* (1965) by the physicist John Desmond Bernal (1901–1971). The biologist Norman Wingate Pirie (1907–1997) wrote a positive review of Bernal's book in the *Scientific American* in 1966 (Pirie 1966). The review was severely criticized in a letter to the editor by the historian of science L. Pearce Williams (*1927) at Cornell University. Williams characterized Pirie's review as "both inaccurate and misleading" (Williams 1966, 8), and criticized Bernal for errors and for repeating

> clichés that are now so out of date as to be laughable if they were not taken so seriously by practicing scientists whose knowledge of the history of science rarely rises above the level of the anecdotal.

Pirie was criticized for his uncritical review. Williams' main point was that

> the history of science is a professional and rigorous discipline demanding the same level of skills and scholarship as any other scholarly field. It is time for the scientist to realize that he studies nature and others study him. He is no more nor no less competent to comment on his own activities and the activities of his fellow scientists than is the politician. Critical political history is rarely written by the politician and the same is true of the history of science. In this case it should be recognized that Bernal has created a majestic myth; it is unfortunate that the reviewer took it for reality.

Thus as scientists, both author and reviewer, were seen as lacking competence in the field of history of science. This dispute has also to be seen in the context of the Cold War. Bernal's largely Marxist perspective and Pirie's esteem for it were attacked on the professional, not on the ideological level.

However, Williams' exclusion of scientists from the realm of history of science was met with opposition by the Wisconsin history of science department. The five Wisconsin professors Erwin N. Hiebert (1926–2012), Robert Siegfried (1921–2014), Aaron J. Ihde (1909–2000), W. D. Stahlman and Victor L. Hilts wrote a letter to the editor in reply to Williams. They argued in favour of cooperation between scientists and historians of science:

> we feel compelled to dissociate ourselves from his [Williams'] strongly implied position that only the professional historian of science can make useful contributions to that field. Such a position would seem to exclude the sympathetic involvement of the professional scientist essential to a meaningful practice of the history of science. (Hiebert et al. 1966, 16)

This debate took place in the very year Unguru arrived at Wisconsin. In the course of time Unguru, however, came to a different conclusion than his Wisconsin professors did, at least in the field of ancient Greek mathematics.

4.2 Mathematicians and the "Bellah Affair" at Princeton

The issue of competence was neither confined to the history of science nor to the letter to the editor section of a scientific journal. In 1973 an academic controversy touching upon the issue of competence and disciplinary sovereignty gained rather wide public attention in the USA. It concerned a dispute on the appointment of the sociologist Robert N. Bellah (1927–2013) as a permanent member of the Institute of Advanced Study (IAS) at Princeton.[15] The *New York Times* made it front page news on March 2, 1973. The *National Observer*, *Time magazine*, *Newsweek*, *Washington Post* and *Science* followed. Bellah's appointment had been suggested by the economist and the IAS-director Carl Kaysen (1920–2010) and the IAS sociologist Clifford Geertz (1926–2006). However, his nomination was rejected in a vote by the majority of the IAS staff. The Board of trustees nominated Bellah nevertheless, because the social scientists at the IAS, i.e. Kaysen and Geertz, were in favour of him. This incident triggered a public debate launched by the dissident IAS majority.

As Bortolini (2011) shows, the fight was basically about academic freedom. However, there were two different interpretations of it. On the one hand, the opponents were fighting for the right of the academic community to choose its colleagues without the interference of the administration. On the other hand, Bellah's supporters were fighting for the right of the discipline (of social sciences) to make its own choices about the appropriate candidate.

In the public, the controversy was often portrayed as a "contest between the "hard" scientists in mathematics at the institute and the "softer" social scientists" (Unkown 1973). In the *Science* article, Shapley (1973, 1209) characterized the controversy as an "overt academic tribal warfare, with pure mathematicians among those most hostile to Bellah [...] and economist Kaysen and the social sciences school defending him." Some IAS mathematicians, in particular André Weil, but also Armand Borel (1923–2003) and Deane Montgomery (1909–1929), were among the main opponents.

Weil, who had been working at the School of Mathematics at the IAS at Princeton since 1958, was among the first to mobilize mathematicians in protest against the Board's decision. He wrote a letter to non-IAS mathematicians and successfully asked for their support. Hence Kaysen and the trustees received many letters supporting Weil and the dissident majority of IAS-members. Weil (and other IAS members) questioned Bellah's competence. They felt that they were in a position to judge Bellah's work even though they were not experts in the field. Weil was said to have rated Bellah's scientific contributions as "worthless" saying that there had been weak candidates before, but that he had never had "the feeling of so utterly wasting [his] time" (Shenker 1973, 43). Weil even warned Bellah in a note that he knew of very few persons in the academic world who would accept an appointment under such circumstances (Weil according to W. Chapman in Bortolini 2011, 10).

[15] I draw here on the detailed sociological analysis by Bortolini (2011).

The supporters, however, believed that a scholar should only be judged according to his or her discipline's standards, and thus argued for the discipline's sovereignty in evaluating their peers. This view gained public acceptance. Other motives behind the attack on the nomination of Bellah were also mentioned (e.g. little respect for Kaysen as director; the role of the Board; the low status of sociology, especially non-quantitative sociology, at the time); and the breaching of confidentiality by making internal documents of the IAS available to the public was addressed, too (Woodward 1973). In the end, Bellah decided to take up his Ford professorship at Berkeley again. And the affair quickly quietened down.

While Unguru was writing his article, the so-called Bellah affair was still an academic topic discussed publicly. Even if one might find Weil's intervention characteristic of him personally, publicly it was the mathematicians as a group who were depicted as trying to interfere in matters outside their discipline, claiming competence for another discipline. And in his 1975 article, Unguru made a point along the same line.

In 1975, a provocative article by the philosopher and sociologist of mathematics Joong Fang (1923–2010) was published in *Philosophia Mathematica* dealing with the Bellah affair. Fang (1975, 129) spoke of 'intellectual' arrogance on the part of the group of mathematicians:

> Such an 'intellectual' arrogance or sheer abberation reflected in the case of Weil et al. has been a norm, however, rather than an exception for the majority of mathematicians who seem to reason that, since they can easily understand mathematics, they can as readily master any academic subject in any area.

This goes to show that considerable resentment existed towards the group of mathematicians within the US-American academic milieu in the early seventies.

4.3 Weil's Slating Review of Mahoney's Fermat Biography

In 1973 Unguru also witnessed another instance of Weil's harsh judgment. This time it was directed against one of his colleagues: Weil wrote a devastating review of Michael Mahoney's Fermat biography, which was published in the *Bulletin of the American Mathematical Society* (Mahoney 1973; Weil 1973). He radically questioned Mahoney's competence to write such a book. At the time Mahoney was an associate professor at Princeton University.

In the first half of the 12-page review, Weil tried to show that Mahoney was lacking "ordinary accuracy", "the ability to express simple ideas in plain English", some knowledge of French and Latin, "some historical sense", "some knowledge of the work of Fermat's contemporaries and of his successors", and, last but not least, "knowledge and sensitivity to mathematics". Weil wondered how one was to review such a book, and thought: "Merely to set things right would require another volume" (Weil 1973, 1144). He went on to give a summary of each chapter, not without continuing to criticize Mahoney's presentation and interpretation. Only the second chapter on Fermat's mathematical career and his controversies with

Descartes and Wallis was judged as somewhat better than the rest. Weil rated it as "the least unsatisfactory" (Weil 1973, 1144). In the end, Weil (1973, 1149) concluded:

> [...] a student of XVIIth century mathematics will find little in that volume that could be helpful to him, and much that can only confuse and mislead him.
>
> Unfortunately, a book on such a subject, published with the imprint of the Princeton University Press, tends to pre-empt the field. Surely Fermat deserved a better treatment. Let us hope he still gets it.

Other critical reviews followed (Itard 1974a, b; North 1974; later also Pedersen 1979). However, none of these was as devastating and ad hominem as Weil's.

But there were also some positive reviews by Carl Boyer (1906–1976), Derek Thomas Whiteside (1932–2008)—both historians of mathematics—and by Alan Gabbey (Boyer 1973; Whiteside 1974a, b; Gabbey 1975; later also Gridgeman 1976). For example, Boyer (1973, 152) appreciated it as

> an altogether exemplary account, for it goes well beyond a catalog of what was done and when, to provide a critical analysis of major aspects in the search for a leitmotif. [...] An outstanding characteristic of this volume is its evenhanded evaluation of Fermat's discoveries. The author does not play the role of a protagonist arguing the case of his hero.

Thomas Kuhn, Mahoney's Ph.D.-supervisor, saw Mahoney's academic future endangered by Weil's review. At the time Mahoney was trying to get tenure at Princeton. Unguru remembered being asked by Kuhn if he could also write a review. Unguru's review was published in *Francia* in 1976 (Unguru 1976). But Unguru had started working on it before his 1975 article was published, as a footnote reveals (Unguru 1975a, 68, fn. 4). In the end, it worked out well for Mahoney.

This episode shows that the themes of the last two subsections were also raised with respect to the history of mathematics: to what extent is a mathematician competent to evaluate the history of his field.

4.4 A Conference on the History and Historiography of Modern Mathematics at the American Academy

In the light of some workshops on the history of recent scientific developments at the American Academy of Arts and Sciences during the early 1970s, the Harvard mathematician Garrett Birkhoff (1911–1996) suggested convening a workshop on the history of modern mathematics. The idea was taken up and a workshop on the historical evolution of modern mathematics was organized by some mathematicians and historians of mathematics: Birkhoff, I. Bernard Cohen (1914–2003) (both at Harvard University), Thomas Hawkins (*1938, Boston University), Kenneth O. May (1915–1977, University of Toronto) and Felix Browder (*1927, University of Chicago). The workshop was held in August 1974 and documented in the second volume of the newly founded journal *Historia Mathematica*. Among its participants were Jean Dieudonné (1906–1992), Boyer, Freudenthal, Ivor Grattan-Guinness (*1941), Dirk Struik (1994–2000), Hiebert, Morris Kline (1908–1992), Gregory Moore, George Mackey (1916–2006), and Hilary Putnam (1926–2016)—but note that Weil was not among them.

The issue of the historiography of mathematics was addressed in a lot of papers. The topic, discussed above, of the different approaches of historians and mathematicians to the history of mathematics was central to the papers by Kenneth May and Judith Grabiner (*1938), who in contrast to Unguru and others at the time, welcomed mathematicians' contributions to the history of mathematics (May 1975; Grabiner 1975).

Grabiner, who at the time was assistant professor of history at the Californian State University at Dominguez Hills, tried to shed light on the different approaches and perspectives of "historians" and "mathematicians". While "the historian is interested in the past in its full richness, and sees any present fact as conditioned by a complex chain of causes in an almost unlimited past", "the mathematician [...] is oriented toward the present, and toward past mathematics chiefly insofar as it led to important present mathematics" (Grabiner 1975, 439). She sketched the history written by mathematicians as technical and of a high mathematical level which results in a "chronologically and logically connected series of papers" with no connection to the authors and the specific historical settings. The approach of the historian was then treated at greater length. It is concerned with the "non-mathematical past" as well as with "what the *total* mathematical past in some particular time-period was like" (Grabiner 1975, 439f., emphasis in the original). So the historian might be able "to construct a total picture of the background of some specific modern achievement" (Grabiner 1975, 441). He or she is also expected to know the general history of the time period in question and be acquainted with historiographical theories and questions in the history of science.

After a brief discussion of the value of history of mathematics, Grabiner (1975, 444) ended her analysis by calling for collaboration between the two approaches that in Grabiner's opinion complement each other:

> We have seen that the mathematician and the historian bring different skills and different perspectives to their common task of explaining the mathematical present by means of the past. Therefore, as this conference by its existence declares, collaboration between mathematicians and historians can be fruitful. The value of such a collaboration will be enhanced if each collaborator understands the unique contributions which can be made by the other.

In a very personal answer to the question "What is good history and who should do it?" May (1975) laid the focus on the different purposes and audiences of texts in the history of mathematics, and thus on the relativity of the answer with respect to the purpose and audience. Like Grabiner, he called for collaboration between historians and creative mathematicians in the study of modern mathematics. May also raised the issue of mistakes and their relativity. He had the impression that a mistake made by a colleague tended to be dismissed lightly whereas the same mistake made by a non-expert was judged as grave. This applied both to mathematical and historical errors, mathematicians and historians:

> Clearly in historical work the danger in missing the mathematical point is matched by the symmetric hazard of overlooking a historical dimension. The mathematician is trained to think most about mathematical correctness without a time dimension, i.e., to think ahistorically. Of course it is interesting to know how a historical event appears when viewed by a twentieth century mathematician. But it is bad history to confuse this with what was meant at the time. The historian concentrates on significance in the historical context and on the historical relations between events. And this is equally interesting to the mathematician who wishes to understand how mathematics *actually* developed. (May 1975, 453, emphasis in the original)

May finished his paper with a speculation that showed his disciplinary affiliations: "It may even be that the best mathematical research is aided by an appreciation of historical issues and results" (May 1975, 453).

5 Concluding Remarks

The analysis of the origin of Unguru's 1975 paper makes for a better understanding of Unguru's contribution: On the one hand, it shows that Unguru was not alone in criticizing the historiography of Greek mathematics. Klein, Szabó and Mahoney had raised important methodological issues that later Unguru focused upon. Thus, Unguru's paper fitted in well with a critical line of reasoning at the time.

As the contextualization in the wider academic background shows, it also took up the more general disciplinary issues of professionalisation revolving around the issue of competence. Unguru thought mathematicians in general unfit for doing history of mathematics, thus rejecting a strictly internalistic history of mathematics. The media coverage of the "Bellah affair" showed the US public that mathematicians boldly claimed competence for issues beyond their field. Weil's slating review could be seen just as another instance of this. The Academy workshop reflected these tensions within the community of historians of mathematics and mathematicians.

On the other hand, Unguru's paper differed significantly from Klein's, Szabó's and Mahoney's: not only the style was different, so was its message. Unguru discarded the standard historiography of Greek mathematics that had a long and rich tradition with the works of Johan Ludvig Heiberg (1854–1928), Zeuthen, Tannery, Heath and others. He put methodological issues on the top of his agenda. Like Weil, he was not diplomatic, but rather out-spoken.

Mahoney's call for a generational change in the history of mathematics in the aforementioned review of Neugebauer's *Vorlesungen*: "As the children of the revolution turn on their fathers, so too one now wishes to examine more closely some of Neugebauer's conclusions about Babylonian mathematics in particular and about methods of the history of mathematics in general," (Mahoney 1971a, 370) perhaps came closest to Unguru's revolutionary statement. However, Mahoney was here much milder in tone and less comprehensive in content. Yet, at the same time Unguru did not belong to this younger generation. He did not go along with all the latest trends in historiography, e.g. an increased interest in non-Western mathematics. Unguru was interested in Greek mathematics and fought for its peculiarity. He did not engage in Neugebauer's project of dethroning Greek mathematics as the origin of mathematics. Thus, seen from this perspective, Unguru's call can also be interpreted as a conservative programme: putting Greek mathematics back on the agenda of history of mathematics.[16]

The analysis also raises questions that need further investigation: Who was doing history of ancient Greek mathematics (in the USA) at the time? What role, if any, did the European (Jewish, emigrant) backgrounds play for their research? How did the controversy affect the study of Greek mathematics and of non-Western mathematics? How was the relation between history of science and history of mathematics at the time? And finally: what influence did Unguru's article have on the historiography of mathematics beyond rhetorical usage?

Acknowledgments I thank my co-editors H. K. Sørensen and V. R. Remmert, as well as K. Chemla and D. E. Rowe for their critical comments on an earlier version of this paper, and C. Friedrich for her generous help with retrieving sources.

References

Artmann, B. (1991). Euclid's elements and its prehistory. *Apeiron, 24*(4), 1–47.

Berggren, J. L. (1984). History of Greek mathematics: A survey of recent research. *Historia Mathematica, 11*, 394–410.

Bortolini, M. (2011). The "Bellah Affair" at Princeton. Scholarly excellence and academic freedom in the 1970s. *American Sociology, 42*, 3–33.

Boyer, C. B. (1968). Vieta and modern algebra. Greek mathematical thought and the origin of algebra by Jacob Klein. *Science, 162*(3858), 1112–1113.

Boyer, C. B. (1973). The Prince of Amateurs: The mathematical career of Pierre de Fermat (1601–1665). Michael S. Mahoney (review). *Science, 181*(4095), 152–153.

Buchdahl, G. (1965). A revolution in historiography of science (review). *History of Science, 4*(1), 55–69.

Dauben, J. W. & Scriba, C. J. (Eds.). (2002). *Writing the history of mathematics: Its historical development. Historical works, historical studies* (Vol. 27). Basel, Boston, Berlin: Birkhäuser.

Elkana, Y., & Segre, M. (1985). History of science in Israel. *Isis, 76*(1), 7–10.

Estes, B. R. (1971). Greek mathematical thought and the origin of algebra by Jacob Klein. *The American Mathematical Monthly, 78*(5), 564.

Fang, J. (1975). "J'Accuse": A politics of mathematics. *Philosophia Mathematica, 12*, 124–148.

Freudenthal, H. (1966). Y avait-il une crise des fondements des mathématiques dans l'antiquité? *Bulletin de la Société mathématique de Belgique, 18*, 43–55.

Freudenthal, H. (1977). What is algebra and what has it been in history? *Archive for History of Exact Sciences, 16*(3), 189–200.

Gabbey, A. (1975). The mathematical career of Pierre de Fermat (1601–1665) by Michael S. Mahoney (review). *The British Journal for the History of Science, 8*(1), 81–84.

Gillies, D. (Ed.). (1992). *Revolutions in mathematics*. Oxford: Clarendon Press.

Glass, H. (2002). *Minderheit zwischen zwei Diktaturen. Zur Geschichte der Juden in Rumänien 1944–1949. Südosteuropäische Arbeiten* (Vol. 112). München: Oldenbourg Verlag.

[16]I thank K. Chemla for drawing my attention to this aspect.

Glass, H. (2005). Das Verschwinden einer Minderheit. Die Auswanderung der Juden aus Rumänien nach 1944. In K. Clewing & O. J. Schmitt (Eds.), *Südosteuropa. Von vormoderner Vielfalt und nationalstaatlicher Vereinheitlichung. Festschrift für Edgar Hösch* (pp. 383–408). München: Oldenbourg Verlag.

Grabiner, J. V. (1975). The mathematician, the historian, and the history of mathematics. *Historia Mathematica, 2*, 439–447.

Gridgeman, N. T. (1976). The mathematical praeface to the Elements of Geometrie of Euclid to Megara (1570) by John Dee; The mathematical career of Pierre de Fermat (1601–1665) by Michael S. Mahoney (review). *The American Historical Review, 81*(4), 840–841.

Hasse, H., & Scholz, H. (1928). Die Grundlagenkrise der griechischen Mathematik. *Kant-Studien, 33*, 4–34.

Hiebert, E. N., Siegfried, R., Ihde, A. J., Stahlman, W. D., & Hilts, V. L. (1966). Letter to the editor. *Scientific American, 215*, 16.

Hilts, V. L. (1984). History of science at the University of Wisconsin. *Isis, 75*(1), 63–94.

Hopkins, B. C. (2011a). *The origin of the logic of symbolic mathematics. Edmund Husserl and Jacob Klein*. Studies in Continental thought. Bloomington, Indianapolis: Indiana University Press. Preface by Eva Brann.

Hopkins, B. C. (2011b). The philosophical achievement of Jacob Klein. *New Yearbook for Phenomenology and Phenomenological Philosophy, 11*, 282–296.

Høyrup, J. (2011). How to transfer the conceptual structure of Old Babylonian mathematics: Solutions and inherent problems. With an Italian parallel. In A. Imhausen & T. Pommerening (Eds.), *Writings of early scholars in ancient Near East, Egypt, Rome and Greece. Translating ancient scientific texts* (pp. 385–417). Berlin: De Gruyter.

Itard, J. (1974a). Ápropos une livre sur Pierre Fermat (The mathematical career of Pierre de Fermat (1601–1665), Princeton Univ. Press, Princeton, N.J., 1973) by Michael Sean Mahoney (review). *Revue d'histoire des Sciences, 27*, 335–346.

Itard, J. (1974b). The mathematical career of Pierre de Fermat (1601–1665). By Michael Sean Mahoney (review). *Historia Mathematica, 1*(4), 470–476.

Kastanis, N., & Thomaidis, Y. (1993). The term "geometric algebra" target of a contemporary epistemological debate [in Greek]. In D. Anapolitanos & V. Karasmanis (Eds.), *Ancient Greek mathematics. Historical and philosophical essays* (pp. 25–52). Athens: Trochalia. (English translation (19 pp.): http://users.auth.gr/nioka/Files/GEOMALGE.pdf; viewed 2012, March 27).

Klein, J. (1936a). Die griechische Logistik und die Entstehung der Algebra. I. Teil. *Quellen und Studien zu Geschichte der Mathematik, Astronomie und Physik, 3*, 18–105.

Klein, J. (1936b). Die griechische Logistik und die Entstehung der Algebra. II. Teil. *Quellen und Studien zu Geschichte der Mathematik, Astronomie und Physik, 3*, 122–235.

Klein, J. (1968). *Greek mathematical thought and the origin of algebra* (2nd ed. trans. by E. Brann, 1992). Cambridge, MA: M.I.T. Press.

Knorr, W. R. (1990). New readings in Greek mathematics: Sources, problems, publications. *Impact of Science on Society, 159*, 207–218.

Lakatos, I. (1976). *Proofs and refutations: The logic of scientific discovery*. Cambridge: Cambridge University Press. (Edited by J. Worrall).

Mahoney, M. S. (1968). Another look at Greek geometrical analysis. *Archive for History of Exact Sciences, 5*(3–4), 318–348.

Mahoney, M. S. (1970). Szabó, Árpád (1969): Anfänge der griechischen Mathematik (review). *The British Journal for the Philosophy of Science, 21*(3), 305–307.

Mahoney, M. S. (1971a). Babylonian algebra: Form vs. content. Essay review of O. Neugebauer: Vorlesungen über die Geschichte der antiken mathematischen Wissenschaften. *Studies in History and Philosophy of Science, 1*(4), 369–380.

Mahoney, M. S. (1971b). Die Anfänge der algebraischen Denkweise im 17. Jahrhundert. *RETE: Strukturgeschichte der Naturwissenschaften, 1*, 15–31.

Mahoney, M. S. (1973). *The mathematical career of Pierre de Fermat, 1601–1665*. Princeton: Princeton University Press.

Mahoney, M. S. (1994). *The mathematical career of Pierre de Fermat, 1601–1665* (revised 2nd ed.). Princeton: Princeton University Press.
Máté, A. (2006). Árpád Szabó and Imre Lakatos, Or the relation between history and philosophy of mathematics. *Perspectives on Science, 14*(3), 282–301.
May, K. O. (1975). What is good history and who should do it. *Historia Mathematica, 2*, 449–455.
Neugebauer, O. (1936). Zur geometrischen Algebra (Studien zur Geschichte der antiken Algebra III). *Quellen und Studien zu Geschichte der Mathematik, Astronomie und Physik, 3*, 245–259.
Neugebauer, O. (1957). *The exact sciences in antiquity* (2nd ed.) (1st ed., 1952). Providence, Rhode Island: Brown University Press.
North, R. (1974). The mathematical career of Pierre de Fermat by Mahoney, Michael Sean (review). *The Mathematical Gazette, 58*(406), 299–300.
Pedersen, K. M. (1979). Mahoney, Michael Sean, The mathematical career of Pierre de Fermat (review). *Mathematical Reviews, 58*(3), 1508f. (MR0490737).
Pirie, N. W. (1966). Science in History, by J. D. Bernal (review). *Scientific American, 214* (March), 131–136.
Poggendorff, J. C. (Ed.). (2004). *Biographisches-literarisches Handwörterbuch der exakten Naturwissenschaften. Ergänzungsband Mathematik. Teilband II*, volume L-Z. Wiley-VCH, Weinheim (Heidi Kühn (Ed.), Sächsische Akademie der Wissenschaften zu Leipzig).
Rowe, D. (1996). New trends and old images in the history of mathematics. In R. Calinger (Ed.), *Vita mathematica. Historical research and integration with teaching* (pp. 3–16). Washington: Mathematical Association of America.
Scriba, C. J. (1970). Greek mathematical thought and the origin of algebra by Jacob Klein (review). *Isis, 61*(1), 132–133.
Shapley, D. (1973). Institute for Advanced Study: Einstein is a hard act to follow. *Science, 179*(4079), March, 1209–1211.
Shenker, I. (1973, March 2). Dispute splits advanced study institute. *The New York Times*, 1, 43.
Szabó, A. (1958). Die Grundlagen in der frühgriechischen Mathematik. *Studi italiani di filologia classica, 30*(N.S.), 1–51.
Szabó, A. (1964a). The transformation of mathematics into deductive science and the beginnings of its foundation on definitions and axioms. Part 1. *Scripta Mathematica, 27*(1), 27–49.
Szabó, A. (1964b). The transformation of mathematics into deductive science and the beginnings of its foundation on definitions and axioms. Part 2. *Scripta Mathematica, 27*(2), 113–139.
Szabó, A. (1969). *Anfänge der griechischen Mathematik.* München, Wien: R. Oldenbourg.
Szabó, A. (1975). The origin of the Pythagorean "Application of Areas". In *Proceedings of the International Congress of the History of Science, Tokyo and Kyoto, Japan*, 1974, August 19–27 (vol. 2, pp. 207–213). Tokyo: Science Council of Japan.
Szabó, A. (1978). *The beginnings of Greek mathematics* (Translation of Szabo (1969) by A. M. Ungar). Dordrecht, Boston, MA: Reidel.
Unguru, S. (1972). Witelo and thirteenth-century mathematics: An assessment of his contributions. *Isis, 63*(219), 496–508.
Unguru, S. (1975a). On the need to rewrite the history of Greek mathematics. *Archive for History of Exact Sciences, 15*(1), 67–114.
Unguru, S. (1975b). On the need to rewrite the history of Greek mathematics. In *Proceedings of the International Congress of the History of Science, Tokyo and Kyoto, Japan*, 1974, August 19–27 (Vol. 2, pp. 169–172). Tokyo: Science Council of Japan.
Unguru, S. (1976). Fermat revivified, explained and regained. *Francia, Forschungen zur westeuropäischen Geschichte, 4*, 774–789.
Unguru, S. (1979). History of ancient mathematics—Some reflections on the state of the art. *Isis, 70*(4), 555–565.
Unkown. (1973, March 19). Ivory tower tempest. *Time Magazine, 49*.
van der Waerden, B. L. (1954). *Science awakening* (Engl. translation by A. Dresden with additions by the author). Groningen: Noordhoff.

van der Waerden, B. L. (1976). Defense of a "Shocking" point of view. *Archive for History of Exact Sciences, 15*(3), 199–210.

Weil, A. (1973). Book review: The mathematical career of Pierre de Fermat (1601–1665) by M. S. Mahoney. *Bulletin of the American Mathematical Society, 79*(6), 1138–1149.

Weil, A. (1978a). History of mathematics: Why and How. In O. Letho (Ed.), *Proceedings of the International Congress of Mathematicians, Helsinki 1978* (Vol. 1, pp. 227–236). Helsinki: Academia Scientiarum Fennica.

Weil, A. (1978b). Who betrayed Euclid? (Extract from a letter to the editor). *Archive for History of Exact Sciences, 19*(2), 91–93.

Weyl, H. (1921). Über die neue Grundlagenkrise der Mathematik. *Mathematische Zeitschrift, 10*, 39–79.

Whiteside, D. T. (1974a). Fermat fulfilled: Michael S. Mahoney. The mathematical career of Pierre de Fermat (1601–1665) (review). *Isis, 65*(3), 398–400.

Whiteside, D. T. (1974b). Michael S. Mahoney. The mathematical career of Pierre de Fermat (1601–1665) (review). *Zentralblatt für Mathematik und ihre Grenzgebiete, 264*, 5–6.

Williams, L. P. (1966). Letter to the editor. *Scientific American, 214* (June), 8.

Woodward, C. V. (1973, March 13). Letter to the editor. *The New York Times*.

Zeuthen, H. G. (1966). *Die Lehre von den Kegelschnitten im Altertum (first ed. 1886)*. Hildesheim: Olms.

Name Index

V.R. Remmert et al. (eds.), *Historiography of Mathematics in the 19th and 20th Centuries*, Trends in the History of Science,
DOI 10.1007/978-3-319-39649-1

MIX
Papier aus verantwortungsvollen Quellen
Paper from responsible sources
FSC® C105338

If you have any concerns about our products,
you can contact us on
ProductSafety@springernature.com

In case Publisher is established outside the EU,
the EU authorized representative is:
Springer Nature Customer Service Center GmbH
Europaplatz 3, 69115 Heidelberg, Germany

Printed by Libri Plureos GmbH
in Hamburg, Germany